SPATIAL INFORMATION SYSTEMS

General Editors

P. H. T. BECKETT
P. A. BURROUGH
M. F. GOODCHILD
P. SWITZER

Statistical Methods in Soil and Land Resource Survey

R. WEBSTER and M. A. OLIVER

OXFORD UNIVERSITY PRESS
1990

Oxford University Press, Walton Street, Oxford OX2 6DP
Oxford New York Toronto
Delhi Bombay Calcutta Madras Karachi
Petaling Jaya Singapore Hong Kong Tokyo
Nairobi Dar es Salaam Cape Town
Melbourne Auckland
and associated companies in
Berlin Ibadan

Oxford is a trade mark of Oxford University Press

Published in the United States
by Oxford University Press, New York

British Library Cataloguing in Publication Data
Webster, R.
Statistical methods in soil and land resource survey.
1. Soils. Surveying. Statistical methods
I. Title II. Oliver, M. A.
631.47
ISBN 0-19-823317-5
ISBN 0-19-823316-7 pbk

Library of Congress Cataloging in Publication Data
Webster, R.
Statistical methods in soil and land resource survey/R. Webster and M. A. Oliver.
(Spatial information systems)
Rev. ed. of: Quantitative and numerical methods in soil classification and survey. 1977.
Includes bibliographical references.
1. Soil surveys—Statistical methods. 2. Soils—Classification—Statistical methods.
I. Oliver, M. A. (Margaret M.).
II. Webster, R. Quantitative and numerical methods in soil classification and survey.
III. Title. IV. Series.
S592.14.W4 1990 631.4'7'072—dc20 90-35409
ISBN 0-19-823317-5
ISBN 0-19-823316-7 pbk

Typeset by Apek Typesetters, Nailsea, Bristol

Printed in Great Britain by Biddles Ltd, Guildford & King's Lynn

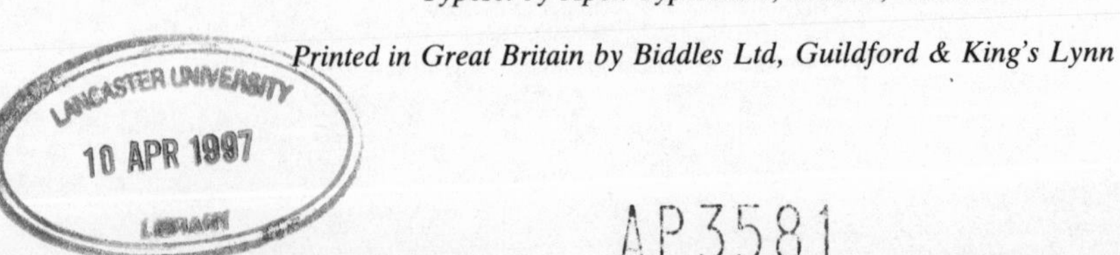

SPATIAL INFORMATION SYSTEMS

Organized societies have collated their knowledge about the earth's surface and of its people, animals, and plants for a very long time. They kept their records on paper as texts, tables, diagrams, and maps, which people could read easily but which were difficult to analyse. Now with the aid of computers these data can be stored automatically and retrieved at will to be displayed in a wide variety of ways. Modern spatial information systems also include many logical and mathematical capabilities for abstraction, generalization, re-expression, recombination, rescaling, aggregation and disaggregation, superposition, summarization, interpolation, and error handling that allow thorough analyses that were previously impossible. These analyses may not only provide new insights into existing situations but, by using fictional data from simulation models, they enable past and future situations to be modelled and explored. These new tools are revolutionizing mapping, and inventory and management of our environment. Today automated information systems are used in many branches of science, in business and commerce, in local and national governments, and in international agencies. The applications range from the utilitarian—the mapping of the networks of telephones, electricity supply, and sewers—to the esoteric and futuristic, such as in modelling the possible effects of climatic change.

The rapid development in automation has outstripped the supply of scientists and technicians trained to use it. Universities and polytechnics the world over are therefore establishing courses on spatial analysis and geographic information systems, both for their own students and to educate staff already in employment. One of our aims is to provide texts for such courses, for third-year undergraduates, for post-graduates, and for practioners in the field. The texts are intended to be interdisciplinary, covering basic principles that can be applied in many fields.

New developments in spatial information systems are reported only in the proceedings of conferences and in scientific journals. A second aim is to provide scholarly monographs, written by experts, to describe both the theory and practice in a well-ordered way. Their authors will gather new knowledge from diverse sources and make it widely available and intelligible.

One of the most striking features of a modern information system is that when its full power is used its results rarely fail to surprise and delight. This brings the danger, of course, that the user will be over-impressed by the sheer technology. A further aim, therefore, is to ensure that users are not only impressed but are also delighted because they have achieved a deeper understanding of the world around them.

P. A. BURROUGH

Utrecht

STATISTICAL METHODS IN SOIL AND LAND RESOURCE SURVEY

There are many ways of assessing soil and other land resources. Survey used to be approached qualitatively. Surveyors divided land into distinct types by observation on the ground and often with the aid of aerial photographs. They then sampled each type at a very few sites to describe it. This was the only way to make an inventory of a large area swiftly. Now planners and land managers want to know in greater detail how the land with which they are concerned varies. In some instances they may seek an optimal classification for their purposes. In others they may wish to choose other ways of assessment, especially where man's activities have overlaid the natural patterns by, for example, pollution. In all these situations the questions arise: 'How many samples should be taken to produce results with a given level of confidence? Does it matter where the sample points are located? How can complex, multivariate data be classified optimally? How can maps of quantitative attributes be made reliably?' This book provides the answers.

If soil and other land resources were truly made up of unique, independent units or entities, then it would not be very difficult to arrange these units in classes, or to demonstrate the various interrelations. But soil and land resources are rarely exact objects: they need to be sampled and treated statistically in order to discover general truths. Moreover, soil and land resources are characterized by many attributes that can show various degrees of co-variation or correlation. In addition, the quantitative properties of soil and land resources cannot be treated as simple stochastic variables because their variation over the earth's surface is governed by interactions at many spatial scales. Richard Webster was one of the pioneers in using statistical methods for studying soil: the first edition of this book, published in the Series Monographs in Soil Science, brought many of these techniques together for the first time. Now he and Margaret Oliver have revised and extended the text to cover recent developments that they and their co-workers have made in the application of spatial statistics to soil and land resource survey. This new material is vital for optimizing interpolation from scattered data and the design of sampling schemes. The theory is applicable not only to soil science, the background from which they write, but also in many other branches of resource survey and mapping.

P. A. BURROUGH

Utrecht

PREFACE

This book, like its predecessor, *Quantitative and Numerical Methods in Soil Classification and Survey*, is addressed to working scientists and advanced students in pedology, engineering, ecology, and geography who study the natural resources at the earth's surface by what are, broadly speaking, survey methods as distinct from experiments. It is written for those who observe, record, and analyse information about those resources with ever-present spatial variation over which they have no experimental control. It describes methods for making surveys quantitative, stressing the need for measurement, sound sampling, sensible and efficient estimation, and proper planning. The traditional techniques of land resource survey all embody classification at some stage, and this book discusses the role of classification, how it can be performed mathematically, and in which situations it is likely to be helpful. There have been major developments over the last few years in geostatistics, which provides a very different approach to local estimation, mapping, and sampling. The book describes the elements of the underlying theory and how they can be applied in practice. Each approach has its place in its proper context. The book explains this. It aims always to help readers choose the most suitable techniques for tackling their problems in context.

The emphasis throughout the book is on soil. This is because as soil scientists that is where our experience lies. Most of the theory and analytical techniques that we describe, however, are equally applicable to other land resources. Similarly, nearly all the illustrative examples are drawn from our own experience. This is not because they are necessarily more suitable than other people's results, but simply that we already had the data and intermediate workings to hand. In some instances they arise from our work with others, and we thank our colleagues for their material and help. In particular, we thank Professor P. A. Burrough for the data from Kelmscot used to illustrate multivariate classification (Chapters 9 and 10), Dr I. M. Buraymah and Mr M. G. Jarvis, who provided the measurements for Figs. 2.4 and 2.5, Dr K. W. T. Goulding for the data for Fig. 6.3, Dr H. E. Cuanalo for the transect data for Sandford (Chapters 12 and 14), Mr W. M. Corbett and Dr S. Nortcliff for data from Hole Farm, the Welsh Plant Breeding Station whose data we used in Chapters 12 and 14, and Dr T. M. Burgess and Dr A. B. McBratney whose collaboration produced many of the results in these two chapters.

Over the years several other colleagues have discussed the application of statistics to soil and land resources with us and have given us a deeper insight. We are especially grateful to Dr F. H. C. Marriott, who took such a strong interest in the first edition, Mr J. C. Gower, always a source of sound common sense, Dr P. H. T. Beckett for stimulating us over many years and for editing our script, and Mrs Joyce Munden for turning our rough diagrams into works of art. Finally, we thank the authors and publishers for permission to reproduce the following figures and tables: Cambridge University Press for Fig. 2.1, Dr A. W. Moore and Messrs Angus and Robertson for Fig. 7.2, Messrs John Wiley and Sons for Table 3.1, and Dr K. Kyuma and the Society of the Science of Soils and Manures of Japan for Table 8.3.

Richard Webster
Margaret Oliver

October 1989

CONTENTS

1

INTRODUCTION

A sensible philosophy controlled by a relevant set of concepts saves so much research time that it can nearly act as a substitute for genius.

N. W. Pirie, *Concepts out of Context*

Aims and means

This book is intended as a practical handbook for earth scientists in many different fields. It is written, however, from our experience as soil scientists, and we have set it in the context of soil survey. Nevertheless, the problems and the needs, in terms of sampling to obtain information and its subsequent analysis, are similar whether the information comes from samples of the atmosphere, the sea or other bodies of water, vegetation, or rocks. In nearly all instances the investigator is concerned with data that have similar properties; they vary continuously in space and they are not predicted readily. The methods of sampling and of analysis included here are general, and geologists, geographers, engineers, and ecologists should all recognize close analogies in their fields and should find the methods equally applicable.

We place considerable emphasis on sampling and estimation from sample information because of the survey context. In the past, sampling has been the weakest feature of resource survey and field research generally. Many of the data so obtained, often with difficulty and at great expense, were of little use because the original sampling was unsatisfactory. Although this is changing, we offer no apology for labouring the importance of sound sampling because everything else depends upon it.

For many years soil survey was regarded as the recognition and subsequent mapping of different types of soil. The situation in geology was similar. Happily that vision is broadening. Increasingly, survey is being seen simply as a way of obtaining information about regions, both large and small, of the earth's surface. With modern computers we can store the information, we can sort it and analyse it, we can collate it with information from other sources, and we can present the results in whatever way seems most appropriate. This is the function of geographic information systems (see e.g. Burrough 1986). In some instances we shall want to display the results of surveys as maps; in others these will be unnecessary or even irrelevant. We do not have to classify the soil or rocks for mapping or to make good use of sample data. In particular, the newer geostatistical techniques provide a sound and effective alternative. Classification, however, may still enable survey to be performed more economically and its findings to be presented more precisely. It can be a convenient tool.

Neither survey nor the analyses performed on the data that accrue are ends in themselves, and the investigator is urged to decide just what he wants to know before considering how to go about finding it out. Only when an investigator is clear about his goals is he in a position to choose the means of achieving them in the most direct way, and to design a survey for that purpose.

Classification and measurement

Rural people have lived in close harmony with their environment for centuries. They learned how to exploit its natural resources and how to manage them without degrading them seriously. In no field has this been more so than with the soil. The husbandman discovered how the soil responded to the way he treated it. He recognized its differences in behaviour from place to place and where it changed from one kind to another in the landscape. So he divided his land into parcels to be managed more or less differently according to these changes in the soil. In other words, he classified the soil, however informally or intuitively. Classification is the practical tool by which man traditionally deals with his environment. It is also a means for communicating information about the environment, or the soil in particular, to neighbours and descendants. Classification of the soil also provides a basis for matching the soil in different places and thus of predicting its behaviour where experience is lacking.

The first soil scientists adopted this approach, both for practical purposes and for more fundamental understanding. At this stage there were big differences to be seen and recorded. The layman's classes and descriptive terms were meaningful, convenient, and adequate. Once the more obvious distinctions had been made, pedologists turned their attention to finer differences. In the practical sphere, agronomists and engineers also needed to describe the soil with which they worked more exactly and consistently, and to predict behaviour more precisely. In both cases the desired consistency and precision could be achieved only by measurement, and so, as in many other branches of science, observation became quantitative. Thus, to describe the soil at some place as 'acid' was no longer adequate: scientists wanted to say how acid, and they devised methods of measuring its acidity in terms of pH. The same applies to the study of many other aspects of the environment.

There is more to measurement than this, however. The soil, for example, is a more or less continuous mantle. The scientist cannot record what it is like everywhere. At best he can measure its properties, whether directly in the field or in the laboratory, on small portions of the soil—that is, from a *sample*. Soil also varies from place to place, often considerably, so measurements from one sampling site cannot describe all soil sensibly. In practice, information about the soil is usually wanted for areas, and surveys are made in many parts of the world to obtain it by sampling at numerous sites. For surveys to be fully quantitative, properties of the soil must be measured at these sites.

There is another kind of data that can be regarded as quantitative. Surveys are often carried out to determine how much land is of a particular type (say, suitable for growing rice) or what proportion possesses some attribute (say, waterlogged). We might attempt to delimit such regions and measure their area by geometry. Alternatively, and more economically, we could inspect the soil at several suitably chosen places and *count* those where the soil possessed the attribute in question. Individual observations would be qualitative, but in total they would assume a quantitative character.

Nor does the matter rest here, for we need to know to which, of all soil or other aspects of the environment of a region, a particular measurement or set of measurements apply. To what extent may a value obtained in one place be used elsewhere? And is it sensible to take an average of several measurements, especially where there are large regional differences? Local averages for smaller areas are likely to be much more reliable. The recognition of such areas means classification, and these are circumscribed by boundaries as on the conventional soil map. We then restrict our use of the data to the classes within which they were obtained. In this way we can predict the values of properties within an area with more confidence. Alternatively, we might recognize neighbourhoods around the localities that interest us and for which we have relevant data. This is the more recent approach to prediction taken in geostatistics. It obviates the need for classification, and so avoids the unnecessary argument and difficult decisions that seem inevitably to accompany classification. It recognizes gradual change where tradition would have us create abrupt boundaries.

Nevertheless, classification still has a place in regional resource survey. When a large body of data is collected in the course of an investigation it often needs simplifying to be intelligible. This can often be achieved by classifying the data. Similarly, when a survey is carried out for planning landuse the sampling sites, though initially described quantitatively, usually need to be grouped. For instance, the farmer cannot vary his management continuously in response to continuous variation in the survey records; he has to treat finite tracts of land in a uniform manner as though they were homogeneous. Likewise, the engineer cannot easily adapt his design to every minor fluctuation in soil character.

Although we replace intuitive classification by measurements for consistent description and for communicating precise information about the soil at particular sites, classification can have an important complementary role in increasing the utility of soil data and enabling us to economize on sampling. Just how important this role is or can be has been a matter of debate, and in recent years classification itself has been the subject of quantitative study. Questions like: how much does my classification improve prediction; how might this population be classified to provide a simple but useful picture; which classification is best for that purpose; to which class should the new individual *I* be allocated; and if classification seems unprofitable is there an alternative, can all be answered quantitatively. Questions such as these are discussed at length in this book.

Statistical methods

Quantitative description of the soil of different areas and its behaviour involves analysis of more or less large bodies of data. Simple statements about the soil must be seen against the background of ever-present variation, which must be taken into account in the analysis. This is the province of statistics, and most of the techniques described in this book are in some sense statistical. Statistics provides accurate and usable mathematical descriptions of the real world, both in the laboratory and in the field.

For many soil scientists with an agricultural background, the subject of statistics often conjures up a picture of carefully designed experiments to compare crop varieties or the effects of fertilizer treatments on yield. The measurements from these are subject to analysis of variance, a powerful yet mysterious process, whose culmination and climax is seen as a test of significance. Success is gauged by the number of stars that can be attached to some *F* ratio to reject the null hypothesis, or to confirm a prejudice! This is unfortunate. Significance tests have a perfectly proper place. In resource survey, however, they will usually be superfluous or even inappropriate to the main purpose of the investigation. In soil survey and soil systematics statistical methods provide means of condensing data from variable material into economical descriptions. They enable us to predict the values of properties from those of others and to estimate the values at unsampled places or larger areas from values measured at the sampling sites. They also enable us to identify relations and structure in our data and to display these. Thus, some tasks are probabilistic while others are not. If a significance test is appropriate we should apply it, since it could prevent our drawing unwarranted conclusions from sample evidence. The precise level of probability at which we test is less important, however, and is to some extent a matter of personal choice.

The book also deals with multivariate methods. Although these can serve very specific purposes, they are also exploratory tools; for identifying relations between soil profiles and classes of soil, for experimental classification, and for suggesting hypotheses that can be tested by other techniques. The methods involve complex and lengthy calculations that have become practicable only over the past twenty years or so with the advent of computers. There are many examples of the application of multivariate analysis in the research literature, and the methods are being applied increasingly in some branches of the earth sciences. Nevertheless, it is somewhat surprising that there are still rather few published accounts of their use in soil science and survey in particular. Most of the studies have been made on familiar data collected for other purposes. This was necessary to give confidence, but it seems that many soil scientists are still afraid. Few handbooks exist for the practitioner, and we hope that books like this one will help soil scientists to overcome their fear.

Scope

A book of this size cannot deal with all the mathematical techniques that a resource surveyor or systematist might need. We have had to select. In doing so we have tried to cover those topics that soil scientists most often ask about and that we have found useful. Although most are described in statistical texts, soil scientists find it difficult, for one reason or another, to link the statistical theory with their work. We have tried to provide that link here, especially with the examples. It seems that few students or practitioners of soil science have studied mathematics since leaving school, and we have borne this in mind constantly in presenting the material. For this reason we have omitted most derivations and proofs. Readers must discipline themselves to mastering symbols, however.

Readers who wish to apply the methods in their investigations might need to consult one or other statistical table. These can be found in many of the standard statistical texts, and it has seemed unnecessary to duplicate them here. The statistical tables compiled by Fisher and Yates, first published in 1938 (see Fisher and Yates 1963), are well known to many soil scientists, and for this reason we have referred to them freely. Lindley and Miller (1953), however, include all the elementary tables, while the *Biometrika* tables (Pearson and Hartley 1966, 1972) are more comprehensive and provide more advanced tests.

A note on terminology and symbols

A number of common English words have a somewhat restricted or special meaning in statistics. Some of these occur frequently throughout this book, and are introduced here.

Variables. The pH and sand content of the soil, the calcium carbonate content of rock, and the gradient of the land surface all vary from place to place, and in some instances from time to time. These are examples of *variables*. Although we may measure their values, it sometimes suffices to record them as only present or absent; for example, we may record a rock as calcareous or non-calcareous. We may choose to distinguish these as qualitative *attributes*. A random variable is often termed a *variate*, and in statistical analysis the term is also used to denote a set of measured values of some variable in which there is more or less random variation.

Individuals. When we record the values of a variable we do so for particular pieces of rock, soil pits, fields, and so on. Each of these objects is an *individual* for the purpose of analysis. In any one study all of the individuals must be of the same general kind.

Population. The whole set of individuals or material under study in a particular instance is referred to as a *population*. A population can be either finite—for example a set of described soil profiles—or infinite—for example the soil of a district, which can be considered as made up of an infinite number of soil profiles.

Sample. A sample is a set of individuals or a collection of material taken from a larger population about which information is wanted. In soil laboratories the term 'sample' often refers to a single bag (disturbed) or core (undisturbed) of soil taken from the field. In the statistical sense it refers more often to several bags or cores, and applies equally to a set of sites where the soil has been or is to be described or measured without any soil necessarily being collected.

Parameter. A *parameter* is a quantity that is constant in the case being considered, though it may differ in other cases. In statistics it is generally reserved for quantities, such as means and standard deviations, of whole populations (q.v.), and is distinguished from estimates of them made by sampling, for which the term *statistic* is used. In computing a parameter is usually a quantity that is held constant for a particular run of a program. 'Parameter' is not synonymous with 'variable' (or 'character'), and the current fashion for using the term in this sense is to be deprecated.

Survey. In the statistical sense a survey simply means a type of investigation in which a situation is observed as it is without alteration other than that unavoidably incurred in sampling. It differs from an *experiment*, in which the investigator changes some feature of the situation on purpose so that he can study effects of the change. The meaning of survey in statistics also differs somewhat from its meaning in soil survey, in which it has implications of mapping.

Classification. This is the act of dividing a population, or agglomerating individuals, into groups. It can also be the set of classes that result from such action. The nouns *class* and *group* are treated as synonyms in this book. *Allocation* is distinguished from classification and is used to mean the assignment of individuals to classes or the *identification* of individuals as belonging to those classes.

The common mathematical symbols will be familiar to readers, but there are several other conventions that might need explaining.

Summation. Summation is one of the most frequent operations in statistics. If there are n values, $x_1, x_2, x_3, \ldots, x_n$, their sum can be written as $\Sigma_{i=1}^{n} x_i$, where

$$\sum_{i=1}^{n} x_i = x_1 + x_2 + x_3 + \ldots + x_n.$$

Mean. The arithmetic mean is also much needed, and is usually signified by placing a bar over the symbol for the variate concerned. Thus, for the n values of variate X, the mean $\bar{x}$, which is read 'x bar', is

$$\bar{x} = \sum_{i=1}^{n} x_i / n.$$

Product. The product of a set of values is occasionally needed and is symbolized by capital Π. The product of n values of X are thus

$$\prod_{i=1}^{n} x_i = x_1 \times x_2 \times x_3 \times \ldots \times x_n.$$

Combination. The number of ways of choosing r items out of n is

$$\frac{n!}{r!(n-r)!}$$

where $n!$ (factorial n) is $1 \times 2 \times 3 \times \ldots \times n$. It is conventionally abbreviated to $\binom{n}{r}$.

Expectation. In the theory of statistics we often have to think of a particular value or set of values as just one realization of a process that could have generated any number of values. The mean or *expected* value of the process is denoted by E.

Several lower-case Greek letters are widely used: μ (mu), σ (sigma), and ρ (rho) indicate the mean, standard deviation, and correlation coefficient of populations respectively, while the sampling estimates of the last two are denoted by Roman equivalents s and r. Alternatively, estimates may be denoted by placing a circumflex, $\hat{.}$, over the Greek symbol. The letter χ, as in χ^2 (chi-square), is used for the distribution with this name. The lower case λ (lambda) refers to latent roots in multivariate analysis. Geostatistics has introduced other conventions, of which γ (gamma) is used to denote the semi-variance and λ the kriging weights. We have used ψ (psi) to signify the Lagrange multiplier.

Matrix notation. Many statistical procedures are most economically and clearly expressed in matrix form, and this practice is followed in Chapters 6–14. A brief account of matrix algebra and symbolism is given in the Appendix.

2

QUANTITATIVE DESCRIPTION OF VARIABLE MATERIAL

> Not chaos-like, together crushed and bruised,
> But, as the world harmoniously confused:
> Where order in variety we see,
> And where, though all things differ, all agree.
>
> A. Pope, *Windsor Forest*

It would be a simple matter to describe the earth's surface if it were the same everywhere. The environment, however, is not like that: there is almost endless variety. Thus when we wish to describe some feature, such as the soil, of an area quantitatively, not only must we be able to describe it at individual observation points, but also we must know how to deal with the differences between points that we shall surely find. In this chapter we shall consider the kinds of observation and measurement that we can make on soil at any one place, and then how to summarize sets of measurements and describe the variation that occurs within any one set.

Scales of measurement

Binary variables—attributes

In the Introduction we saw that qualitative terms are often quite adequate to describe clearly distinct states of soil. There are places where the soil changes abruptly from sand to clay, and others where it changes equally abruptly from red to black. For all practical purposes, only two states of the soil need to be recognized in each case. So, for example, sand and clay can be regarded as points on a *binary* scale and given the values 0 and 1. Characters that are either present or absent, for example the presence or absence of litter, calcium carbonate, manganese concretions, or earthworms, can also be recorded on binary scales. If the character is present then it is given the value 1, otherwise it takes the value 0. In these circumstances be sure that 0 means absent. Do not use it to mean 'inapplicable', 'irrelevant', or 'forgotten': other symbols should be used for these. Binary characters are also known as *attributes*, and division of a population according to the presence or absence of an attribute constitutes the simplest form of classification. Binary scales are sometimes known as *dichotomies*.

Multi-state characters and ranked variables

Properties such as the shape of structural aggregates are also described in qualitative terms; granular, blocky, prismatic, platy, and so on. When there are more than two states they constitute a *manifold*, or a *nominal* scale. They are also known as *multi-state characters*, and when there is no sense in which they are ordered they may be termed *unordered* multi-state characters. For computing purposes it is usually convenient to identify each state by a number, but one must avoid mistaking them subsequently for quantitative values.

Other multi-state morphological characters include dry, slightly moist, moist, wet, waterlogged, and non-plastic, slightly plastic, plastic, which describe the moisture state and plasticity of the soil respectively; the terms to describe the degree of structural development are also multi-state. Each of these sets is ordered, however: they are *ordered multi-state characters*, or *ranked* variables.

Continuous variables

Properties are recorded in a fully quantitative way when they are measured and assigned values on a continuous scale with equal intervals. Examples include the thickness of horizons (dimension: length), organic carbon content (proportion), shear strength (dimensions: mass $\times$ time^{-2} $\times$ length^{-1}), and water held at 50 mbar (proportion). The Munsell colour scales may be regarded as fully quantitative since they have been arranged as far as possible in steps of equal perception. Under standard lighting, however, the scales are not linear, and the CIE (1971) has provided formulae for converting Munsell codes to linear form. We mention this again shortly.

It is sometimes important to distinguish between those scales that have an absolute zero and those for which the zero is arbitrary. Most scales on which soil properties are measured do have absolute zeros. Some properties, like those mentioned above, rarely attain zero in nature: others, such as the proportion of stones and calcium carbonate, often do. Measurements made on scales that have an absolute zero are said to be *ratio* measures. It is meaningful to compare values as multiples or ratios of one another. For example, soil that contains 4 per cent of calcium carbonate may be considered twice as calcareous as soil that contains only 2 per cent. Some scales have only arbitrary zeros. Colour hue and longitude are examples. In practice, both are represented in soil survey by small segments of circular scales. The zero of the pH scale is also arbitrary. Such scales are usually known simply as *interval* scales.

We need not distinguish between interval and ratio scales for most purposes. Data recorded on either are amenable to all kinds of mathematical operation and to many forms of statistical summary and analysis. They are the most tractable and informative of all data. Most soil properties can be measured on such scales, either in the field or in the laboratory. Measurement usually requires more effort than qualitative description, but the investigator who seriously seeks a quantitative description or analysis should be prepared to make the effort. It is likely to be worth while.

Coarse stepped scales and estimates

Although quantitative scales are continuous, there are often practical limits to the minimum interval on a scale. In the laboratory this is usually determined by the sensitivity of an instrument or the accuracy of its calibration. In either case, the minimum interval is negligibly small for practical purposes. However, refined instrumentation is often too expensive for routine field survey, and properties measured in the field are often recorded in coarse increments on short scales. Soil colour provides an example. On the Munsell scales, colour value (lightness) extends from 1 to 9, and the test charts provide steps of 1; chroma (intensity) ranges from 0 to 8 in steps of 1 or 2. Hue (spectral composition) has a longer scale but, where the greenish and bluish greys are absent, it rarely extends over more than five increments, e.g. from 10R to 5Y with intermediate values 2.5YR, 5YR, 7.5YR, 10YR, and 2.5Y. In high latitudes the range is usually less than this. It is possible to interpolate between the panels of the Munsell charts, but such interpolation needs good lighting and very careful perception, and so is not common practice. The pH of soil can be measured reasonably quickly with universal indicator, but a surveyor is unlikely to be able to judge to better than 0.5 of a pH unit by this means. Although records of pH and colour are quantitative and can be treated legitimately by most mathematical and statistical techniques, results might be somewhat crude.

There are also other properties observed in the field for which time-consuming measurement is often replaced by rough estimation, but on scales related purposefully to measurable quantities. Examples include root density, carbonate content, and soil strength. The steps of their scales are not necessarily equal in terms of measured values, but they have been chosen as the best compromise between increments of equal practical significance and increments whose limits can be detected consistently. Such assessments need to be treated with some caution, but there are many instances where data so derived can be treated as fully quantitative, especially where the central limit theorem applies (see Chapter 3).

Counts

An attribute can sometimes be given quantitative character by counting. Instead of recording roots as present or absent, or giving a rough estimate of their frequency, we can count those cut by a particular cross-section of the soil. Similarly, it is common practice to count grains of particular species of heavy mineral in studies of soil provenance. Counts can be treated by many of the methods used for continuous variables, though care is needed to ensure that the treatment and results are sensible.

Circular scales

Although hue and latitude and longitude are circular, treating them as linear has no serious consequences when only a small part of each scale is used in any particular study. The situation is different when a whole circle or a large part of it is represented. In land resource survey this most often arises in records of the direction faced by the land surface—its slope azimuth or aspect. Aspect may be recorded quite properly in degrees from 0 to 360°, or coded 1 = north north-east, 2 = north-east, . . . , 16 = north. Absurd results will almost surely follow if such records are treated as though they were linear. The orientations of stones and sand grains in soil are also directional. Special methods are needed to summarize and analyse directional data. They are not mentioned further here, and readers can find details in Mardia (1972).

If a large range of colour hue is encountered then the problem can be overcome in another way. Imagine a somewhat distorted cylinder standing vertically on its base. Its vertical axis represents the colour value. Its radius represents chroma, and the direction in the horizontal plane represents hue. The system is a mixture of Cartesian and polar coordinates. It can be converted to a fully Cartesian system on orthogonal axes, and the CIE (1971) system mentioned above does this, at the same time linearizing the scales. Melville and Atkinson (1985) describe the conversion at some length in the context of soil recording, and McBratney and Webster (1981) illustrate it with an example.

Representing variation: frequency distributions

Soil varies continuously, and measurements of almost any property made at different places will differ more or less. Likewise, replicate measurements made in the laboratory on a single sample (bag) of soil will vary. This *experimental* or *observational error* can often be diminished to negligible proportions by mixing and sub-sampling the soil carefully and by using well maintained instruments and good laboratory technique. It cannot be eliminated completely. Variation from place to place is often substantial, and is a major source of uncertainty in soil survey. Many of the difficulties of applying mathematics to the study of soil are caused by such variation. It is one of the main reasons why this book was written. In the rest of this chapter we shall see how to treat the variation present in sets of measurements.

The histogram

One of the first steps in studying a set of measurements is to divide it into a manifold. To do this, values are chosen at equal intervals as limits of successive classes within the observed range. The number of individual values falling within each class is counted and is the *frequency* for that class. The frequencies for all classes constitute the *frequency distribution* for the set. A graph can be drawn of the distribution, with frequency on the ordinate and the variate values on the abscissa, and with contiguous bars representing the frequencies of the classes. Such a graph is a *histogram*. Several examples are shown in Fig. 2.1; the first pair is derived from the values in Table 2.1.

Choosing the class interval is a matter of judgement and experience of the particular property and locality. Generally speaking, the fewer observations there are, the fewer the classes and the wider the class interval needed to give a reasonable picture. With 30 observations five or six classes will probably serve best, increasing in number to around ten for 60–100 observations, and perhaps as many as 25 with 300 or more observations. Little is to be gained by having more than about 30 classes, however many observations there are. The locations of the intervals, i.e. the class limits, are a matter of convenience. In some studies it is helpful if the class limits are round numbers. However, there must be no ambiguity about the assignment of individuals with values falling on a borderline. Such ambiguity is avoided best by assigning all individuals with borderline values to the classes immediately below or immediately above. A similar result can be obtained by defining class limits to one further place of decimals than that at present in the recorded measurements. When the recorded values have already been rounded it may be better to assign individuals lying on a particular borderline alternately to the class above and the class below.

Although histograms are commonly drawn by plotting the number of observations as the ordinate, it is the *area* beneath the horizontal top of each bar that actually represents the class frequency. When class intervals are equal the bars of any histogram drawn this way automatically have their areas proportional to the class frequencies. If class intervals are not equal then the heights of the bars must be calculated so that the areas of the bars are proportional to the frequencies. A histogram is not appropriate for a discontinuous variable such as a count. For such variables frequencies are best represented as isolated vertical bars.

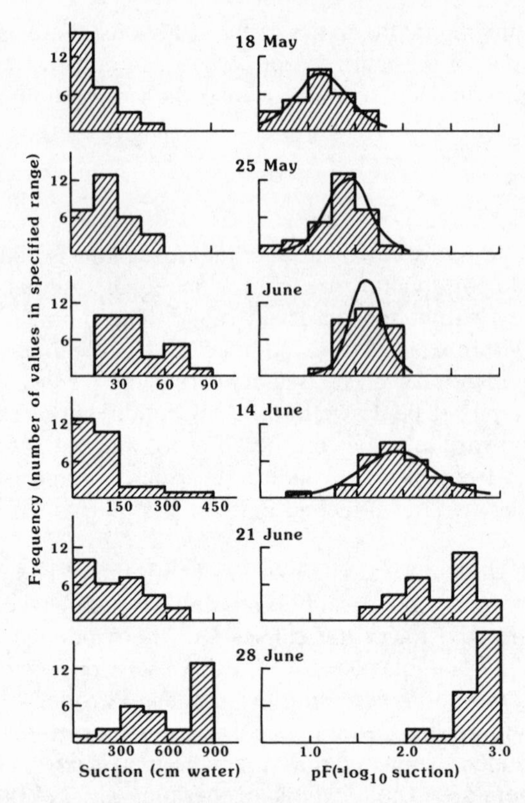

FIG. 2.1 Histograms of suction in centimetres of water (left) and pF = $\log_{10}$ cm (right). Curves of the normal distribution are fitted to the first four pF histograms. (From Webster 1966)

TABLE 2.1 Values of soil matric suction at 38 cm in Oxford Clay on 18 May 1963 in centimetres of water and as pF (= $\log_{10}$ cm water)

cm water	pF	cm water	pF	cm water	pF
53.3	1.73	5.7	0.76	9.8	0.99
19.3	1.29	6.9	0.84	22.4	1.35
15.5	1.19	4.4	0.64	34.9	1.54
13.0	1.11	4.5	0.65	40.2	1.60
10.5	1.02	20.9	1.32	15.0	1.18
11.9	1.08	12.0	1.08	22.5	1.35
10.7	1.03	16.1	1.21	11.2	1.05
23.3	1.37	8.5	0.93	30.1	1.48

Source: Webster (1966)

Other ways of displaying frequency distributions are by *frequency polygons* and cumulative frequency diagrams. The latter are often used to present the results of particle size analysis. A frequency polygon differs from a histogram in that the ordinates at the centres of the classes are joined to their neighbours by straight lines instead of being stepped. Its disadvantage is that the area under the graph is no longer exactly proportional to the sum of the frequencies.

Valuable though frequency tables and histograms are, they need summarizing further for most practical purposes. We shall now consider the two most important characteristics of distributions of soil properties, representing the position and spread of the values on the measurement scale respectively.

The arithmetic mean

The position of a distribution can be represented by its centre or average, and there are several measures that may be used, each appropriate in certain circumstances. The most useful is the *arithmetic mean*, or simply the *mean*. Suppose we have a set of N observations, $x_1, x_2, \ldots, x_N$; then the mean μ is defined as

$$\mu = \frac{1}{N}\sum_{i=1}^{N} x_i. \tag{2.1}$$

As a measure of position the mean has several advantages: it takes account of all the observations; it can be treated algebraically to compute, for example, totals and to combine averages from several sets of observations; the mean of a sample is an unbiased estimate of the population mean, a point that will become clearer in the next chapter.

Median and mode

Two other measures of the position of a distribution are the *median* and the *mode*. The median is the central value of the variate when the observations are ranked in order from smallest to largest. There are, therefore, as many observations with values less than the median as there are with values greater than it. Mathematically it is less tractable than the mean. When a property is recorded on a coarse scale the median is likely to provide only a rough estimate of the true centre.

Its main advantage is that it is not disturbed by the size of extreme values. Suppose soil is sampled at several places to determine its potassium status. If a bag of fertilizer had been spilled at one of the sampling points or if there had been a bonfire then the potassium value there might be one or more orders of magnitude larger than the rest. Its inclusion would influence the mean greatly, but could leave the median unaffected. In some instances extreme values are suspect because the instrument used for measurement is insensitive in that part of the range, or they are unobtainable because the true values lie beyond its range. Again, the median is not upset in this situation whereas the mean is. This situation is illustrated in Fig. 2.1 and Table 2.2, taken from a study in which soil matric suction was measured at weekly intervals by tensiometers placed at 38 cm in soil on a small plot of land (Webster 1966). On the logarithmic pF scale of measurement the distribution is symmetrical, and the median closely approximates the mean for the first four sets of readings. However, the tensiometers could not measure suctions more than about 710 cm water, pF 2.85, and as the soil dried out an increasing number of them failed to provide reliable values. This is illustrated in the last two histograms on the right-hand side of Fig. 2.1. The values bunch at the high end of the scale. The centre of the true pF values on 21 June was probably nearer to the median, pF 2.41, than to the mean, pF 2.33. On 28 June the median is a much more realistic estimate of the true average suction than the mean, though by then even that might have been too small.

The mode is the most frequent or the typical value. It implies that the distribution has a single peak. Some distributions have more than one peak and may be termed *multi-modal*. Although we may identify intuitively what we think is the mode, its numerical value is difficult to determine in practice. If the class interval can be made very small and the class frequencies still increase and decrease smoothly, then the mid-value of the most frequent class would probably be a good approximation. Such a procedure is likely to work only if many observations have been made. The alternative is to fit to the frequency values an ideal frequency curve whose peak would then be the mode of the distribution.

For any symmetrical distribution with a single peak, the mean, median, and mode coincide. In asymmetric distributions the median and mode lie further from the longer tail of the distribution than the mean, and the median lies between the mode and the mean. The formula

$$\text{mode} - \text{median} \simeq 2(\text{median} - \text{mean})$$

provides a rough guide to their relative positions. The mean and median of the suction values in cm water given in Table 2.2 illustrate the difference:

mean	16.6 cm
median	11.9 cm

The mode is difficult to judge on so few data. The histogram shows it to lie somewhere between 0 and 15 cm. It seems to be less than 11.9 cm, the median, but the value of 2.5 cm given by the above formula is obviously too small.

TABLE 2.2 Means, standard deviations, coefficients of variation (CV), and medians of matric suction expressed as pF (= $\log_{10}$ cm water) recorded by 30 tensiometers at 38 cm depth in the soil in 1963

Date	Mean	Standard deviation	CV(%)	Median
18 May	1.13	0.29	25.5	1.11
25 May	1.41	0.23	16.2	1.39
1 June	1.58	0.18	11.1	1.59
14 June	1.91	0.38	20.0	1.90
21 June	2.33	0.35	15.1	2.41
28 June	2.73			2.85

Dispersion

As with the position of a distribution, so there are several measures for describing its dispersion or spread. They include the *range*, the *interquartile range*, the *mean deviation*, and the *standard deviation*. Of these, the standard deviation and its square, the *variance*, are without doubt the most valuable. The variance of a set of observations or finite population, $x_1, x_2, \ldots, x_N$, is often defined as

$$\sigma^2 = \frac{1}{N}\sum_{i=1}^{N}(x_i - \mu)^2, \tag{2.2}$$

where μ is the mean as above. In Chapter 3 and later in the book we shall define the variance slightly differently by replacing N in the denominator of this expression by $N-1$. The reasons for this will be made clear then.

The variance and standard deviation, like the mean, are based on all the observations; they can be treated algebraically and are the least affected by sampling fluctuations. They are both special cases of important quantities in the theory of errors. Suppose u is some particular value of a property about which we have N measurements; then the quantity

$$s^2 = \frac{1}{N}\sum_{i=1}^{N}(x_i - u)^2 \tag{2.3}$$

is the *second moment* of the distribution about u. If u is μ then the second moment about the mean is the variance. Its square root s, the *root-mean-square deviation*, is the *standard deviation*. The merit of these quantities will become more apparent when we apply theoretical distributions to data. Until then we can gain a 'feel' for the standard deviation by matching the histograms in the right-hand side of Fig. 2.1 with the values of the standard deviation given in Table 2.2. In the first four distributions, and in many other reasonably symmetrical distributions, we shall not go far wrong if we assume that about two-thirds of the observations lie within one standard deviation of the mean.

Coefficient of variation

The standard deviation expresses dispersion in the same units as those in which the variable is measured, and it takes no account of the position of the distribution in relation to the zero of its scale. Sometimes we wish to express variation in relative terms. Suppose we have measured some soil property at several points in two different regions. The standard deviations of the two sets of measurements, σ_1 and σ_2, are equal, but there is a two-fold difference in their means; $\mu_1 = 2\mu_2$. Relative to its mean, the soil of the second region is twice as variable as that of the first. The *coefficient of variation*, CV, which equals the standard deviation divided by the mean, expresses this. It is usually given as a percentage:

$$\mathrm{CV} = 100\frac{\sigma}{\mu}\%.$$

It is useful for comparing variation in different sets of observations of the same property, but care is needed to ensure that its use is appropriate. In particular, it is unlikely to be helpful for scales with arbitrary zeros.

Skew and kurtosis

The mean and variance of a set of observations describe their position and spread. Sometimes we wish to describe other characters of a distribution, in particular its degree of symmetry, or *skew*, and its peakedness, or *kurtosis*. These are derived from the third and fourth moments about the mean respectively.

Considering skew first, we compute the third moment:

$$m_3 = \frac{1}{N}\sum_{i=1}^{N}(x_i - \mu)^3, \tag{2.4}$$

For symmetrical distributions m_3 is zero, since the positive deviations balance those that are negative. A distribution with its peak nearer the small end of the range and large values extending far beyond the mean has a positive value of m_3. It is said to be *positively skewed*. If a distribution has a long tail at the small end of its scale then m_3 is negative and the distribution has negative skew. Skew is usually measured by a dimensionless quantity, $\sqrt{\beta_1}$ or γ_1, defined as

$$\sqrt{\beta_1}=\gamma_1=\frac{m_3}{m_2^{3/2}}, \tag{2.5}$$

where m_2 is the second moment about the mean, i.e. the variance.

Soil variables often have skew distributions, and the left-hand sides of Figs. 2.1–2.4 all show examples of positive skew. The suction data for Wytham are more skewed (γ_1 for 18 May is 1.51) than the values of aluminium at Ginninderra ($\gamma_1=0.87$, Fig. 2.2), as are the data for copper (Fig. 2.3) and electrical conductivity (Fig. 2.4).

The fourth moment about the mean is

$$m_4=\frac{1}{N}\sum_{i=1}^{N}(x_i-\mu)^4. \tag{2.6}$$

The importance of m_4 relates especially to the normal distribution, which we shall discuss next, and for which the ratio

$$\beta_2=\frac{m_4}{m_2^2}=3. \tag{2.7}$$

The quantity γ_2 is then defined to equal β_2-3, so that $\gamma_2=0$ for normal distributions. Distributions that are more peaked than normal (*leptokurtic*) have positive values of γ_2, and those that are flatter than normal (*platykurtic*) have negative values.

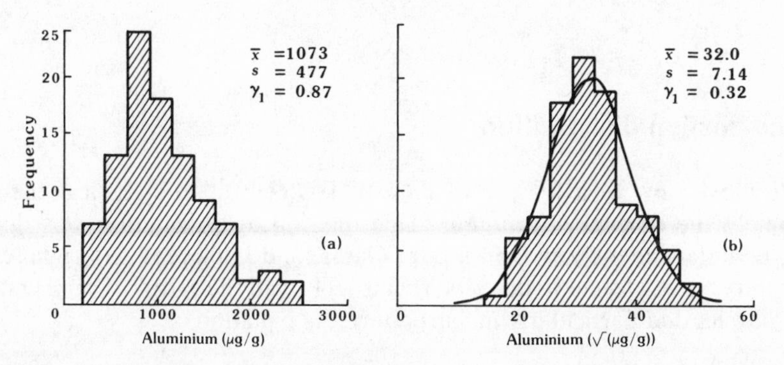

FIG. 2.2 (a) Histogram of extracted aluminium in parts per million. (b) Histogram of extracted aluminium transformed to $\sqrt{}$(p.p.m.). The normal curve is fitted to (b). (Data from study by Webster and Butler 1976)

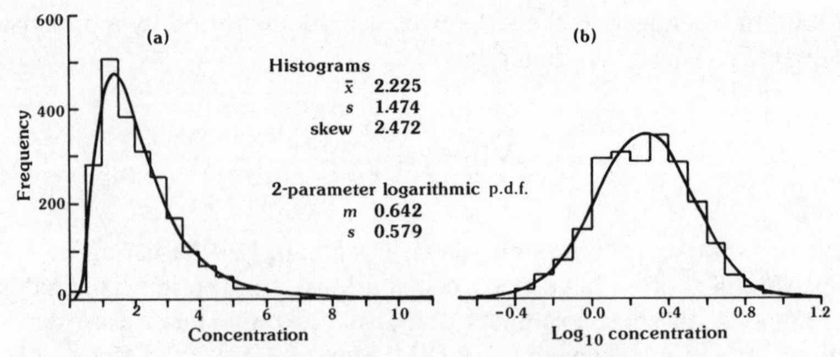

FIG. 2.3 (a) The two-parameter lognormal curve fitted to the histogram of easily extracted copper in the topsoil in part of south-eastern Scotland (McBratney *et al.* 1982). (b) Histogram of the copper measurements transformed to their common logarithms

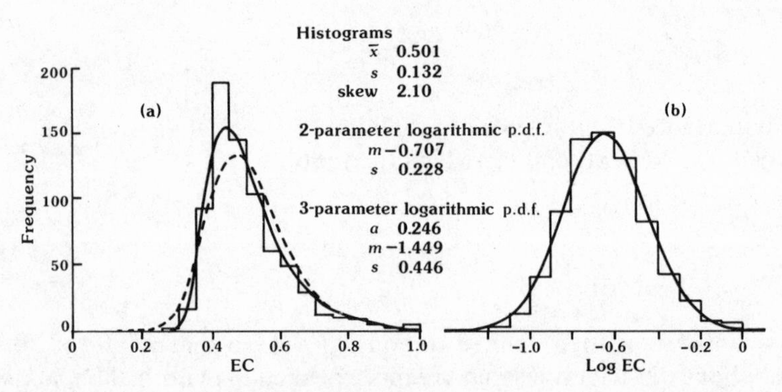

FIG. 2.4 Histograms of electrical conductivity in the topsoil of irrigated land in the Gezira of Sudan (Buraymah and Webster 1989), (a) on the original scale and (b) on the logarithmic scale

The normal distribution

We move now to a feature of data that is central to statistical theory, namely the *normal distribution*. This theoretical distribution was discovered independently by De Moivre, Gauss, and Laplace in the eighteenth century. It was found to describe remarkably well the errors of observation in physics and particularly in astronomy. Its equation is

$$y=\frac{1}{\sigma\sqrt{(2\pi)}}\exp\left\{-\frac{(x-\mu)^2}{2\sigma^2}\right\}, \tag{2.8}$$

where π and exp have their usual meaning, and μ and σ are the mean and standard deviation of the particular variable. The shape of the distribution is characteristically that of a vertical cross-section through a bell. Several graphs of the normal distribution are shown in Figs. 2.1–2.5. They show the kind of graphs that we might expect to obtain if we have very many measurements of the properties in question, had made the class intervals very small, and then had joined the class frequencies to form frequency polygons.

The normal curve has a number of features of interest. It is continuous and symmetrical, with its peak at the mean of the distribution. It has two points of inflexion, one on each side of the mean at a distance equal to the standard deviation σ. The ordinate y at any given value of x is a measure of the probability of that value of x occurring, and is known as the *probability density* at x. Since the peak of y is at $x=\mu$, the mean is the most likely value of x. The area under the curve is equal to 1, the total probability of the distribution, i.e. the probability that x takes any of its possible values. The area under any portion of the curve, say between x_1 and x_2, represents the proportion of the distribution lying in that range. Some useful figures to remember are: slightly more than two-thirds of the distribution lies within one standard deviation of the mean, i.e. $\mu-\sigma$ and $\mu+\sigma$; about 95 per cent of it lies in the range $\mu-2\sigma$ to $\mu+2\sigma$; and 99.74 per cent of the distribution lies within three standard deviations of the mean. When considering a particular set of data it is natural to think of frequencies rather than probabilities, in which case the above probabilities or proportions are simply multiplied by N, the number of observations.

Very many variables, not only of soil, but also of other natural materials, plants, and animals, and also manufactured goods, are distributed in a way

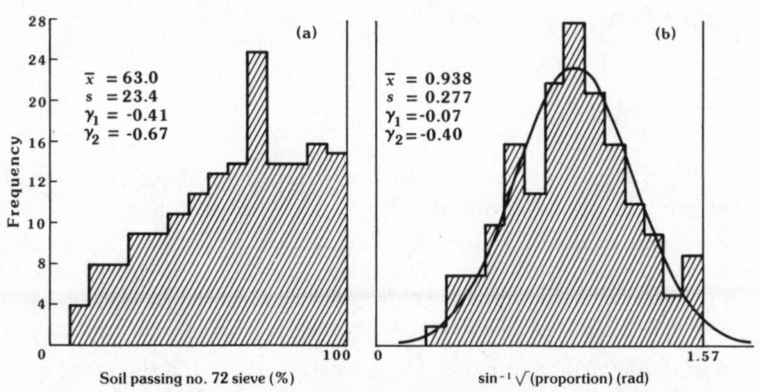

Fig. 2.5 (a) Histogram of soil material passing through no. 72 sieve as a percentage. (b) Histogram of soil material passing through no. 72 sieve transformed to arcsine. The curve of the normal distribution is fitted to the right-hand graph. (Data for Berkshire kindly provided by M. G. Jarvis)

that approximates closely the normal curve. Many measurements that are not normally distributed can be made so by transforming their measurement scales. The matric suction data of Table 2.1 are just such. The original measurements in cm water are very skew, but their common logarithms (pF) match the normal distribution closely. Other transformations will normalize some other non-normal measurements. Even for variables that are far from normally distributed and cannot be normalized by transformation, their sample averages tend to become more nearly normal as the size of the sample is increased.

For these and other reasons the normal distribution has been intensively studied. It is very well understood, and although it never describes exactly any real distribution of a natural variable, it can be applied usefully and reliably to many tasks involving estimation, prediction, and comparison of soil in one or more regions. Further, tables of its ordinates and integrals have been accurately prepared, and are available in many standard texts. Thus the investigator is likely to have the strong backing of statistical theory and practice when he undertakes an analysis of the properties of interest.

Other distributions

The normal distribution is one of several distributions that we shall come across in this book. In the next chapter we shall see that proportions are distributed somewhat differently; they follow the *binomial distribution*. Certain mathematical quantities, when calculated for random samples, are distributed in their own characteristic way and are known by their own names. They include *Student's t*, χ^2 (chi-square), and the *F ratio*. These distributions have been worked out and are published in sets of tables; see for example Fisher and Yates (1963).

Transformations

Sets of observations that are not normally distributed give rise to difficulties that are absent with normally distributed sets. We have already seen that there is some doubt as to which measure of the centre we should use when the distribution is skew. Other results that depend on normality cannot be applied, or can be applied only roughly. Comparisons between means of different sets of observations (see Chapter 4) are especially unreliable if the variable is skew, for the variances within the sets can differ substantially from one set to another. Our estimate of the proportion of a population lying in any given range could be erroneous if we assumed normality when there was appreciable kurtosis. These difficulties can often be overcome by transforming the scale of measurement so that the distributions on the new scale are normal. Further, it is often found that the same transformations also make the variances of samples less dependent on the mean, and therefore they are often made mainly to stabilize the variance.

Skewness is the most common departure from normality in measurements of soil properties. Matric suction measured in cm water, or more usually nowadays in bars, is just one example. If the skew is only slight then no transformation need be made, and results of a statistical analysis are unlikely to mislead. For more pronounced skew, however, some transformation may be desirable, and the investigator has several options.

Square root transformation

Moderate skew can be removed by taking square roots of the measurements. Fig. 2.2 shows an example of this.

Logarithmic transformation

A more powerful transformation than taking square roots can be obtained by taking logarithms, either to base 10 or base e. Care is needed to ensure that there are no zero or negative values in the data. If there are, then a small value should be added to each measured value to ensure that the smallest is just positive. Fig. 2.1 shows how well matric suction values can be normalized in this way. Note that means calculated on transformed values are transforms of the geometric means of the original data. Thus, if

$$\bar{g} = \left\{ \prod_{i=1}^{N} x_i \right\}^{1/N},$$

then

$$\log \bar{g} = \frac{1}{N} \sum_{i=1}^{N} \log x_i. \qquad (2.9)$$

If a transformation to logarithms produces a normal distribution then the variable is said to be lognormally distributed. Its parameters in the logarithmic scale are the mean μ and the standard deviation σ. However, it may be possible to approximate a normal distribution more closely by shifting the origin of the scale on which the measurements, x, were made by an amount a. This gives rise to the general three-parameter distribution in which $\log(x-a)$ is normally distributed. The parameter a defines the location of the distribution. The full probability function is then

$$y = \frac{1}{\sigma(x-a)\sqrt{(2\pi)}} \exp\left[-\frac{1}{2\sigma^2}\{\ln(x-a)-\mu\}^2\right]. \tag{2.10}$$

The formula is sometimes written in the form

$$y = \frac{1}{\sigma(x-a)\sqrt{(2\pi)}} \exp\left[-\frac{1}{2\sigma^2}\left\{\ln\frac{x-a}{b}\right\}^2\right], \tag{2.11}$$

where $b = \exp(\mu)$. The three parameters a, b, and σ can be thought of as representing the location, size, and shape of the distribution. In these formulae a must obviously be less than any actual value of x. Aitchison and Brown (1957) describe the distribution fully.

Fig. 2.3a shows an example of the two-parameter lognormal curve fitted to the histogram of easily extractable copper in the topsoil in part of south-eastern Scotland (McBratney *et al.* 1982). It fits well. Transforming the measurements to their common logarithms gives the histogram in Fig. 2.3b. It is now symmetric, and a normal curve fits well. In this instance no shift of origin could improve the fit.

Fig. 2.4a is the histogram of electrical conductivity in the topsoil of irrigated land in the Gezira of Sudan (Buraymah and Webster 1989). Here both two- and three-parameter lognormal curves are fitted. The two-parameter curve fits moderately well (dashed curve), but shifting the origin by 0.25 mS/cm improves the fit significantly (solid curve). Fig. 2.4b shows the histogram on the logarithmic scale.

Many properties of soil and other naturally occurring materials seem to be approximately lognormally distributed. Matric suction, electrical conductivity, and the concentrations of micronutrients are just a few. Exchangeable cations, available nutrients, hydraulic properties, and microbial activity are also widely reported to be distributed in this way. Many chemical elements appear to be distributed lognormally in rock also. Nobody has proposed a satisfactory explanation for this fact (see Ahrens 1965, and Shaw 1961, for discussion). The lognormal distribution has a special place in exploration geochemistry and leads to interesting problems in estimation. We shall return to it briefly in the next chapter.

Reciprocals

Still more powerful is the reciprocal transformation, though this is rarely needed in soil studies. If the transformation is made then the mean of the reciprocal data is the reciprocal of the harmonic mean of the original data.

Angular and logit transformation

Data that consist of proportions are clearly confined in the range 0–1, or 0–100 if expressed as percentages. Their distributions are often compressed near 0 and 1. The aim of the angular transformation is to spread the distribution out near the ends of its range. If the proportion is p then the transformation is

$$\phi = \sin^{-1}\sqrt{p}, \tag{2.12}$$

i.e. the angle whose sine is $\sqrt{p}$. For this reason the transformation is also known as the *arcsine* transformation. The circumstances in which this transform is likely to help are fairly clear.

The transformation is appropriate where the proportions are based on counts, and follow the binomial distribution (see Chapter 3). It is widely used in biology. In soil studies it is likely to apply to proportions of mineral species based on counts of sand grains. However, it can also be applied empirically to other measures expressed as proportions, such as particle size fractions. Fig. 2.5 shows an example.

When the observed values fall in the range 30–70 per cent there is very little to be gained by the transformation, and it is unlikely that there will be much gain when only a small proportion of the observations fall outside this range. However, it should be noted that the angular scale is still constrained between 0 and $\pi/2$. If observations are concentrated close to 0 or 1 then the *logit* transformation will usually spread them out to approximate normality more nearly. If p is an observed proportion and $q = 1 - p$, then its logit is

$$Y = \ln(p/q), \tag{2.13}$$

and its scale ranges from $-\infty$ to $+\infty$. As with the logarithmic transformations, if any value is exactly 0 or 1 then some small value should be added to or subtracted from all the values to keep the logit within a manageable range. McBratney and Webster (1983b) used logits to analyse particle size distributions.

Faced with all these options for transformation, the scientist might well wonder which, if any, he should use. To some extent it is a matter of good judgement that comes with experience. The beginner should be prepared, therefore, to compare histograms of original observations with those transformed in several ways and to fit normal curves to them after calculating their means and standard deviations. He will not go far wrong if he transforms to a scale on which his data 'look' normally distributed, and does not transform data when he has doubts about them. Bartlett (1947) provides further guidance in a useful review.

Correcting for positive skew is often necessary. The best transformation for a single set of observations can be judged as above. When several sets of observations have been made, especially if they have been on soil of different types, or from different regions, or on which different treatments have been applied, there is an additional way of forming a judgement. Differences between the groups of soil will usually be reflected in differences among the mean values of the variable concerned, and if there is skew then the variances of sets will also tend to differ. If the skew is positive then the larger the mean, the larger the variance is likely to be, and the most appropriate transformation can be determined from the relation between the means and variances. Jeffers (1959) suggests a neat approach, as follows. On double logarithmic graph paper plot the variance of each set of observations against the mean for that set and note any trend. If the plotted points fall approximately on a line through the origin with a slope of 1, i.e. 45°, then the variance increases in proportion to the mean and the original values should be transformed to their square roots. If the trend passes through the origin and has a slope of 2 then the standard deviation is proportional to the mean, and the logarithmic transformation should be used. If the slope of the line is 4, i.e. if the variance increases as the fourth power of the mean, then the data should be transformed to their reciprocals.

Transformation enables many sets of measurements to be analysed and the results to be applied with confidence. However, the results are not so readily understood as are those from data that do not need transformation. It is helpful to transform means back to the original scales, and sometimes to express standard deviation as two values, the back transforms of $\mu - \sigma$ and $\mu + \sigma$. Confidence limits (see Chapter 3) should be treated similarly.

Transformation might smack of 'cookery'. In one sense there is an analogy: transformation renders data more digestible, and hence more useful than they would otherwise be. But as Moroney (1956) remarks wisely, the natural scientist regards it as entirely proper if he makes the transformation himself, for example from hydrogen ion concentration to pH. It is equally proper when a transformation is done for the purpose of analysis.

3

SAMPLING AND ESTIMATION

Truth lies within a little and certain compass, but error is immense.

Henry St John, *Reflections on Exile*

We have already seen that we cannot measure any property of the soil everywhere. To find out what the soil of a region is like we must be satisfied with measurements made on part of it, that is, on a sample. However, sample information is of little practical value unless it can be used as a reliable description of the region as a whole. We want information that is truly representative of the region, and a means of sampling that will ensure this, bearing in mind that soil is very variable.

In the 1930s and 1940s statisticians devoted much attention to the theory of sampling, which now provides a base for sound sampling in many spheres. The results of this work are recorded in several standard texts. Those by Cochran (1977) and Yates (1981) are especially recommended. In this chapter we consider how to apply this classical theory to obtain a good sample from an area, how to use the information from it, and some of the difficulties that arise in soil survey. Later, in Chapters 12–14, we shall introduce the more advanced Theory of Regionalized Variables developed largely by Matheron (1965, 1971) and colleagues. There we shall describe the nature of spatial variation in soil and how it affects sampling and estimation.

The population

When we want to learn something about soil we have to decide first which soil: which of all possible soil is to be included in the investigation? The kinds of question that must be asked and answered are:

1 Is the study to be of the whole soil profile, or just the topsoil, or the material below 45 cm?
2 What is the area to be covered in the investigation: the experimental station or the neighbouring locality? the administrative district or the national territory?

3 Is the study to be of one or a few particular kinds of soil only? brown earths, or brown earths and podzols? soil developed from granite? soil used for agriculture? or 'undisturbed' soil?
4 Is the study to be restricted to soil material already collected or to a set of data from an earlier investigation?

If the answer to the last question is yes then, of course, there is no sampling problem. The investigator should know how the sample was obtained, since this might affect his conclusions. The other questions might seem equally obvious and easy enough to answer, but a little reflection will show that they are not so, especially questions 1 and 3. For example, we can well understand an agronomist's interest in the topsoil, but when we come to measure its properties we need to know more precisely what constitutes topsoil. Is it to be the A horizon, or the plough layer, or that part of the soil that would be cultivated if cultivation were undertaken? And what constitutes the 'whole soil profile'? Despite years of debate, this question is almost impossible to answer in many instances. If we wait for it to be resolved on pedological grounds we might never begin our investigation. We might decide to consider all the soil down to rock. In some areas this would be feasible, but in others it would not. We might decide instead to consider that part of the soil containing plant roots: the bottom of the soil will be that depth below which there are no roots. This too might be unsatisfactory if we have some soil under grass and some under forest. So we might decide quite arbitrarily to take the top metre, or two, or three metres. What is needed in each case is a workable definition of the soil to be included in the study, a rule that can be applied easily and is reasonable and appropriate in the circumstances.

Once the investigator is quite clear what soil he wishes to include in the study and can define it operationally, that soil becomes the *population* for the purpose of the study.

Question 2 involves somewhat different problems and pitfalls. For example, suppose an assessment of the potential productivity of the soil is wanted in an area of several thousand square kilometres. It turns out to be impossible, however, to carry out trials uniformly over such an area and to prevent depredations, whether by insects, game, or local people, far from base. So the investigation takes place on local experiment stations instead. Or information about the soil of a national territory is desired, but because the area is large and manpower scarce, only a small portion can be visited, or has been surveyed when the data come to be analysed. Or a disease such as Rhizomania breaks out and access is denied to part of the region. These are factors that should become apparent at the planning stage, though some, like the outbreak of disease, might occur during execution. The result in each case is that the population actually sampled is not the population that the investigator originally intended to study. The latter is sometimes known as the *target population* to distinguish it from the actual population.

Although the soil exists as a continuous mantle, we have to regard it as made up of a number of discrete elements or individuals for sampling purposes. We shall actually observe or measure some of these individuals in the field or in the laboratory. What constitutes an individual is more or less arbitrary and is determined largely by convenience. The size of the individual might be chosen less arbitrarily as the volume of soil occupied by the roots of a single plant, or the volume of soil deformed under a given load. Individuals are often sampling cores from the topsoil or from the whole profile with finite size in three dimensions. They might be pedons, though the pedon (the volume of soil beneath a metre square of the soil surface) is usually too large for practicality, and its extended definition to embrace cyclic variation is statistically quite unsatisfactory. Or they might be faces of pits with length and depth and just sufficient thickness for their morphology to be ascertained. In one sense it is immaterial what the individuals are, for in any large area there are so many that we may regard their number as infinite, and this simplifies the statistical treatment. However, the degree of variation that we observe between individuals depends very much on their size. Variation among 2 cm auger cores could include substantial effects caused by burrowing small animals. Such variation is likely to be smoothed out if the individuals are 1 m cubes. In general, for a given region, the larger the individuals on which the measurements are made, the more variation is encompassed within them, which is effectively hidden, and the less there is between them to be revealed by analysis. It is important, therefore, to specify at the outset the dimensions of the individual—its size, shape and orientation, known as its *support*, and to adhere to that decision throughout a survey. The support should always be stated when reporting results.

In a particular study we are not usually interested in the whole soil so much as certain characters of it; its pH or colour or clay content, for example. Therefore, we may regard the soil population as a set of individuals each possessing a limited number of characters.

Sampling

Having defined the population for the purpose of a study, the next task is to choose a method of sampling it. One way is for someone familiar with that soil to select 'typical' or 'average' occurrences. Measurements can then be made on these. Such sampling is sometimes known as *purposive sampling*, and is often used to describe individual classes of soil. When a population is very variable and resources allow the soil at only one or two sites to be examined, this method can provide a more accurate description than others. It relies heavily on personal judgement. There is no way of knowing just how good that judgement is, or of communicating the expert's confidence in his choice. Further, there is a strong risk that an expert's choice will be biased to some extent; that is, it will give preference to some part of the population at the expense of the rest. Bias is almost always present in human judgement, and it cannot be avoided either by training or by conscious effort. Yates's (1935) work on bias at Rothamsted showed how serious this risk is. Selection of the 'typical' is not a safe way to sample.

Sampling from 'convenient' spots is even worse. Soil surveyors are well aware how atypical the soil is near to gates and other places of access to agricultural land. Furthermore, if parts of an area are avoided because they are too inconvenient to reach then the sampled population is not the target population.

The only sure way of avoiding bias is by *random* or *probability* sampling. In common parlance 'random' tends to mean haphazard or casual. For example, a participant in a card trick might be asked to 'take a card at random'. A soil sample chosen casually is quite as likely to be biased as one chosen purposively, and will not do. Random sampling means choosing individuals from a population in such a way that all members of that population have an equal chance (probability) of being chosen.

Although it is easy to state this principle, it is far from obvious how to achieve random sampling in practice, and paradoxically it can be achieved only by following firm rules.

To sample soil randomly we need two things: first, a means of identifying individuals in the soil population; second, a means of selecting individuals once we have decided how to identify them. If the population consists of pots of soil in a greenhouse or fields in a parish, each pot or field is identified by a unique number. These numbers are then matched to a sequence of random digits, which may be read from one of several published tables of random digits (e.g. Fisher and Yates 1963; Rand Corporation 1955), or may be generated by machine for the purpose. There are several well tried computer routines for producing pseudo-random numbers. Thus, if we wish to choose at random 10 pots out of a total of 240 in a greenhouse, we number the pots from 1 to 240. Then we can take a row of digits in a table and read off digits in sets of three until we have ten unique three-digit numbers less than or equal to 240 and excluding 0. The pots with these numbers then constitute the random sample.

When the population is the soil of an area and the individuals are sampling cores or profile pits this technique is too cumbersome. An area of 100 km^2, for example, contains 10^{12} disjoint 10 cm × 10 cm elements that we might choose; and there is in principle an infinite number of profile faces that we could expose. The most convenient alternative is to regard the individuals as points in a plane and to identify them by reference to a rectangular grid whose origin and orientation are known. Points to be included in a sample can then be chosen by selecting pairs of numbers from a table of random digits and using these as the coordinates of the points on the reference grid. The result is a *simple random sample*. Other ways of obtaining unbiased samples that are more efficient, give more even coverage, and are preferred for mapping are described later.

The degree of resolution, that is the number of significant digits in the coordinates, requires some thought. Referencing to 10 cm is needed to give equal probability of selecting 10 cm × 10 cm elements once the position of the grid is fixed. For areas of more than 1 ha this is impracticable and unnecessary. Resolution to one-thousandth of the length of either axis of the grid is more than adequate for any one area, since this will give nearly one million possible sampling points in the field. If a surveyor cannot locate himself to better than 100 m there is no point in his obtaining more precise coordinates than this. Nor is it necessary to locate points with great accuracy, provided bias is not introduced at this stage and an accurate map is not to be constructed from the observations.

It can be very convenient to identify sampling points by reference to a standard geographic grid. The British National Grid is one such. It is an orthogonal square grid with principal divisions at 1 km intervals on most British maps. Latitude and longitude may also be used, though if the area is very large the parts nearest the poles will be more densely sampled than those nearer the equator, and some compensatory adjustment will be needed either to the sampling procedure or to the results of the survey. Once it is decided to use a particular grid and degree of resolution, most potential sampling points are excluded from the sample, but this is scarcely likely to matter.

Location in practice

Having decided on a sampling scheme and specified where each observation is to be made, we then have to find each point in the field. This can be done accurately by triangulation, but as already noted such accuracy is rarely necessary and, provided we avoid bias, we may use more rough and ready methods. The following procedure will usually suffice. Choose from a map or air photograph some easily recognizable feature that is near the sample point to be reached, and go there. Estimate from the map the direction and distance to the sample point. Then, using a compass to give direction, measure out the estimated distance. If the ground surface appears uniform the sample point may be located by pacing. However, if the ground is variable, especially close to the sample point, then beware of introducing bias by shortening or lengthening paces to avoid areas that may seem atypical, unpleasant (e.g. swamp or thicket), or difficult to dig or record (e.g. stony ground). It is safer to use a chain or tape in these circumstances.

Inevitably snags arise, and the investigator must be prepared in advance to deal with them. *Ad hoc* decisions made in the field as they arise can easily lead to bias. One of the most common difficulties occurs when a sampling point happens to be on a road, in a river, or in a farm building. What should the surveyor do? The answer depends on the nature of the investigation. If the aim of the survey, even in part, is to assess *how much* soil there is or what proportion of an area is usable soil then such sample points must be accepted and the findings there recorded. If the survey is concerned solely with, say, agricultural soil then the investigator has a choice. He may simply ignore the point of may follow a predetermined rule for choosing a substitute near by. Such a rule might be: when the selected site is not on agricultural soil choose as substitute a site x metres away on a bearing θ degrees. The values of x and θ may themselves be chosen from a table of random numbers and replaced by a new pair when once used. Alternatively, x may be fixed and θ chosen as 0, 90°, 180°, 270° in turn. The distance x should be kept to a small proportion of the average distance between neighbouring sampling points. These refinements are of little advantage if simple random sampling is used, and the first course is the safest. But they are valuable when even coverage is desired for mapping.

A surveyor may go astray more readily when a sample lies in a hedgerow or gateway, or near a ditch that has recently been cleared or deepened. It is easy to regard such situations as atypical and to omit them from the sample. The matter must be resolved by deciding in advance whether such sites are part of the target population. If they are then they are accepted when they occur in the sample; otherwise they are rejected.

Another difficulty arises when a landowner refuses to grant access to his land. If only a few selected points cannot be visited for this reason then they may be omitted from the sample without serious consequence. If several landowners deny access then the investigator should consider whether there is some relation between the kind of land and owners' attitudes to survey. This is by no means as silly as it might seem at first. If there is any such relation then the population actually sampled will not be the target population, and the fact should be recorded. The same holds if any large tract of land cannot be visited for this or other reasons.

Yet other problems arise when a farmer is willing to co-operate, but with restrictions. The investigator might reasonably comply with a request not to dig up the front paddock where the horses are kept. To keep out of the standing crops, on the other hand, could bias results, and the investigator should consider returning there after harvest. Numerous small snags such as these can occur during a survey and prevent the sample from being truly random. The course taken over many of them is unlikely to affect the outcome seriously provided the investigator is always on the look-out for bias.

Estimation and confidence

Quantitative data from a sample can be summarized just like those from any other set of individuals. Thus, we can calculate the mean and variance and higher-order moments. Provided that the sampling is sound, we can also use these values to *estimate* the corresponding parameters in the population. Using Greek letters for population parameters and Roman ones for the sample values, $\bar{x}$ estimates μ and s^2 estimates σ^2. The estimates apply strictly to the population that was actually sampled, and if that differs from the target population then any inference extended to the target or other population rests solely on the judgement of the investigator.

In a sample of variable material such as soil there are inevitably differences between the individuals included in the sample and those excluded. As a result, sample estimates differ more or less from the true population values. In a random sample such differences are attributed to sampling error, and they are distinguished from errors arising from bias. Sampling error can usually be reduced by increasing the size of the sample, so that the estimate converges on the population value. Similarly, in the absence of bias the average of sample estimates converges to the true value as the number of samples is increased.

We must now make a formal distinction between two kinds of bias. That mentioned so far is the systematic error that arises from faulty selection. When present it is a constant source of inaccuracy that is unaffected by the size of the sample. The same kind of bias can result from slack laboratory technique and the use of poorly calibrated instruments. The second kind of bias is that associated with an estimator. An estimator is said to be biased if its average value taken over all possible random samples differs from the population parameter being estimated. An unbiased estimator is one that gives the true value on average, and is generally regarded as desirable.

Since sampling error is inevitable, we need to be able to assess it and to take it into account when applying sampling estimates to populations. To do so we introduce a further formality, namely that of expectation, which we denote by E. The mean value of a population is the expected value of any member of it, thus

$$\mu = \mathrm{E}[x].$$

Likewise, the population's variance is the expected squared deviation from the mean:

$$\sigma^2 = \mathrm{E}[(x-\mu)^2].$$

As above, any estimate of the population mean based on the mean of a sample, $\bar{x}$, will be in error, and its variance is

$$\sigma_{\bar{x}}^2 = \mathrm{E}[(\bar{x}-\mu)^2]. \tag{3.1}$$

This is known as the *estimation variance*, and its square root, the standard deviation of the sample mean, is the *standard error*. They represent the variation in $\bar{x}$ that would occur if repeated samples of size n were taken. For a simple random these quantities themselves are estimated readily from the sample statistics by

$$s_{\bar{x}}^2 = \hat{\sigma}_{\bar{x}}^2 = s^2/n, \tag{3.2}$$

where

$$s^2 = \frac{1}{n-1}\sum_{i=1}^{n}(x_i-\bar{x})^2. \tag{3.3}$$

Finite population correction

The above expression for standard error assumes that the sample is drawn from an infinite population. If the population is finite—say, pots of soil in a greenhouse—and the sample is more than about one-twentieth of the population then a correction should be applied. For a sample of size n drawn from a population of N individuals, the sampling fraction is

$$f = \frac{n}{N}.$$

The standard error is then corrected by multiplying it by $(1-f)$. The quantity $1-f$ is known as the *finite population correction*, or f.p.c. Its effect is to diminish the calculated standard error somewhat. The populations of concern in soil survey are usually infinite, or practically so, and in this case the f.p.c. can be ignored. We shall ignore the f.p.c. henceforth.

Degrees of freedom

Equation (2.2) for the variance of a finite population contains N alone in the denominator. By analogy we might have expected to replace it by n when calculating the variance of a sample, but in equation (3.3) we use

$n-1$, and this needs explanation. The variance of a population is the expected squared deviation from the population mean μ. When sampling, however, we do not know μ: we have only our sample estimate $\bar{x}$, which is more or less in error. Further, calculation of the mean as

$$\bar{x}=\frac{1}{n}\sum_{i=1}^{n}x_i$$

ensures that $\Sigma_{i=1}^{n}(x_i-\bar{x})^2$ is less than it would be if $\bar{x}$ were replaced by any other value, and in particular is less than $\Sigma_{i=1}^{n}(x_i-\mu)^2$. Thus, if the variance is calculated from sample data using n instead of $n-1$ in (3.2) then it will underestimate σ^2: it is a biased estimator of the population variance. Equation (3.3) ensures that σ^2 is estimated without bias.

It might help to grasp the difference between the variance of a finite population and a sample estimate if we consider what happens in the limit when N, the size of the population, equals 1. A population consisting of a single measurement clearly has zero variance: $x_1=\bar{x}$, $(x_1-\bar{x})^2=0$, and the whole expression $(x_1-\bar{x})^2/N$ is zero. If we have a sample of size 1 the fact that its variance is zero tells us nothing about variation in the parent population. The formula

$$\frac{1}{n-1}\sum(x_1-\bar{x})^2$$

reduces to 0/0, which expresses this lack of information.

The quantity $n-1$ is known as the number of *degrees of freedom* in the estimate of the variance. This concept arises, as above, because the variance is calculated from the squares of deviation from the sample mean, which is itself calculated from the same sample data, and so leaves one fewer independent quantities than there are individuals in the sample. Using degrees of freedom also simplifies the mathematics in more complex analyses of data and in tabulating sampling distributions. We shall use it in the remainder of this book for calculating both population variances and their sampling estimates.

Central limit theorem

Before considering how we can use the standard error, we shall discuss another important feature of sampling. When samples are drawn at random from a population their means tend to be distributed in a more nearly normal fashion than do the individual values. Further, however the original population is distributed, provided that its variance is finite, the distribution of the sample mean approaches normality as the size of the sample increases. This result is known as the *central limit theorem*. Its importance lies in the fact that it allows a large body of statistical theory to be applied to practical problems even though the underlying population distributions are far from normal.

Confidence limits

The standard error measures the precision attained in a survey. We can use it to determine the range within which the true population mean can be said to lie with any desired degree of confidence. Assuming that the sample is large and its mean $\bar{x}$ is normally distributed, the confidence limits for the population mean are

$$\bar{x} - z\sqrt{\frac{s^2}{n}} = \bar{x} - zs/\sqrt{n} = \bar{x} - zs_{\bar{x}}, \text{ the lower limit,}$$

and (3.4)

$$\bar{x} + z\sqrt{\frac{s^2}{n}} = \bar{x} + zs/\sqrt{n} = \bar{x} + zs_{\bar{x}}, \text{ the upper limit.}$$

The quantity z is the value of the normal deviate for the desired level of confidence, and can be obtained from tables. Frequently used values are

Confidence (%)	75	80	90	95	99
z	1.15	1.28	1.64	1.96	2.58

For example, the probability that the interval $x - 1.64s/\sqrt{n}$ to $x + 1.64s/\sqrt{n}$ embraces the true mean is 0.9, and we can be '90 per cent confident' that the true mean lies in this range.

When n is small confidence limits may still be calculated in this way if a precise estimate of the population variance, σ^2, is already available, as it might be from a previous study. However, when the only estimate of σ^2 is s^2, the sample variance, then z must be replaced by Student's t. The quantity t is the deviation of the estimated mean from the population mean expressed as a ratio of $s/\sqrt{n}$; thus,

$$t = \frac{\bar{x} - \mu}{s/\sqrt{n}}. \qquad (3.5)$$

We cannot calculate t because we do not know μ, but the sampling distribution of t has been worked out, and values can be obtained from published tables for n up to 120. Thus, to find the 90 per cent confidence limits for a mean estimated from a sample of size 10, for example, we find the corresponding value of t with $n-1=9$ degrees of freedom. It is 1.83, and confidence limits are therefore

$$\bar{x}-1.83s/\sqrt{10} \quad \text{and} \quad \bar{x}+1.83s/\sqrt{10}.$$

The question naturally arises: how small is 'small'? It has at least two aspects. The first concerns whether to use t instead of z. For n greater than about 60, z and t are so nearly equal that it is largely immaterial which is used. The t distribution holds exactly only when the individual values are normally distributed, though moderate departure from normality does not affect it seriously. The second aspect is that if the original population is not normally distributed then the distribution of the sample mean might not be sufficiently close to normal to justify basing confidence limits on it. The sample must be large enough for probabilities calculated from the normal distribution to apply. There is no easy answer to this aspect of the question, however. In most instances involving measurements on soil the problem can be avoided by applying a normalizing transformation to the original data beforehand.

Sometimes only one confidence limit is of interest. For example, a mining company might sample the soil of an area with a view to ore extraction (for example, bauxite for aluminium). There would be some critical concentration, say c, of the metal less than which extraction would be uneconomic and greater than which extraction would be profitable. If the sample mean exceeded c then the company would wish to know the probability that the true mean exceeded this value. Thus it would calculate the value for z or t to satisfy

$$c=\bar{x}-zs_{\bar{x}} \quad \text{or} \quad c=\bar{x}-ts_{\bar{x}}, \tag{3.6}$$

and would determine the corresponding level of confidence from the appropriate table. If this exceeded 99 per cent then it might proceed to extract. If not, then the company would probability carry out further sampling to increase the precision of the estimate. A public health authority, on the other hand, might be more interested in the toxicity of a metal in soil. It too could have some critical concentration of metal that served as a warning. In this instance, however, the authority would become increasingly alarmed the nearer the upper confidence limit approached the critical value.

Note, when calculating single confidence limits, that the probability that the true mean lies beyond one limit is half that of its lying beyond both limits. Thus, in the example above with $n=10$, the probability that the true mean is less than $\bar{x}-1.83\sqrt{(s^2/10)}$ is 5 per cent.

Estimating from the lognormal distribution: special problems in geochemistry

We now return briefly to a matter that we had to leave in Chapter 2, namely the lognormal distribution and its implications. We have already seen some of the advantages of normal distributions: they are mathematically tractable, they lead to estimates of confidence, and there is no doubt about the centre or average because the mean, median, and mode coincide. We may now add another advantage: sample means and variances are efficient, in the sense of being precise estimators of the population parameters. We often transform data that are not normally distributed in order to gain these advantages. Having done our computation, however, on what scales should we report our results, and how should we transform results back to the original scales if we have to?

The soil chemist who uses the pH scale to measure acidity probably never considers transforming his results to hydrogen ion concentration. The mean pH is a quantity he can grasp easily. In the case of the matric suction the mean pF also makes sense, and taking logarithms enabled it to be estimated efficiently. In geochemistry, however, where the lognormal distribution is so widespread, much more care is needed. In an agricultural context such as that for which Fig. 2.3 was obtained, the mean of the logarithms of the concentration of available copper and its exponent, the geometric mean, may still be satisfactory: they estimate the median and provide an index of deficiency, sufficiency, or excess. In mining and exploration geochemistry, however, the mean of the logarithms or the geometric mean are not the estimates that are wanted. Here the investigator and his client usually want to know how much metal or other mineral an ore body contains. This quantity is the straight sum of the amounts in any complete set of subdivisions of the ore body. Before extraction the mine planner will want an estimate of the quantity, and this must be based on the arithmetic mean of the amounts present in samples. In other words, the value from which the total quantity of mineral can be calculated should be an unbiased estimate of the expectation. Clearly, the geometric mean is not such, and if we are to normalize our data in the interests of efficiency then we must convert the results back to the original scale in a way that avoids bias. For the lognormal distribution the appropriate conversion is

$$x^* = \exp(\hat{\mu} + \tfrac{1}{2}\hat{\sigma}^2) \tag{3.7}$$

where x^* denotes our estimate of the arithmetic mean by the method, $\hat{\mu}$ is the estimated mean of the natural logarithms, and $\hat{\sigma}^2$ is the estimated variance of the natural logarithms.

Let us take the data for copper in the topsoil of south-east Scotland, Fig. 2.3, as an example. The mean and variance in the logarithms are estimated as $\mu = 0.6244$ and $\sigma^2 = 0.3443$ respectively, giving $x^* = 2.217$. This is close to the arithmetic mean of the measurements, $\bar{x} = 2.225$, which, with more than 2000 data and assuming unbiased sampling, will be close to the true mean for the region. The geometrical mean, however, is only 1.867.

Unfortunately, for strongly skewed distributions the problem is still not entirely resolved, because the presence of a few erratic values in the long tail of the distribution can influence the estimate of the mean very strongly. As a result the usual estimate is inefficient: either the estimate will have a large standard error, or a very large sample must be taken. For the lognormal case Sichel (1966) provides a solution. The geometric mean is multiplied by a factor that depends on the size of the sample and on its variance (on the logarithmic scale). These factors are tabulated (Sichel's Table A), and further tables (Sichel's Tables B and C) enable confidence limits to be gauged. Sichel's tables and worked examples are quoted by David (1977), who discusses at length the application of the lognormal distribution to ore estimation. Readers who wish to estimate concentrations of constituents that are lognormally distributed or approximately so should consult these two works.

Chi-square and confidence limits of variance

In many investigations the mean values of the properties studied are clearly the important ones, and variances are of interest only in so far as they provide measures of confidence. In soil survey and classification the variances themselves are often of interest, and we then wish to know how precisely we have estimated them. The sampling distribution of variance is closely related to the chi-square (χ^2) distribution, which is introduced now.

Let $z_1, z_2, \ldots, z_m$, be m values drawn at random from a normally distributed population with a mean of zero and unit variance. The quantity χ^2 is then the sum of their squares; that is,

$$\chi^2 = \sum_{i=1}^{m} z_i^2. \tag{3.8}$$

When sets of m values are drawn repeatedly and independently from such a population the values of χ^2 are distributed in a characteristic way, the chi-square distribution with m degrees of freedom. This distribution is continuous. Its shape depends only on the number of degrees of freedom, which is also the mean of the distribution. For few degrees of freedom it is very skew: since χ^2 is a sum of squares and cannot be less than zero, occasional large absolute values of z give it a long upper tail. It becomes less skew, however, as the number of degrees of freedom increases. For m exceeding 30 the distribution of $\sqrt{(2\chi^2)}$ is close to normal with a mean of

$\sqrt{(2m-1)}$ and variance of 1. The chi-square distribution has many applications, and it is tabulated in several standard texts.

The relation between the distribution of χ^2 and variance will be clear if we consider a sample of size n from a normal population with known mean μ and variance σ^2. The variance in the sample is

$$s^2 = \frac{1}{n}\sum_{i=1}^{n}(x_i - \mu)^2. \tag{3.9}$$

If we divide through by σ^2 we have

$$\frac{s^2}{\sigma^2} = \frac{1}{n}\sum_{i=1}^{n}\left(\frac{x_i - \mu}{\sigma}\right)^2. \tag{3.10}$$

The term in brackets is a random normal deviate, hence

$$s^2/\sigma^2 = \frac{1}{n}\chi^2$$

and

$$\chi^2 = \frac{ns^2}{\sigma^2} \tag{3.11}$$

with n degrees of freedom. Usually μ is estimated by $\bar{x}$, and s^2 is an estimate with $n-1$ degrees of freedom. In these circumstances

$$\frac{(n-1)s^2}{\sigma^2} = \chi^2. \tag{3.12}$$

This quantity is distributed as χ^2 with $n-1$ degrees of freedom, and we can use the expression to determine confidence limits for variances, provided the original population is normal.

The published chi-square tables contain the values of χ^2 that are exceeded with given probabilities. Table IV of Fisher and Yates (1963), for example, lists χ^2 for 14 probabilities P ranging from 0.99 to 0.001.

If we read values of χ^2 given for $P = 0.05$ then the probability that a randomly chosen value lies between these two is 0.90, i.e. $0.95 - 0.05$. If now we rearrange (3.12) we have

$$\sigma^2 = (n-1)s^2/\chi^2, \tag{3.13}$$

and by inserting the 0.95 and 0.05 values of χ^2 we obtain a 90 per cent confidence interval for σ^2 as

$$\frac{(n-1)s^2}{\chi^2_{0.05}} \leqslant \sigma^2 \leqslant \frac{(n-1)s^2}{\chi^2_{0.95}}. \tag{3.14}$$

Incidentally, since the mean value of χ^2 is equal to $n-1$, the number of degrees of freedom, equation (3.13) shows that s^2 is an unbiased estimator of σ^2.

As an example, consider the pF values in Table 2.1 as a sample from the plot of land in which the instruments were inserted. The sample variance is 0.08297 and there are 26 degrees of freedom. The values of χ^2 at $P=0.05$ and $P=0.95$ are respectively 38.90 and 15.38; 90 per cent confidence limits for σ^2 are therefore

$$26 \times 0.08297/38.90 = 0.0554$$

and

$$26 \times 0.08297/15.38 = 0.1402.$$

Their square roots are the corresponding limits for the standard deviation; thus, with $s=0.29$, σ lies between 0.235 and 0.374 unless we have been unlucky enough to obtain a sample with unusually small or unusually large variance, the chances of which are 1 in 10.

The standard error in planning

When planning a survey an investigator naturally wants to known what effort is likely to be required—how big a sample should he take? When he wants to estimate the mean value of some soil property the above methods can often help. The investigator must make two decisions. He must decide what degree of confidence he wishes to place in the estimate, and how wide a confidence interval he can tolerate. He also needs a prior estimate of the population variance σ^2. Assume that the investigator is happy with a 90 per cent level of confidence and sets limits l on either side of the mean. Then

$$l = 1.64\sigma/\surd n, \tag{3.15}$$

which can be solved to find n. His estimate of σ^2 is often little better than a guess, so no great accuracy can be expected.

The relation between standard error and sampling effort is illustrated in Fig. 4.1 below. Precision can be increased substantially by modest increases in the size of small samples. Once the sample size is more than 10–15, however, the standard error declines very slowly in response to increases in sample size.

Area sampling schemes

Simple random sampling

Fig. 3.1 is an example of a simple random sample of 49 points. The coordinates of the points were taken from the first four columns of Fisher and Yates's (1963) table of random digits. The first pair of digits in each row identifies the x-coordinate and the second pair the y-coordinate for each point. Notice how the points seem to cluster, and that two parts of the area, bottom centre and the upper half just right of centre, are comparatively sparsely sampled.

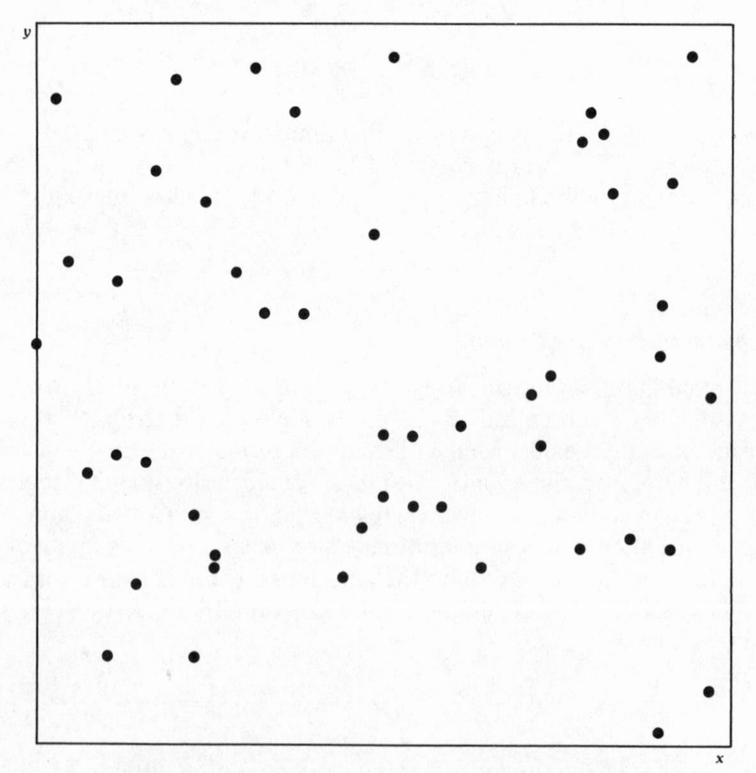

FIG. 3.1 A simple random sample of 49 points

Uneven coverage often makes simple random sampling inefficient (in a sense to be explained later) for describing the soil of a region, and several techniques are available for choosing points with a more even distribution.

Stratified random sampling

In this method the total area is divided into cells (they are usually squares, but they can be elongated rectangles, triangles, or hexagons), within each of which one or more points is chosen at random. Each cell can be regarded as a stratum for sampling, and the method is one example of a more general technique known as *stratified* random sampling. Fig. 3.2 is an example with two points in each of 25 squares chosen using columns 1 for x and 3 for y from p. 134 of Fisher and Yates's (1963) tables. Cover is still somewhat uneven. Several groups of points appear to be clustered, and there seems to be one substantial gap (just below centre).

Provided all cells are equal in area and contain the same number of

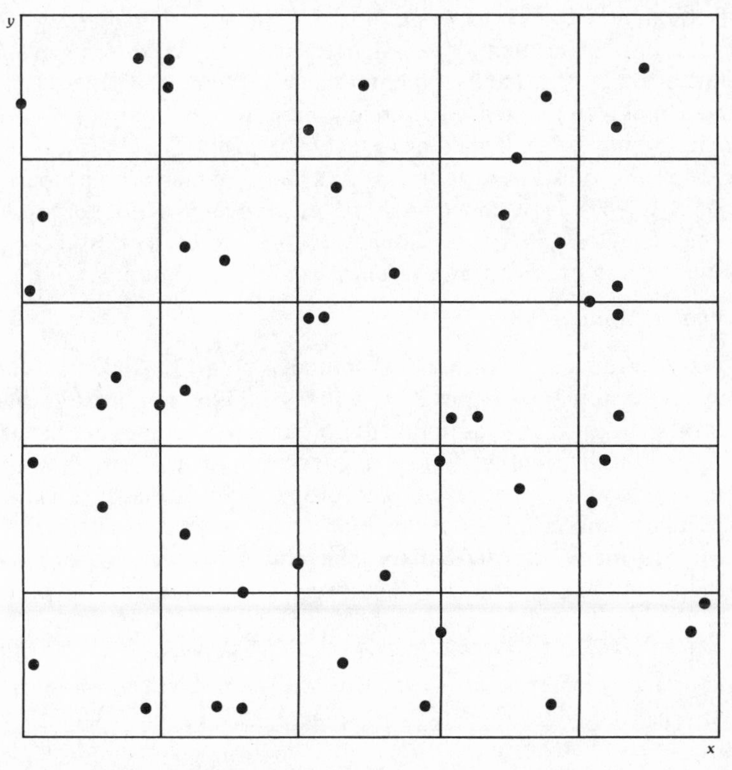

FIG. 3.2 A stratified random sample of 50 points with two points chosen randomly within each of 25 squares

sample points, the mean $\bar{x}$ of all observations estimates the population mean without bias, as before. However, the standard error is now estimated from the pooled within-cell variance according to

$$s_{\bar{x},st} = \frac{1}{m}\sqrt{\left(\frac{\sum_{k=1}^{m} s_k^2}{n_k}\right)}, \tag{3.16}$$

where m is the number of strata (cells), s_k^2 is the variance of n_k observations within the kth stratum, and the subscripts $\bar{x}, st$ refer to the mean of the stratified sample. There must be at least two points per cell to estimate it.

The term $(1/m)\,\Sigma_{k=1}^{n} s_k^2$ represents the pooled variance *within cells*, which we may denote by s_w^2, and replaces s^2, the variance for the whole sample, in the formula for the standard error. Thus, the sampling error in a stratified sample derives solely from the variation *within* the strata. Differences among strata make no contribution. Generally speaking, the smaller the area of land, the less the soil varies within it. So when an area is divided into cells as above, s_w^2 is usually less than s^2, and $s_{\bar{x},st}$ is less than $s_{\bar{x}}$ for the same-sized sample. This represents a gain in precision that results from stratification, and may be expressed as $s_{\bar{x}}^2/s_{\bar{x},st}^2$, the *relative precision* of the method. It also means that fewer individuals are needed to achieve a given standard error by stratified sampling than would be required in a simple random sample. In this sense stratified sampling is the more efficient, and its efficiency may be expressed as $(n_{\mathrm{random}}/n_{\mathrm{stratified}})$.

Experience suggests that, of the total variance of many soil properties in areas of 10 to 1000 ha, as much as a quarter, or even half, can occur within a few square metres. Surveyors should therefore expect modest rather than dramatic gains in precision or efficiency from stratification.

Systematic sampling

Completely even coverage, and potentially greater efficiency, can be obtained by systematic sampling, in which sampling points are located at regular intervals on a grid, as in Fig. 3.3. Systematic samples are the easiest to select, and if the sampling grid is aligned with the map grid they are the most easily located and indexed. The preparation of maps is also easier from regularly spaced observations than from irregularly scattered ones. However, as many statistical texts take pains to warn, there could be periodicities in the population. If these were to coincide with the period of the grid, or were some simple multiple or fraction of it, then dire results would surely follow. The risk of introducing bias in this way when sampling an orchard or plantation is clear. But the period of variation and its direction are equally obvious, and a grid can be chosen with quite unrelated spacing and orientation. A field that has been underdrained may possess a less obvious pattern, and sampling it systematically could be hazardous. It is less often realized that if a systematic sample is small, with

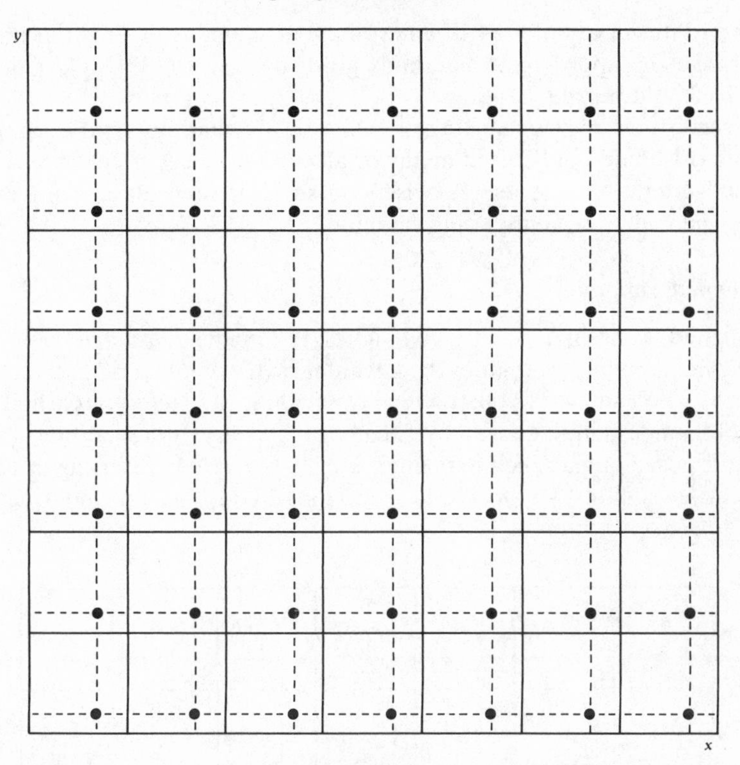

FIG. 3.3 A systematic sample of 49 points. The coordinates of one point are chosen at random; all others are at fixed intervals equal to the cell width in both horizontal and vertical directions

say 20 members, its period does not need to be at all closely tuned to the period of variation in the population or its harmonics for bias to occur. But when a survey extends over much larger areas than the single field, orchard, or plantation, and the sample is large, the taboo on systematic sampling seems quite unjustified. No regular soil variation with a period of more than a few tens of metres or of more than local extent has so far been identified. Bias can also occur when systematic sampling is applied to an area across which there is some general trend.

There is another disadvantage of systematic sampling. On its own the method gives no entirely valid estimate of sampling error, since the sampling points are not located at random within the strata. Nevertheless, there is much empirical evidence to show that systematic sampling is often considerably more precise than simple random sampling and somewhat more precise than stratified random sampling. The early theoretical studies by Yates (1948) and Quenouille (1949) supported this experience in certain circumstances, but the Theory of Regionalized Variables (Matheron 1965) was needed to provide a full explanation and the means of analysis. We return to this in Chapter 14. If an estimate of sampling error is required at this stage the approximate methods given by Yates (1981) and Cochran (1977) should be adequate.

Clearly, it is impossible to say categorically that systematic sampling should or should not be used in any or all circumstances. Each case should be judged on its merits. A wise course is to consult a sympathetic statistician when contemplating sampling.

Unaligned sampling

Unaligned sampling, or, to give it its full name, *stratified systematic unaligned sampling*, combines the advantages of a regular grid and randomization. The sample is constructed by dividing the survey area first into cells (again usually square) by means of a coarse grid. A fine grid is superimposed on each cell in turn as a reference system. Starting in one of the corner squares, say top left, a horizontal coordinate x and a vertical

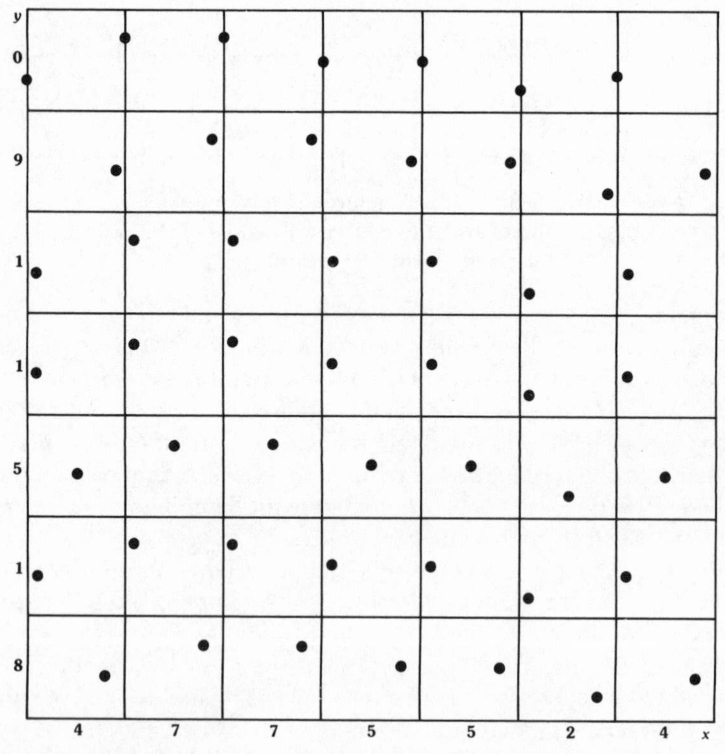

FIG. 3.4 An unaligned sample. The horizontal coordinates in each is constant for any one row and is given in the left-hand margin. The vertical coordinate is constant for any one column as is given along the bottom of the figure

coordinate y are chosen randomly and the point located by means of the reference grid. The reference grid is moved to the next cell of the top row, the *horizontal* coordinate is retained but a new random value for y is chosen, and the point in this cell is then located. The procedure is repeated for all the remaining cells in the top row, keeping the x coordinate constant but choosing y afresh each time. For example, in Fig. 3.4 a square area is divided into 49 square cells. The reference grid for each cell is 10×10 with its origin at the bottom left corner. The x and y coordinates of the point in the top left cell are 0 and 4. The x coordinate of all cells in the top row is 0, but y changes along the row and has values $4, 7, 7, 5, \ldots$. Sample points are chosen similarly in the left-hand column, but this time holding the vertical coordinate constant and varying x. In the first column of Fig. 3.4 $y=4$ and x is $0, 9, 1, 1, \ldots$. The position of the point in each remaining square is then determined by the x coordinate of the point in the left-hand square of its row and the y coordinate of the point in the uppermost square of its column. There is thus a constant interval both along the rows and down the columns, and this ensures even covereage without alignment.

This method provides unbiased estimates of means. As with systematic sampling, there is no simple way of estimating the sampling error, though Quenouille (1949) and Das (1950) have shown that unaligned sampling can be more precise than either stratified random or regular grid sampling in some circumstances.

Unequal sampling

In the methods described above all points initially have equal probability of being chosen. However, if one part of an area is known beforehand to be more variable than the rest then intuitively it makes sense to give this part more attention. The corollary is even more evident: to sample a seemingly uniform and endless plain at the same intensity as would be needed for an area of intricately dissected and contrasting sediments makes no sense. Clearly, it is desirable to be able to vary the sampling intensity. This can be done, retaining all the advantages that probability sampling confers, provided that the different probabilities of selection are known for each part of the area. This is another case of stratified sampling. In Chapter 4 we shall discuss other kinds and uses of stratification.

Suppose we divide an area on its visual appearance (on geology, or relief, or some combination of these) into several strata, $k=1, 2, \ldots, m$, which we sample with different intensities. To combine data from several strata we assign a weight w_k to each stratum such that

$$w_k = \frac{\text{area of stratum } k}{\text{total area}}.$$

Then the mean, μ, for the whole area is estimated by

$$\bar{x}=\sum_{k=1}^{m} w_k \bar{x}_k, \tag{3.17}$$

where $\bar{x}_k$ is the mean for the kth stratum. Its standard error is given by

$$s_{\bar{x},st}=\sqrt{\left(\sum_{k=1}^{m}\frac{w_k^2 s_k^2}{n_k}\right)}, \tag{3.18}$$

where s_k^2 is the variance and n_k is the number of sample points in the kth stratum.

Sampling for proportions

A sample survey can often be undertaken to determine the proportion or amount of land in a region that possesses some attribute or belongs to some particular class of soil. For example, what proportion of the soil is calcareous or how much land is suitable for irrigation, or for growing wheat, or whatever is of interest? Questions of this kind are especially important in economic planning. They can be answered by a sample survey, provided, of course, that for each sampling point the surveyor can decide whether the soil possesses the attribute of interest.

Let us assume that an area has been sampled at random, n sites have been recorded, and a sites possess some attribute X. Then $p=a/n$ is the proportion of observed sites possessing X. This is an unbiased estimate of the proportion, P, of the area possessing X. We shall define $Q=1-P$ as the proportion lacking X, and $q=1-p$ as its sampling estimate.

Since we can regard an attribute as a variable that takes only the values 1 (when the attribute is present) and 0 (when it is absent), the proportion p is equivalent to the sample mean

$$p=\frac{a}{n}=\frac{1}{n}\sum_{i=1}^{n} x_i=\bar{x}. \tag{3.19}$$

It can be shown algebraically that the variance of p is PQ/n, and its estimate is

$$s_p^2=\frac{pq}{n-1}. \tag{3.20}$$

Its standard error is therefore s_p.

Cochran (1977) points out that the variance of a proportion is usually given as pq/n, but that this is biased. However, as we shall see, n usually needs to be large, and the difference between results calculated by the two formulae is then so small that it can be ignored safely.

Binomial distribution

When samples of size n are drawn from a population in which the proportion possessing a given attribute is P, the probability Pr_a of obtaining any given number of individuals a possessing the attribute can be calculated as

$$\mathrm{Pr}_a = \frac{n!}{a!(n-a)!} P^a Q^{n-a}. \tag{3.21}$$

The probabilities for all possible values of a are the successive terms in the expansion of $(P+Q)^n$, in which the first part of each is the familiar binomial coefficient. The distribution of a, and therefore of p, is said to follow the *binomial distribution*.

Confidence limits

Confidence limits for P can be obtained from sample evidence by calculating the probabilities of each value of a from the above formula. Alternatively, the probabilities can be obtained from published tables (National Bureau of Standards 1950; Romig 1952; Harvard Computation Laboratory 1955). Note that since a must be integral the exact level of confidence cannot be chosen in advance.

If n is at all large the method is cumbersome even when the probabilities are read from tables, and it is usual to determine approximate limits from the normal distribution instead. This gives limits for P as

$$p \pm \left\{ z\sqrt{\left(\frac{pq}{n-1}\right)} + \frac{1}{2n} \right\}, \tag{3.22}$$

where z is the normal deviate for the chosen probability. The normal distribution is continuous whereas the binomial distribution is not, and the term $1/2n$ represents a correction for continuity.

The central limit theorem guarantees good approximation to the normal distribution when n is large or when p is between about 0.3 and 0.7. Table 3.1, taken from Cochran (1977), indicates when the normal approximation may be used safely. It shows especially the importance of taking

TABLE 3.1 The smallest value of np and sample size for use of the normal approximation

P	np, or $nq = n(p-1)$, whichever is the smaller	n
0.5	15	30
0.4	20	50
0.3	24	80
0.2	40	200
0.1	60	600
0.05	70	1400

Source: Cochran (1977)

a large sample if P is near 0 or 1, when the binomial distribution is very skew.

When the normal approximation is likely to mislead confidence limits should be calculated from the binomial distribution or read from Table VIII of Fisher and Yates (1963).

Size of sample for estimating proportions

We come now to a question of some importance when estimating proportions, namely, how big a sample should we take? For a continuous variable confidence limits were $l = \bar{x} \pm z\sigma/\sqrt{n}$. Assuming that the normal approximation holds, the limits for a proportion are similarly

$$p \pm z\sqrt{\left(\frac{PQ}{n}\right)}. \tag{3.23}$$

As before, we decide our level of confidence, and hence z, and the degree of error we can tolerate. We do not know P, and initially have to make a guess. Having done so, we can then solve for n.

We now consider the consequences. Suppose that a very attractive dam site (from an engineering point of view) has been identified, and that water from the dam, if built, could irrigate some 2000 km^2 of land by gravity. Not all the land that it commands has soil that is suitable for irrigation, and the extent of suitable soil will decide whether building the dam would be a sound investment. So we need a fairly precise estimate of the area of irrigable soil in the region. The suitability of soil for irrigation is determined largely by characters such as its clay content, permeability, and exchangeable sodium. Sites must be visited to measure these or to collect soil for laboratory analysis. Sampling and laboratory analysis are expensive, and therefore we shall wish to know the minimum number of sites that will provide a reliable estimate of the extent of irrigable soil.

We start with the best prior estimate of P that we can obtain. Perhaps someone with a general knowledge of the region says that about 30 per cent of the soil is likely to be suitable, i.e. 60 000 ha. We then choose a tolerance and the probability that the true area lies within the chosen tolerance. Suppose our tolerance is 10 000 ha on either side of the estimate, i.e. that confidence limits are 60 000 ha $\pm$ 10 000 ha, and that we accept a chance of 1 in 10 that the true area lies outside these limits. Thus, our estimate of P is 0.3, $\hat{Q}=1-\hat{P}=0.7$, the permitted error, l, is 0.05, and z at the 90 per cent level is 1.64. Inserting these in (2.23) and rearranging, we have

$$n=\frac{z^2\hat{P}\hat{Q}}{l^2}=\frac{1.64^2\times 0.3\times 0.7}{0.05^2}=226.$$

This is a fairly large number. But a planner could well want more confidence in the result—say 95 per cent instead of 90 per cent. If so, n would be 323. Or, if he wanted narrower confidence limits—say, ±5000 ha—then 904 sites would need to be recorded.

Table 3.2 gives some examples of the size of sample required to estimate P within limits $P\pm 0.05$ for different values of P and three levels of confidence. It shows how greater confidence can be bought by increased

TABLE 3.2 Size of sample required to estimate P within limits $P\pm 0.05$ for three levels of confidence, calculated from the normal approximation

P or $(1-P)$	n		
	80%	90%	95%
0.5	164	269	384
0.4	157	258	368
0.3	136	226	323
0.2	(105)	(172)	245
0.1	(59)	(97)	(139)
0.05	(31)	(71)	(101)

Values in parentheses should be replaced by values from the binomial distribution

sampling, and also that a smaller sample will suffice to estimate P near 0 (or near 1) than is needed with P near 0.5 for the same tolerance. This is equivalent to stating that the standard error of P is largest when $P=Q$ (or $p=q$) and decreases as P approaches either 0 or 1. In fact, n varies in proportion to PQ, as is obvious from equation (3.23). If nothing is known about the size of P to start with then setting $P=0.5$ will ensure an adequate sample size, and if P happens to be in the range 0.3 to 0.7 then it will not entail serious over-sampling.

When we want an estimate of an area, however, it is more appropriate to express our tolerance in terms of a percentage of the true value. A tolerance of ±10 000 ha might be acceptable if the true area is about 60 000 ha. If the area were only 10 000 ha we should almost certainly regard the same confidence interval as too wide to be helpful, and would want the interval to be much smaller, perhaps 1000–2000 ha. Table 3.3 shows how expressing tolerance as a constant percentage of the true value affects sample size. Confidence limits are set as 10 per cent of the likely value on either side of the proportion. The sample size now depends not on PQ but on PQ/P^2, i.e. Q/P, and clearly rises dramatically as P decreases. A very large sample is needed to estimate the area of a scarce soil type with reasonable precision if it is chosen in a simple random fashion.

Here we note two points. First, a single observation of an attribute carries little information; it can take only one of two values, and this is partly why samples for estimating proportions must be so large. Second, it clearly pays to have good prior estimates of the proportions of land possessing the attributes of interest when attempting to determine areas. A sample survey designed to estimate the area of a soil type covering 10 per cent of a region could cost several times too much if the type turned out to cover 30 per cent.

TABLE 3.3 Size of sample required to estimate P within limits $P \pm 0.1P$ for three levels of confidence, calculated from the normal approximation

P or $(1-P)$	Confidence limits	n		
		80%	90%	95%
0.5	±0.05	164	269	384
0.4	±0.04	245	331	576
0.3	±0.03	382	627	896
0.2	±0.02	655	1075	1536
0.1	±0.01	1474	2420	3456
0.05	±0.005	3112	5109	7296

Increasing efficiency

Sampling for attributes can often be made more efficient, and therefore less costly, by using one of the alternatives to simple random sampling described earlier. We saw that several sampling designs in which the sample points were more evenly spread gave more precise estimates of the means of continuous variables than simple random sampling. The same applies to estimates of proportions.

We shall consider first the advantages of stratified random sampling. In the simplest case we divide the region to be sampled into m square cells by a grid, randomly choose an equal number of points in each cell, and observe whether each point possesses the attribute of interest, X. Using the same notation as before,

$$p = a/n \tag{3.24}$$

is an unbiased estimate of P, and if a_k points out of n_k in the kth cell possess the attribute then

$$p_k = \frac{a_k}{n_k} \quad \text{and} \quad q_k = 1 - p_k, \tag{3.25}$$

and

$$p = \frac{1}{m}\sum_{k=1}^{m} p_k = \frac{1}{m}\sum_{k=1}^{m}\frac{a_k}{n_k}. \tag{3.26}$$

The variance of p is, as before, a function of the variance within cells, and is estimated by

$$s_{p,st}^2 = \frac{1}{m^2}\sum_{k=1}^{m}\frac{p_k q_k}{n_k - 1}, \tag{3.27}$$

and the standard error is its square root.

If the cells are not of equal area then the formula becomes

$$s_{p,st}^2 = \sum_{k=1}^{m} w_k^2 \frac{p_k q_k}{n_k - 1}, \tag{3.28}$$

where w_k is the ratio of the area of the kth stratum to the total area.

Within small cells the soil will tend to be all of one type, and either to possess the attribute everywhere or to lack it everywhere. Thus $p_k q_k$ is likely to be smaller for most cells than pq for the whole population, and as a result $s^2_{p,st}$ is likely to be smaller than s^2_p, the variance for the simple random case.

The early survey of soil and fruit in the Vale of Evesham (Osmond *et al.* 1949) provides an example. The report suggests that the best types of soil for all-round fruit production are heavy but not poorly drained, namely the Worcester, Evesham, Chadbury, and Hipton Hill series. Assuming the soil series map to be correct, we can estimate the proportion of the region surveyed that is good for fruit growing and the area that it covers. The map was divided into 2 km × 2 km cells. Some 1 km^2 squares near the edge of the map were not included in the 4 km^2 squares and were grouped into 4 km^2 cells of different shapes. Four points were then chosen at random in each cell, i.e. one point per km^2. The results were as follows:

$$n = 344$$

$$a = 163$$

$$p = 0.474;$$

i.e., the total area is 34 400 ha, of which 16 300 ha are estimated as good for fruit growing. The standard errors are

$$s_{p,st} = 0.022$$

$$s_{\text{area},st} = 760 \text{ ha},$$

giving $\hat{P}$ and $\hat{A}$ and their 90 per cent confidence limits as

$$\hat{P} = 0.474 \pm 0.036$$

$$\hat{A} = 16\,300 \pm 1245 \text{ ha}.$$

Had the same values of a and p been obtained from a simple random sample we should have had

$$s_p = 0.027$$

$$s_{\text{area}} = 926 \text{ ha},$$

with correspondingly wider confidence limits than those actually obtained.

Sampling was repeated with the map divided into cells 2 km × 1 km and two points chosen per cell. The results, summarized in Table 3.4, are very similar to those for the 4 km^2 cells. Both stratified samples provide modest improvements over simple random sampling for the same effort.

If we regard the standard error of 0.027 as acceptable we can determine roughly the extent to which sampling could be reduced by inserting $s=0.027$ for the tolerance, l, in equation (3.28) and rearranging. Since $w_k=n_k/n$, the value of n is obtained from

$$n=\frac{1}{l^2}\sum_{k=1}^{m} w_k n_k \frac{p_k q_k}{n_k-1}.$$

For the 2 km × 2 km strata this gives a sample size of 230, and for the 2 km × 1 km strata, 207, and the efficiencies relative to simple random are approximately 1.5 and 1.7 respectively. These represent worthwhile savings. These estimates are somewhat optimistic, because with fewer sampling sites the cells would be bigger and therefore the variances within them would also tend to be larger. However, the difference between the variances within the 2 km^2 and 4 km^2 cells is small, and the estimates of efficiency are likely to be good enough for planning purposes.

Comparison of sampling designs was taken one stage further for the Vale of Evesham. The soil map was resampled eight times using unaligned and regular systematic schemes. The points in the unaligned design were chosen by random coordinates within the kilometre squares of the National Grid. A kilometre grid with a fresh random origin and orientation was laid

TABLE 3.4 Estimated proportions and areas, standard errors, and 90% confidence limits for two stratified random samples of the Vale of Evesham

	Stratification	
	4 points/2 × 2 km^2	2 points/1 × 2 km^2
Count	163	166
Proportion	0.474	0.483
Area (ha)	16 300	16 600
Standard error		
Proportion	0.022	0.021
Area (ha)	760	712
90% confidence limits		
Proportion	±0.036	±0.034
Area (ha)	±1245	±1168
Efficiency compared with random	1.5	1.7

TABLE 3.5 Estimated proportion and area, standard errors, and 90% confidence limits for unaligned and systematic sampling of the Vale of Evesham

	Design	
	Unaligned	Systematic
Count[a]	162.5	164.25
Proportion[a]	0.472	0.477
Area (ha)[a]	16 250	16 425
Standard errors		
Proportion	0.012	0.0093
Area (ha)	424	319
90% confidence limits		
Proportion	±0.020	±0.015
Area (ha)	±695	±523

[a] Average of eight

over the map to obtain each systematic sample. The results are summarized in Table 3.5. They show that unaligned sampling is about three times as precise as the stratified random scheme for the same sampling effort, and that systematic sampling is about five times as precise. Furthermore, both are even more precise than simple random sampling. With the proviso already noted, we should expect comparable improvements in efficiency from these two sampling designs. For the record, we note that sampling at 1 km intersections of the National Grid gave a count of 163: hardly an indication of bias.

In a similar study carried out by Berry (1962), stratified sampling with only two observations per cell reduced the standard error much more than in this one. Berry also found that systematic and unaligned sampling performed better still. The improvement was similar to that obtained in the Vale of Evesham.

Systematic or unaligned sampling will usually be the most efficient schemes for estimating proportions, and they are recommended except where a precise estimate of sampling error is needed.

In the next chapter we discuss other forms of stratification that should increase efficiency further, and in Chapter 14 we describe how to plan systematic sampling using more advanced theory.

4

GENERALIZATION, PREDICTION, AND CLASSIFICATION

Classification is easy:
it is something you just do.
F. C. Bawden

The descriptive measures and sampling methods discussed in the last two chapters enable us to generalize with known confidence about a soil population, whether it is a set of soil specimens or profiles, or the soil of an area. Thus, if we have measured the pH of the topsoil at 100 sites in a region and found that on average the pH is 5.6 and the standard deviation is 0.3 we can say that in general the soil at those sites is moderately acid. We can lose the detail and yet retain in the mean a valuable picture of the soil's acidity. Assuming normality, some 95 per cent of the observations fall within 2 standard deviations of the mean, i.e. in the range pH 5.0–6.2. Provided the sites were chosen probabilistically, we may say the same about the soil of that region with considerable confidence. With 100 sites the standard error of the mean is 0.03 pH units, giving very close confidence limits. It is obviously useful to be able to generalize in this way.

If, on the other hand, the observed pH ranged from 4.5 to 9.7 with a mean of 7.0 and standard deviation of 2.0, it would be much less sensible to attempt to generalize. It is undoubtedly true that the soil is neutral on average; but this hides the fact that an appreciable proportion of the soil is acid, and a similar proportion alkaline, and that the associated properties of soil at opposite ends of the range of pH will be very different. In these circumstances some division of the population is needed. Distinguishing just two classes, acid and alkaline, might be enough to allow useful generalization about each class. Adding a third, say neutral, could well allow generalization to be even more useful.

Exactly the same philosophy applies in sampling the soil profile. If the pH or other properties of interest are much the same all the way down then their values may sensibly be averaged for the profile. If they are not, and there is often an important change in some property with depth, then it is better to keep values separate and average them only for particular horizons.

Prediction

In addition to generalizing from sample data we may wish to predict. Prediction in the context of soil survey usually means predicting the state of the soil at some place or places where the soil has not been observed previously and often where it cannot be observed for one reason or another. Soil is continuous, but we can observe it, at least in depth, at only a finite number of sampling points. Thus, clearly it is important to be able to predict its properties elsewhere. The techniques described in Chapters 3 and 14 enable us to do this.

Following from Chapter 3, the expected value of a property at any place is the mean, μ. If we knew it then this should be our predictor, and the variance on our prediction, the estimation variance σ_E^2, would be the population variance, σ^2. In practice we have only our sampling estimate of μ, x. We may use this without bias to predict provided we have sampled properly in the first place. The variance on our prediction, however, is now amplified by the estimation variance of the mean, σ^2/n, giving

$$\sigma_E^2 = \sigma^2 + \sigma^2/n. \tag{4.1}$$

If the distribution is normal we can set confidence limits for our prediction. These are often termed *prediction limits* to distinguish them from the confidence limits of a parameter. In general, the symmetrical prediction limits on a prediction are $\bar{x} \pm z\sigma_E$ where z is the standard normal deviate for the confidence we choose; i.e., approximately 1.0 at 66 per cent, 1.28 at 80 per cent, 1.64 at 90 per cent, 1.96 at 95 per cent, and so on.

Often we are concerned with only one prediction limit, say a critical strength of the soil, and we wish to know the likelihood that the soil strength is less than this value. We can estimate this likelihood if we know μ and σ or have good estimates of them. The value of z corresponding to the critical value, c, is calculated as $z = |c - \mu|/\sigma$, and the probability of obtaining a value larger than or equal to z can be found from tables, e.g. Table III of Fisher and Yates (1963).

As mentioned in the last chapter, our value of the standard deviation is itself usually an estimate based on a sample, and so is subject to error. Prediction limits are wider than they would be if we knew σ precisely, therefore, especially when samples are small. So as before, we replace z by Student's t with the appropriate number of degrees of freedom. With samples larger than 60 there is little difference, but for smaller ones there is, and t should be used instead of z.

Consider the second illustration now, where the mean and standard deviation were 7.0 and 2.0 respectively. Our best prediction of the pH is 7.0 but the confidence intervals are wide. Thus, even at the 80 per cent level of confidence, the lower and upper limits are approximately 4.4 and 9.6. Intervals as wide as this are clearly unhelpful. The interval must be narrowed to achieve worthwhile prediction, and this can often be done by classifying the soil.

In some instances not only is the spread of observed values large, but the frequency distribution of the variable of concern has more than one distinct peak. The population is then almost certainly heterogeneous, and the sample almost certainly consists of a mixture of two or more soil types. The population cannot be represented adequately by a single mean, nor can knowledge of its standard deviation lead to confident prediction. Each type of soil needs separate recognition, usually with a separate estimate of its mean, and perhaps of its standard deviation.

So even when we measure soil properties, instead of describing them qualitatively, soil classification can have an important role in generalization, prediction, and, as we shall see later, estimation. In the remainder of this chapter we shall consider kinds of classification and how they enhance the value of data or the efficiency of the survey.

Kinds of soil classification

Dissection

When a single soil property is all-important classes of soil, each with strictly limited variation, can be created by dividing its range at certain fixed points. These critical values can either be determined by technological need or be chosen arbitrarily at convenient points to achieve some desired number of groups. The process is often known as *dissection*. Dissection implies neither that the soil is heterogeneous nor that the divisions are in any sense 'natural'; the only consideration is that the classification be useful for the purpose in hand.

An example arises when using alkaline soil, especially for irrigated agriculture. Ease of management and plant growth depend critically on the exchangeable sodium percentage (ESP) of the soil, that is, on the amount of exchangeable sodium as a proportion of all exchangeable cations. Soil with more than about 15 per cent ESP deflocculates readily and is difficult to make and keep permeable. It needs ameliorative treatment before it can be used for agriculture. Soil with ESP in the range 7.5–15 per cent is usable but needs careful management, and in particular, care is needed to ensure that its ESP does not increase when irrigated. Where the ESP is less than 7.5 per cent the exchangeable sodium has no appreciable effect on the soil. The critical dissection values are thus 7.5 and 15 per cent.

Fertilizer recommendations are often based on measurements of the available nutrients in the soil. Relations between measurements and responses are rough, and there is little justification for having more than four or five levels of recommended treatment for any one crop, and sometimes no more than three, corresponding to no fertilizer, a moderate dressing, and a heavy dressing. For example, Cooke (1975) divides the scale of soluble phosphorus (in parts per million) into four groups: <5, 5–10, 10–15, and >15, and indicates for each group suitable fertilizer dressings to grow cereal crops. The limits of each class are not critical, but are round numbers that divide the range conveniently.

If use of the soil depends largely on two or three properties then classes can be created by dividing all their measurement scales at critical or convenient values.

When a survey is undertaken to divide land for different types of use and management, or to resolve competing claims for its use, several properties of the soil may be recorded. All are thought to be relevant, but none are known to be critical. Normally the number of kinds of land use or management envisaged will be few, and it is convenient if the soil can be classified into a few groups, but ones based on all the measurements. Clearly, if every scale were divided, even into two, the number of groups would be unmanageable. Much effort has been devoted to finding an alternative. It has involved the search for close relations between sets of variables in the hope that division on one scale might effectively divide others also. Until recently there was no generally satisfactory technique. The advent of computers stimulated research on this problem. There are now numerical methods for classifying soil from multiple observations without creating a plethora of groups and in ways that reflect relationships in the data and that are generally useful. Some of these are described in Chapters 10 and 11.

General purpose classification

Much soil classification is not at all explicit. At the local level soil is often divided into classes intuitively. The observer makes decisions on what he can see on the surface and in profile and on any supplementary information that he might gain in the laboratory. The way he classifies soil can be influenced by what grows on the land and by any apparent relations between soil and its environment. Here the scientist is simply extending the layman's approach. He forms in his mind 'notional types', and as his experience grows these become his standards or nuclei around which his classes of soil grow. We may call the result 'typological' classification. Alternatively, the observer might approach a new area or set of profiles with a fairly clear idea of the types he wishes to recognize and attempt to divide the new soil into groups. Butler (1980) discusses this at length.

Classification based purely on intuition can be very fluid and is not easily communicated. Organizations with large staffs often feel the need to define standard classes using criteria that everyone can apply consistently. The classes might initially be recognized intuitively, but their definitions become explicit. Such classifications can be called 'definitional'. The classes are usually disjoint, and in a number of schemes soil profiles are grouped at different levels, or categories, in a hierarchical fashion like the Linnaean classification of plants and animals. The soil classification of the US Department of Agriculture (Soil Survey Staff 1975) is the most meticulously defined of those. Others include those by Avery (1973) for England and Wales, and by Northcote (1971) for Australia.

Definitional classifications dissect the scales of the properties on which definition is based. They rarely produce a disjoint division of the scales of other properties, however: there is almost always some overlap in the latter, and when used to predict these other properties they are not necessarily better than intuitive classifications.

It is often claimed that both intuitive and definitional classifications are 'natural', because they take account of many properties of the soil, or because they divide the population at natural discontinuities in the soil, or because they are generally useful, or some combination of these. For much the same reasons they are also called *general purpose classifications*. In this chapter we shall use this term for such classifications and shall distinguish them from dissection.

Horizons, profiles, and areas

Soil in the field is classified in three main ways: into horizons, into classes of profile, and into parcels (areas) or groups of parcels. The recognition of distinct horizons is a classification of soil within a profile. Its purpose is to enable a more precise description of the soil profile by describing parts of it in detail rather than describing the whole soil more generally. The horizons may be labelled A, B, C, and so on to emphasize homologies in different profiles.

Likewise, the soil of a region can be divided into parcels within any one of which the soil seems reasonably similar and different from its neighbours. Separate parcels that possess similar soil are grouped to form classes that are usually known as *mapping units*.

Both horizons in a profile and mapping units in an area are separated by boundaries in space. The classification of soil profiles is a little more abstract, in that profiles in the same group can be near or far away on the ground. Their spatial proximity is immaterial in principle, though in practice classes of profile are often recognized because the particular range embraced happens to occur in sizeable tracts of country. Profile classification often derives from mapping exercises. In other instances surveyors seek to delineate areas containing previously recognized types of profile. The mapping unit is then an attempt to display the extent of that particular class of profile. In either event there is a duality in the classification, and the same names or labels are usually given to classes of profile and to their corresponding classes of area. In general discussion the two kinds of class need not be distinguished. But in more critical work, especially that concerned with sampling and quantitative appraisal, the investigator must be quite clear which kind of classification he means.

General purpose soil classifications are most appropriate for resource inventories and surveys for multiple land use. The classes are intended to convey a wide variety of information. However, a similar approach is often taken for more specific purposes, and the reasons are primarily economic, and sometimes logistic. A few examples will make this clear.

1. Much agricultural land in Britain can be or has been improved by underdraining to remove surface water from impermeable soil. A drainage engineer who has measured the hydraulic conductivity of soil that needs draining can design a suitable scheme for that soil. Since individual drainage schemes are small (only about 5 ha on average) and the likely economic benefits also small, the cost of measuring hydraulic conductivity (which is large) cannot usually be justified (Thomasson 1975). An economically feasible alternative is to classify the soil on its appearance and on cheap measurements, paying special regard to characters that seem related to hydraulic conductivity. The hydraulic conductivity is determined at a few sites on each class of soil, and it is predicted elsewhere according to the classes of soil there.

2. Establishing productive forest on land where there is none or planting with exotic species is expensive, and it takes a long time to discover whether the species will grow successfully. It could be risky and potentially very wasteful to plant the whole of a large area without prior indication of the performance of the species to be used. Records of previous experience are therefore assembled. If these are insufficient to allow a firm decision then trials are carried out on small sample areas. Provided the species are suitable for the climate, their growth is likely to depend on the soil. So the results of experience and experiment are indexed according to the type of soil on which they were obtained, and are used to forecast the performance of the same species on the same soil types elsewhere. The soil classification is again the means by which information from a few sites is extrapolated to many. The same philosphy is often applied to agricultural development. General soil survey is followed by field trials on representative sites of any soil type that is obviously not unsuitable. These are then followed by

planning land use to produce crops on the kinds of soil on which they grew successfully in the trials.

3. When land is accessible only with difficulty, either because it is remote or in the military context because it is held by an enemy, or when large areas must be covered by a small staff, the assessment of soil conditions relies heavily on the interpretation of aerial photographs or digital imagery. In most instances interpretation involves, first, a classification of the image into relatively homogeneous regions and, second, sampling on the ground at representative sites of the same class. Photo interpretation has been especially successful in civil engineering for predicting the foundation qualities of soil for roads and airfields and for identifying construction materials (Brink *et al*. 1982).

The problem common to all three examples is this. The soil is known to vary too much for all of it to be treated similarly. The properties that determine how it can be used are known—hydraulic conductivity, nutrient status and pH, bearing capacity—but they cannot be measured everywhere. Classification is seen as a solution whereby generalization and prediction are based on sample evidence for each class separately. Variation within any one class is, or is hoped to be, less than that in the whole population and small enough for generalization or prediction to be useful. Success depends on the extent to which this is so. Yet until recently very few scientists tested their classifications formally to see whether they served the purposes for which they were intended. Kantey and Williams (1962) were among the first to demonstrate such success quantitatively. The same criterion of success, namely utility for specific purposes, applies to any general purpose classification. Although a person may create a classification that he intends to be generally useful, it is difficult to imagine a general purpose user. Any individual user will usually have a specific interest to which a few specific soil properties relate.

So, what are we looking for in an effective classification? First, we want classes that differ from one another in respect of the properties that we wish to predict, and the more so the better. If they do not then classification is of no benefit. Second, we want classes that are internally homogeneous. No class of soil is ever completely uniform, but the less variation there is within the classes, the better we can predict values of the properties using the classification. It is with these criteria in mind that we shall judge the effectiveness of a classification (Webster and Beckett 1968).

Effects of classification: analysis of variance

Dissection clearly and predictably limits the range of the variable or variables concerned. The ranges of other variables within each of the classes created by dissection are much less predictable. Similarly, though the variation of some properties within classes of a general purpose classification will be less than that in the population as a whole, the extent of their variation is far from predictable. For other properties variation may be no less within classes than it is in the whole population. The variation present must therefore be determined empirically.

Just as we estimated the variance and standard deviation of whole soil populations, so we can for each class separately. For a class i, sampled at n_i places with sample mean $\bar{x}_i$, the variance σ_i^2 is estimated by

$$s_i^2 = \frac{1}{n_i - 1} \sum_{j=1}^{n_i} (x_{ij} - \bar{x}_i)^2. \tag{4.2}$$

If s_i^2 is small enough we can usefully generalize about that class and use $\bar{x}_i$ as the predictor for points within it with variance $s_i^2 + s_i^2/n_i$.

Now, consciously or otherwise, when an investigator classifies soil at a particular level he tends to create classes within which the degree of variation is much the same. Thus, all soil series recognized in a survey project, however much they differ in their mean values, might reasonably be expected to have approximately the same standard deviation of any one variable (transformed if necessary). Experience confirms this. When this is so we can obtain a precise estimate of the within-class variance, σ_W^2, by pooling the individual ones. This is done by taking their averages, weighted by their corresponding degrees of freedom, from which the standard deviation is derived. For k classes this is equivalent to

$$s_W^2 = \frac{\sum_{i=1}^{k} \sum_{j=1}^{n_k} (x_{ij} - \bar{x}_i)^2}{\sum_{i=1}^{k} (n_i - 1)}, \tag{4.3}$$

where the sample of the kth class contains n_k observations.

In addition to the total and the within-class variances, σ_T^2 and σ_W^2, there is a third variance that is often of interest: it is the variance among class means. These three variances and the relations between them can be used to evaluate and compare classifications. They can all be estimated by *analysis of variance*, which is at once one of the most powerful and elegant techniques in statistics, and was developed by R. A. Fisher in the 1920s.

The basis of analysis of variance is that variances are additive and that the total variance is the sum of the variances contributed from two or more independent sources. The aim of the analysis is to estimate these separately. We know already how to estimate the total variance and the pooled within-class variance, two of the quantities we need. Let us

TABLE 4.1 Analysis of variance

Source	Degrees of freedom	Sum of squares	Mean square
Between classes	$k-1$	$\sum_{i=1}^{k} n_i(\bar{x}_i-\bar{x})^2$	$\frac{1}{k-1}\sum_{i=1}^{k} n_i(\bar{x}_i-\bar{x})^2=B$
Within classes	$N-k$	$\sum_{i=1}^{k}\sum_{j=1}^{n_i}(x_{ij}-\bar{x}_i)^2$	$\frac{1}{N-k}\sum_{i=1}^{k}\sum_{j=1}^{n_i}(x_{ij}-\bar{x}_i)^2=W=s_W^2$
Total	$N-1$	$\sum_{i=1}^{k}\sum_{j=1}^{n_i}(x_{ij}-\bar{x})^2$	$\frac{1}{N-1}\sum_{i=1}^{k}\sum_{j=1}^{n_i}(x_{ij}-\bar{x})^2=T=s_T^2$

Notes
N is the total size of the sample.
k is the number of classes, the ith class containing n_i observations.
$\bar{x}$ is the mean for the whole sample.
$\bar{x}_i$ is the mean for the ith class.
B, W, and T are convenient symbols for the three mean squares.

suppose, as will happen sometimes for a particular variable, that the classes do not differ in their means. In these circumstancers $\sigma_T^2=\sigma_W^2$, and s_T^2 and s_W^2 both estimate the same population variance. We can also obtain a third estimate of the variance from the sample means of the classes. As we saw in the last chapter, each mean is an estimate of the population mean with a variance σ_T^2/n_i, where n_i is the number of observations in the ith class. So the sum of squares of deviations of the class means from the general mean, each multiplied by the sample size and divided by the number of degrees of freedom, gives us our third estimate, say B. Thus,

$$B=\frac{1}{k-1}\sum_{i=1}^{k} n_i(\bar{x}_i-\bar{x})^2. \tag{4.4}$$

This is usually laid out as in Table 4.1.

If, however, the classes differ in their means, μ_i, then σ_T^2 and σ_W^2 are not equal, and B estimates σ_W^2 plus a contribution from the variation between classes. Put another way, the variance among the sample means,

$$V_B=\frac{1}{k-1}\sum_{i=1}^{k}(\bar{x}_i-\bar{x})^2,$$

estimates the variance among the true class means plus the sampling variances of the means, $s_W^2 n_i$. Thus, if all n_i are equal ($n_i = n$),

$$V_B = s_B^2 + s_W^2/n,$$
$$B = nV_B = ns_B^2 + s_W^2$$

and

$$s_B^2 = (B - s_W^2)/n. \tag{4.5}$$

If the n_i are not equal then n is replaced by n_0 (Snedecor and Cochran 1980), where

$$n_0 = \frac{1}{k-1}\left(N - \frac{\sum_{i=1}^{k} n_i^2}{N}\right) \tag{4.6}$$

and

$$s_B^2 = (B - s_W^2)/n_0, \tag{4.7}$$

where N is the total number of observations.

The meaning of s_B^2 is not always easy to appreciate, and in some instances is a matter for debate. It depends very much on the nature of the sampling in relation to the classification. Two distinct situations can be recognized.

Fixed effects. The first occurs when we have classified the soil in an area into several classes, every one of which is sampled. The aim is usually to estimate the means μ_i for each class $i = 1, 2, \ldots, k$. The deviations between the class means and the general mean are fixed quantities: they are determined by the classification. And though s_B^2 estimates

$$\frac{1}{k-1}\sum_{i=1}^{k}(\mu_i - \mu)^2,$$

it is not usually of much interest.

Random effects. In studies of classification and sampling schemes the attributes of the schemes themselves are of more interest than those of the individual classes. Schemes are devised so that replicate samples are obtained for some classes, but it may be largely immaterial which. The actual classes included depend on chance, and the differences between the class means and the general mean $\mu_i - \mu$, are subject to random variation. In this event s_B^2 estimates the variance among a larger population of means, usually denoted by σ_B^2 and termed a *component of variance*, which is of considerable interest.

Intraclass correlation

The size of σ_B^2 in relation to σ_W^2 can be used as a measure of the effectiveness of a classification. The relation can be expressed as the intraclass correlation, ρ_i, defined as

$$\rho_i = \frac{\sigma_B^2}{\sigma_W^2 + \sigma_B^2} \tag{4.8}$$

and estimated by

$$r_i = \frac{s_B^2}{s_W^2 + s_B^2} = \frac{B - s_W^2}{B + (n-1)s_W^2}. \tag{4.9}$$

The last expression enables r_i to be calculated swiftly from the analysis of variance table. The term derives its name from the fact that it expresses the 'correlation' among individuals within the same class.

Clearly, r_i has a theoretical maximum value of 1 when each class is uniform ($s_W^2 = 0$). In practice, there is always some variation within classes in a measured variable, so r_i never attains 1. The minimum value of r_i is less certain and its interpretation more problematic. The minimum value of σ_B^2, and hence of ρ_i, is zero. However, their sampling estimates s_B^2 and r_i are often negative, and this disconcerts investigators. In the analysis of variance it appears as $B < W$, and in soil survey the cause is simply sampling error in situations where the differences between classes are small.

A simpler but closely related way of expressing the effect of classification is by the ratio s_W^2/s_T^2, sometimes called the *relative variance* and used by Beckett and Burrough (1971). Its complement, $1-(s_W^2/s_T^2)$, can be regarded as the proportion of variance accounted for by classification. When both the total sample size and the number of groups are large its value is very similar to that of r_i.

Example

Some years ago the Royal Engineers wished to know whether classifying soil or land according to features visible on aerial photographs would improve their ability to predict engineering properties of soil. This example is drawn from a study of the situation (Beckett and Webster 1965*a*, *b*). The land in south-central England was classified largely on its air-photo appearance. Seventeen of the classes were sampled by choosing sites within each class in as nearly random a way as circumstances would allow. Several soil properties were measured at each site. The following is the analysis of variance for the plastic limit of the soil at 13 cm taken from a single 10 cm auger core at each site.

The mean plastic limit for the region was estimated as 38.1 with a variance per site of 245.7 and hence a standard deviation of 15.7. Confidence limits for prediction are somewhat wide. The variance was then partitioned for the classification; the details are given in Table 4.2. The F ratio is well in excess of that at the 0.001 level of probability, and we conclude that there are highly significant differences between classes (see the section on significance below). Table 4.3 gives the mean values for each class. Classes 9.1 and 9.2, river flood plains, have organic-rich soil with the largest values; classes 12, 13.1, and 13.2 lie on the Corallian formation with predominantly sandy soil and the smallest plastic limits. Confidence limits for prediction are obtained for each class from the class mean and the pooled within-class standard deviation, which is 11.5.

The estimate of the between-classes component of variance is 121.1, and the intraclass correlation is therefore

$$r_i = \frac{121.1}{121.1 + 132.4} = 0.477.$$

This compares with the simple estimate of variance accounted for,

$$1 - (s_W^2/s_T^2) = 0.461.$$

This is a fairly typical result. About half the variance in the physical properties of soil in a region can be attributed to differences between classes in a fairly simple classification of soil based on profile appearance,

TABLE 4.2 Analysis of variance for the plastic limit at 13 cm

Source	Degrees of freedom	Sum of squares	Mean square	F
Between classes	16	15 837	989.8	7.48
Within classes	105	13 900	132.4	
Total	121	29 737	245.7	

TABLE 4.3 Sample size and mean plastic limit for 17 classes

Class	Sample size	Mean	Class	Sample size	Mean	Class	Sample size	Mean
1	6	27.0	8.2	6	38.0	13.1	5	23.6
2	8	36.4	8.3	11	44.1	13.2	4	19.8
3	6	26.3	9.1	6	67.9	13.3	11	33.9
4	17	32.8	9.2	6	60.5	14	3	29.0
5	10	41.5	10	6	36.3	15	6	39.7
8.1	7	51.9	12	4	25.0			

physiography, and geology. A few other results from the same study, originally reported by Webster and Beckett (1968), are given in Table 4.4. Experience suggests that chemical properties that are modified by farming are less easily differentiated by such simple classifications, and that the between-classes components of variance are likely to be only about one-tenth of the total variance.

The example has another feature that is not obvious but could be important. One aim of the survey was to obtain a reasonably precise estimate of the mean for each class sampled, and in the context of the survey all classes were equally important, whatever their extent. From the air-photo analysis and prior fieldwork, some of the classes seemed likely to be more variable than others. These classes also happened to include the ones of smallest extent, the spring lines (class 2) and small valley floors (classes 8.1, 8.2, and 8.3), in addition to the river flood plains (classes 9.1 and 9.2). Further, to obtain good estimates of their class means, some of these very variable classes were sampled more than the more uniform

TABLE 4.4 Means, $\bar{x}$, within-class variances, s_W^2, and intraclass correlations, r_i, of topsoil properties for a classification of 1000 km^2 in south-central England

Property	$\bar{x}$	s_W^2	r_i
Clay (%)	37.2	90.2	0.61
Mean matric suction in summer (pF$_1$)	2.66	0.07176	0.66
Mean matric suction in winter (pF$_1$)	1.82	0.00402	0.61
Mean soil strength in winter (cone index)	138	510	0.70
Liquid limit (%)	68.0	309.8	0.48
Organic matter (%)	9.8	9.48	0.28
pH	7.09	0.326	0.33
Available P (%)	0.031	0.00114	0.09
Available K (%)	0.013	0.0000939	0.06

ones. These variable classes contribute disproportionately therefore to the pooled within-class variance in the above analysis, both because of their small extent and because of their above-average sampling. If the classes had been represented more nearly in proportion to their extent then class 2 would have been eliminated and classes 8.1, 8.2, and 8.3 would have been represented by only two sites each. The total variance of this reduced sample remains much the same, but the pooled within-class variance is substantially less, and the intraclass correlation is correspondingly larger, as follows:

Total variance	253.7
Within-class variance	99.32
Between-classes component	167.5
Intraclass correlation	0.63

Except in the classes mentioned that are more variable than average, the confidence intervals are narrower than would have been judged from the first analysis, since the within-class standard deviation is now only 10.0.

Short cuts

The number of degrees of freedom (d.f.) between classes plus the d.f. within classes sum to the total d.f.:

$$(k-1)+(N-k)=N-1. \tag{4.10}$$

Likewise, the entries in the sums of squares columns for between classes and within classes sum to the total. This can provide a useful check on the arithmetic. However, when computing is done on a small calculator it is normal practice to calculate the sums of squares for the total and between classes and to derive the within-class sum of squares as the difference. This saves time.

It is also easily shown that

$$\sum_{i=1}^{k}\sum_{j=1}^{n_i}(x_{ij}-\bar{x})^2=\sum_{i=1}^{k}\sum_{j=1}^{n_i}x_{ij}^2-\frac{1}{N}\left(\sum_{i=1}^{k}\sum_{j=1}^{n_i}x_{ij}\right)^2 \tag{4.11}$$

and

$$\sum_{i=1}^{k}n_i(\bar{x}_i-\bar{x})^2=\sum_{i=1}^{k}n_i\bar{x}_i^2-\frac{1}{N}\left(\sum_{i=1}^{k}\sum_{j=1}^{n_i}x_{ij}\right)^2. \tag{4.12}$$

This enables further time to be saved, since deviations from the general mean no longer need to be calculated separately. They are taken into account in the second term on the right in the above equations. This term is known as the *correction for the mean*.

Many textbooks recommend these short cuts. However, there are two hazards. The first is that the investigator loses track of the meaning of the terms he is calculating. The very slickness of the computation blinds him. The second hazard is that the terms on the right-hand sides of equations (4.11) and (4.12) are often very large quantitites with small differences between them. When the arithmetic is performed by computer these quantities can easily be rounded and large errors introduced into their differences (see Nelder 1975). Thus these methods are best avoided when using a computer.

Significance

We come now to an aspect of analysis of variance, and of statistics generally, namely tests of significance. Many scientists are puzzled by these and use them out of context.

In the last chapter we saw that sample estimates of a population mean were not all equal. They deviated more or less from the true mean. Provided sampling was unbiased, we regarded this as sampling error. When several samples are taken there is variation from sample to sample, and as we have seen this is measured by the mean square B in the analysis of variance. If the samples are drawn from different classes and the term B is larger than s_W^2, i.e. if $s_B^2>0$, the question arises: how should we decide whether this result is a chance effect of sampling or a reflection of real differences between the classes?

The problem is resolved by adopting a *null hypothesis*. We begin by assuming that all class means are the same—that there are no differences among them, that $\sigma_B^2=0$ if we are dealing with random effects, and that if s_B^2 is larger than zero it represents sampling error. When s_B^2 is small this assumption is reasonable, but the larger s_B^2 is, the less reasonable the assumption becomes. Eventually we decide that it is too large to attribute to chance and that it represents real differences among classes. We then reject the null hypothesis, and the differences between means are said to be *significant*. The criterion for deciding whether to accept or reject the null hypothesis is the ratio $F=B/W$ in the analysis of variance. Its sampling distribution was worked out by R. A. Fisher, actually as $z=\ln \sqrt{F}$, and it is tabulated in many statistical textbooks. The calculated F is compared with the tabulated values for the same degrees of freedom, $k-1$ and $N-k$, and the probability of obtaining that value of F or a larger value is thereby obtained. Conventionally an F ratio or value of s_B^2 is regarded as significant if the probability P of its occurring by chance is less than 0.05. An investigator who feels that this is too stringent a test may choose a somewhat larger value of P, say 0.1. If he does then he runs the risk of judging classes to differ when they do not. On the other hand, he may choose a smaller value of P, say 0.01. In that event the risk of chance variation's being regarded as real variation between classes is less, but real differences are less readily detected.

There are published tables of F values for probabilities 0.25, 0.1, 0.05, 0.025, 0.01, 0.005, and 0.001. The precise level of probability is to some extent a matter of choice, and in making that choice an investigator should consider the consequences of his judgement. Is it a matter of life or death, as it might be if a land-settlement scheme fails or a military convoy gets bogged in soft ground? Or is it a situation in which a mistaken judgement could readily be corrected or a fairly painless recovery be made from ill-advised action? The proper course in all cases is to report the probability associated with the estimate of σ_B^2 or the F ratio and number of degrees of freedom. The user of the information can then make his own judgement.

Significance tests are rarely edifying in studies of soil classification. When comparing recognizable classes of soil the null hypothesis is nearly always highly implausible—there are nearly always some differences in whatever property we happen to measure. So, however small and trivial such differences are, we can find them significant if we take a large enough sample. Even when a significance test is essential at the end of an investigation, and we quote the value of P, that is not the only important result. The class means, s_W^2, s_B^2, and r_i, matter more.

Statistical method embraces many kinds of test for significance. The F ratio in the analysis of variance leads to just one. The feature common to all is that we have results for a sample, but wish to make inferences about a population. To do that the sample must have been drawn without bias from the target population. If $s_B^2 > 0$ for a sample then that is a fact; but the extension to the population, i.e. $\sigma_B^2 > 0$, is a matter of judgement. Life is uncertain, the real world is an uncertain place, and significance testing is one of the means whereby we can make decisions against this background of uncertainty.

Homogeneity of variances

We noted that classification at some particular level of generalization tends to result in classes each having much the same variance. This is the intention. However, in an actual survey it is not always possible to achieve this. Even detailed maps of the soil of some areas show parcels of varying complexity. If the classes are not equally variable with respect to the property of interest it is as well that we know, since we shall wish to apply different confidence intervals to them.

Clearly, we can sample each class and determine its variance. We shall obtain different values for each class, and as with the differences between means, we may test for significance. If we have only two classes we calculate their variance ratio,

$$F = s_1^2/s_2^2.$$

If s_2^2 is the larger variance then we invert the expression. The resulting value of F is compared with tabulated values to find its probability. Most published tables, e.g. Table V of Fisher and Yates (1963), are for 'one-tailed' tests. The probabilities given assume that we know in advance which variance is the larger. In the present situation, where either σ_1^2 or σ_2^2 for which we have estimates may be the larger, the test is 'two-tailed', and the level of significance that $\sigma_1^2 \neq \sigma_2^2$ is double the published values.

With more than two classes we can use a test due to Bartlett (1937). For k classes we compute

$$M = \ln s^2 - \sum_{i=1}^{k} (n_i - 1) \ln s_i^2 \tag{4.13}$$

and

$$C = 1 + \frac{1}{3(k-1)} \left\{ \sum_{i=1}^{k} \frac{1}{n_i - 1} - \frac{1}{N-k} \right\}, \tag{4.14}$$

where s^2 is the total variance in the sample and s_i^2 the variance of the ith class. The quantity M/C is distributed approximately as χ^2 with $k-1$ degrees of freedom, and calculated values can be compared with those tabulated for the desired level of probability.

The F test for two variances and Bartlett's test are sensitive both to real differences of variance and to departures from normality. Analysis of variance is much more robust. Even when significant differences are found among variances, the investigator is often quite justified in proceeding with analysis of variance on the assumption that the variances are equal.

Estimation

When classification divides a population into classes within each of which the variance is less than the variance in the population as a whole we can use the classification to generalize more usefully and predict more precisely than we could without it. Nevertheless, we may still be concerned about a whole population. For example, we may wish to know the total amount of bauxite in a district, or the total area of land suitable for irrigation. Classification can be helpful here too.

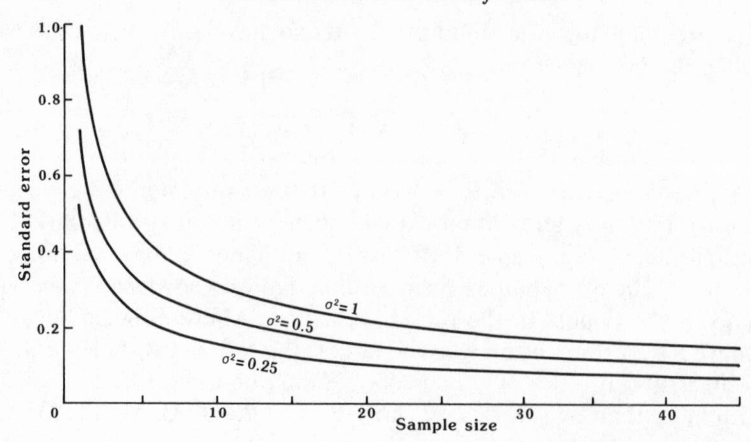

FIG. 4.1 Graphs of standard error for samples of varying size drawn from populations with variances of 1, 0.5, and 0.25

We saw in the last chapter that the standard error (s.e.) of a mean and its associated confidence intervals vary inversely as the square root of the sample size, n. In principle, therefore, we could reduce the s.e. to any desired value simply by increasing n. Fig. 4.1 shows that very large samples are needed to diminish the s.e. to much less than one-fifth of the population standard deviation.

We also saw in Chapter 3 how to improve the precision of an estimate by stratifying a population. If an area were divided into cells and the soil of each cell sampled separately the sampling error would depend on the variation within cells, which is generally less than that in the whole area. Sampling strata need not be regular cells, but can be created by any classification of soil. When the variation within such classes is less than in the population as a whole population parameters can be estimated more precisely or more efficiently by sampling the classes independently. If, for example, the variance within strata is half that in the population as a whole then, other things being equal, the sample need be only half the size to achieve the same standard error. If simple division of an area into regular cells achieves this then the advantage is pure gain. If, on the other hand, we have to make subsidiary observations to recognize different soil classes and delineate their boundaries then the effort involved must be set against the advantages of the smaller variance that results.

In many instances investigators do not know what precision to aim for, nor can they forecast how effective a soil classification will be. They cannot decide easily what effort to commit either to classification or to sampling, and they have to do what seems reasonable in the circumstances. Graphs like those in Fig. 4.1 are then useful aids. Assuming the population variance is 1, it might seem unreasonable to carry out expensive trials at 25 sites to reduce the standard error to 0.2, and so in the absence of other information an investigator might settle for s.e. = 0.29 from 12 sites. If with modest effort he could recognize soil classes within which the variance was 0.5 then he might consider it worth devoting that effort to classification and still sample at 12 sites to give an s.e. only slightly more than 0.2. To estimate the sampling error from the sample there should be some replication within classes. When sampling is costly and classification is cheap the options are represented mainly in the left of the figure. When sampling is cheap and classification is expensive attention should be focused on the upper curve but can extend to the right and beyond.

Experience to date suggests that the variances within the classes of a typical general purpose soil classification of an area range from about half the total variance for physical and mechanical properties to less than one-tenth for some chemical properties (Beckett and Webster 1971). It is unlikely that the within-class variance will be reduced to a quarter of the population variance without very substantial effort. Some attributes are closely associated with particular classes of soil; for example, some types of soil are well suited to agriculture, others quite unsuited. A survey intended to determine the proportion of land suitable for agriculture could be carried out efficiently after preliminary stratification.

Is classification worth while?

We have seen some of the effects and potential advantages of classification and how to measure them. We have also noted that classification does not always confer advantages, and it will be salutary to end this chapter with some guidance for action.

First, the soil within classes must be less variable with respect to the properties of interest than is the whole population. Otherwise prediction can be no more precise and generalization no more useful than if no classification had been made: the classification is irrelevant.

Second, the classification must be easy and cheap to create and to use. If it costs more to identify the class of soil than to measure the soil property of interest then it is better to measure the soil directly than to attempt prediction.

Many situations are not clear-cut. Classification allows somewhat better generalization, and somewhat more precise prediction and estimation, than are possible without it. However, the effort needed to classify the soil or identify types of soil is substantial, even though it is less than that required for direct measurement. The investigator must attempt to balance the costs of classification against the benefits that he can expect from it. If in doubt he will probably do best to limit classification to the most obvious differences and to increase sampling as far as resources allow.

5

RELATIONS BETWEEN VARIABLES: COVARIANCE AND CORRELATION

A poor relation is the most irrelevant thing in nature

Charles Lamb, *Poor Relations*

Chapters 2, 3, and 4 dealt with the quantitative description, estimation, and prediction of properties of the soil taken one at a time. In many investigations two or more properties of the soil, or of the soil and its environment or its productivity, are measured. They may all be intrinsically interesting; some may be the cause of others, some might be of little interest in themselves but are recorded in order to predict the values of others that are difficult or expensive to measure. Rarely are any two such properties independent in the statistical sense: they are almost always related to some extent. Technically, they are correlated.

Correlation is an advantage if we want to predict values of one property from another. This is the field of regression, which is the subject of the next chapter. A statistical relation may point to a causal one, either directly or through some intermediary, but not necessarily. On the other hand, correlation among recorded variables in a survey represents redundancy, some of which we may wish to remove for further analysis. Multiple correlation of this sort has been a major problem in soil systematics, and we deal with it in Chapter 8.

A soil property can also be related to itself in the sense that values at different places are related to one another. The property is then said to be *autocorrelated* or *spatially dependent*, and this form of relation is described in detail in Chapter 12.

Here we describe the basic linear form of relation between variables and how to measure it. It is a prelude to the chapters that follow on regression and multivariate analysis.

The scatter diagram

We begin with a graphic representation to help our understanding. Suppose that we have measured two properties of the soil at each of several sites, which for present purposes we shall consider to be the population.

TABLE 5.1 The clay + silt (material <0.06 mm diameter) and plastic limit of topsoil at 40 randomly located sites in west Oxfordshire (%)

	Clay + silt	Plastic limit		Clay + silt	Plastic limit
1	48.7	25	21	66.6	22
2	49.4	29	22	75.6	46
3	41.0	24	23	78.5	78
4	73.8	26	24	88.9	71
5	84.3	34	25	84.2	70
6	73.0	35	26	69.7	49
7	74.7	33	27	81.7	86
8	58.4	38	28	37.6	22
9	68.2	46	29	23.5	5
10	68.2	30	30	42.3	25
11	72.1	27	31	26.2	5
12	76.5	33	32	36.9	19
13	68.9	30	33	50.6	21
14	72.1	33	34	34.1	24
15	85.6	66	35	70.5	34
16	42.9	5	36	56.4	34
17	80.6	76	37	46.8	25
18	53.8	39	38	65.2	43
19	80.6	43	39	42.9	30
20	60.4	21	40	39.0	16

The population is said to be *bivariate*. We can represent the two variables by perpendicular axes on a graph. The axes define a two-dimensional space that we shall term a *character space*. Any individual can then be located in that space by rectangular coordinates equivalent to its observed values. Table 5.1 contains the values of clay + silt content and plastic limit of topsoil measured at 40 randomly located sites in West Oxfordshire. The data are displayed in Fig. 5.1.

The graph, a *scatter diagram*, shows how the 40 sites are distributed in the two dimensions and illustrates the relations between individuals. It also throws light on a quite new aspect—the relation between the characters. We consider this next.

Relations between variables

Examine Fig. 5.1. In general, sites with a large proportion of clay + silt also have large values of plastic limit, while those that have a small proportion of clay + silt tend to have small values of the plastic limit. The two sets of measurements depend to some extent on one another; they co-vary. We can express the degree of their dependence by their *covariance*. Let the observed values of the two properties X_1 and X_2 on n individuals be $x_{11}x_{12}$, $x_{21}x_{22}, \ldots, x_{n1}x_{n2}$. Then the variances of X_1 and X_2 are estimated as before, by

$$s_1^2 = \frac{1}{n-1}\sum_{i=1}^{n}(x_{i1}-\bar{x}_1)^2 \tag{5.1}$$

$$s_2^2 = \frac{1}{n-1}\sum_{i=1}^{n}(x_{i2}-\bar{x}_2)^2, \tag{5.2}$$

where $\bar{x}_1$ and $\bar{x}_2$ are the means of X_1 and X_2. In addition, the covariance is

$$c = \frac{1}{n-1}\left\{\sum_{i=1}^{n}(x_{i1}-\bar{x}_1)(x_{i2}-\bar{x}_2)\right\}. \tag{5.3}$$

The expression in braces is the sum of products about the means. It is clearly analogous to the sum of squares in the expressions for the variance of a single variate, equations (5.1) and (5.2).

The covariance of the percentage of clay + silt and the plastic limit in Table 5.1 is 265.31. The value is difficult to interpret as a measure of the relation because it is a function of the two variances whose values depend on the scales on which the properties are measured. It can be made more

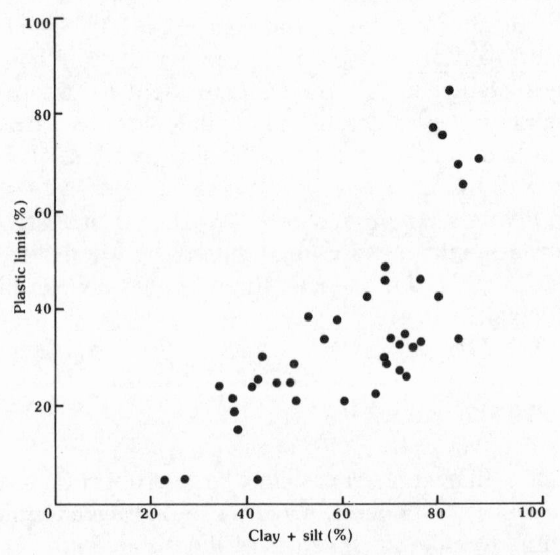

FIG. 5.1 A scatter diagram of the data given in Table 5.1

meaningful, however, if it is brought to a standard form by dividing it by the geometric mean of the variances; thus,

$$r = \frac{c}{\sqrt{(s_1^2 s_2^2)}}. \tag{5.4}$$

The c of Table 5.1 becomes

$$r = \frac{265.31}{322.62 \times 386.05} = 0.752.$$

The quantity r is known as the *product moment correlation coefficient* of the two variates, or usually just their *correlation coefficient*. The correlation of 0.75 between the clay + silt percentage and the plastic limit is moderate, as we shall see.

The correlation coefficient can neither exceed $+1$ nor be less than -1. If r is $+1$ then all the points lie on a straight line, conventionally rising from left to right on a graph, showing perfect positive correlation between the variables. If r is -1 then the points lies on a straight line falling from left to right: the variables are then perfectly negatively correlated. If r is zero there is no correlation, and the covariance, c, is also zero. Perfect correlation is very rare in real populations of material like soil: actual values of r fall between these extremes.

Fig. 5.2 illustrates some intermediate values of r. The variances on the two axes have been made equal, so that if correlation were perfect the points would lie on a line at 45° to the axes, dashed in the figures. It can be seen that r measures the extent to which the points deviate from such a line. The more they deviate, the nearer r is to 0, whereas the more closely they approximate to a straight line, the closer r is to 1 or -1.

Figure 5.2d shows a configuration for which r is sensibly zero. The two variates do not depend on one another. However, $r=0$ does not always mean that the two variates are independent. The points could lie, for example, on a curve or close to two perpendicular straight lines. Thus, when we speak of two variates' being uncorrelated we mean strictly only that they are not linearly related. It is always worth drawing a scatter diagram to check this.

If the variances on the two axes of a scatter diagram are not equal then the principal axes of the configuration will not be 45° bisectors. This can be seen in Fig. 5.3, in which a particular set of data is plotted with a fixed vertical scale but a varying horizontal scale. The correlation coefficient, r, is 0.752. Fig. 5.3a shows the configuration on axes with equal variance.

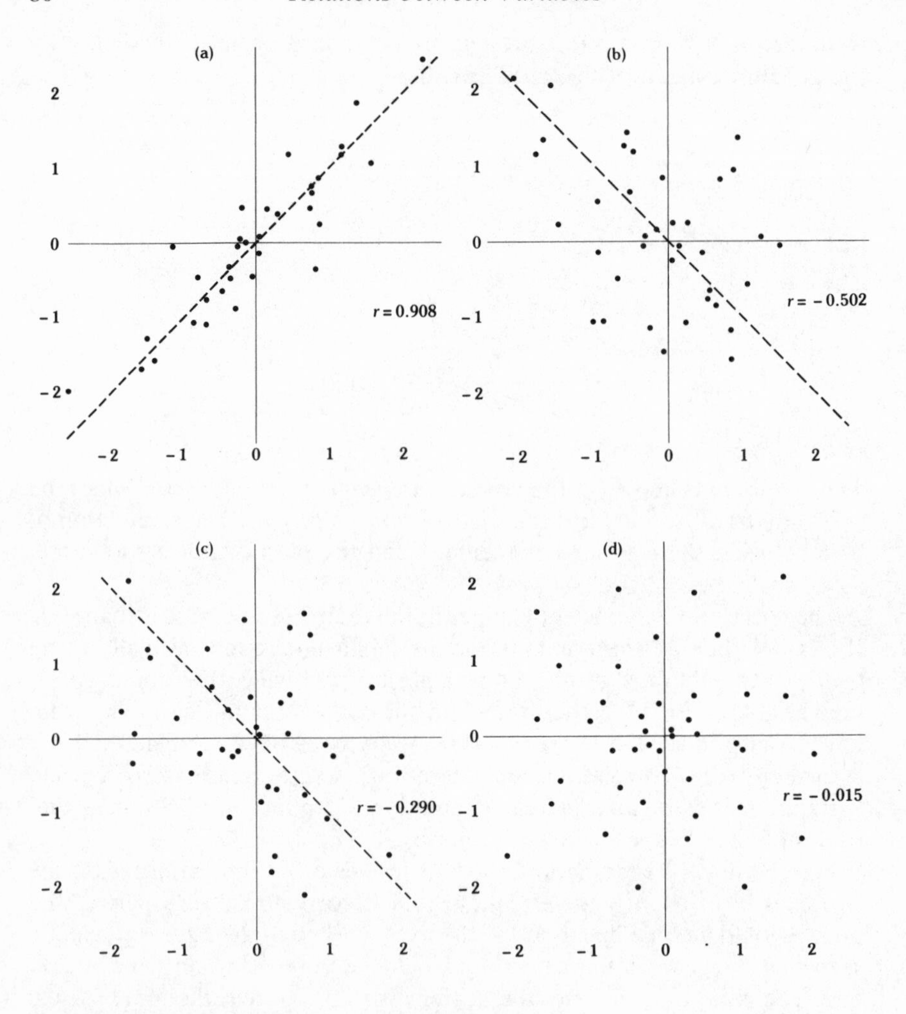

FIG. 5.2 Graphs showing scatter of points for a range of correlation coefficients. The variances are equal on both axes in all four graphs

In Fig. 5.3b the horizontal scale has been halved, thereby reducing the variance s_1^2 to a quarter of its original value and the covariance c to a half. The horizontal scale is multiplied by 2.5 in Fig. 5.3c: the variance is multiplied by 6.25 and the covariance by 2.5. Thus, by compressing and expanding the scale of measurement the configuration is alternately compressed and stretched. It is worth noting here that if one or other axis of Fig. 5.2d were compressed then the circular configuration of points would be compressed to an ellipse with two distinct principal axes, one almost vertical and the other almost horizontal. This is characteristic of uncorrelated variates.

Actual values of correlation coefficients

Numerous examples of the use of the correlation coefficient can be found in the literature of soil research. In many investigations theory would predict relations, and so large coefficients could be expected. In routine survey, however, where data are recorded for a variety of purposes there

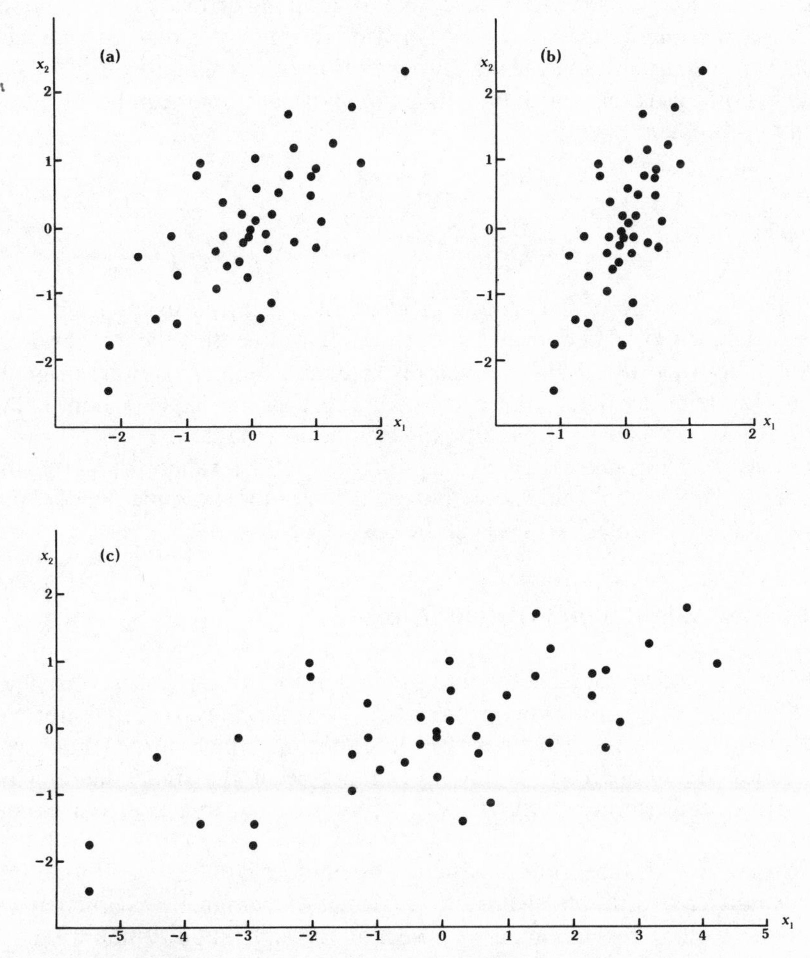

FIG. 5.3 Scatter diagrams showing the effect of change of measurement scale on a configuration of points. The correlation coefficient is 0.72

is no reason for such strong correlations. Published values of correlation coefficients between properties from such surveys (e.g. McKeague *et al.* 1971; Moore *et al.* 1972; Webster and Butler 1976) lie mainly in the range 0.3 to -0.3, and a few exceed 0.5 in absolute value. Indeed, if two properties are strongly correlated only one of them need be recorded. So neither view of correlation gives an entirely accurate picture of what to expect.

Estimation and significance

A value of r determined on data from a sample may estimate the correlation coefficient in the parent population, which is usually denoted by ρ. Since r is an estimate, it is subject to sampling error. If it is not equal to zero we might wish to know whether it reflects a real dependence between the variables in the whole population or is a sampling effect. The null hypothesis is that $\rho=0$, and, assuming normality, this can be tested by computing Student's t as

$$t=\frac{r(n-2)^{1/2}}{(1-r^2)^{1/2}}, \qquad (5.5)$$

where n is the size of the sample. The result is referred to the t table for $n-2$ degrees of freedom. If t exceeds that for the chosen level of probability then the null hypothesis is rejected: the correlation is significant. Inserting $r=0.752$ and $n=40$ for the data in Table 5.1 into this expression, we have $t=7.03$, a highly significant result.

Tables of r for several levels of probability and a range of degrees of freedom have been calculated and published, for example Fisher and Yates's (1963) Table VII, and can be consulted directly.

The bivariate normal distribution

We have already seen something of the importance of the normal distribution in both theory and practice for single variables. The bivariate normal distribution, and its extension to situations in which more than two variables are of interest, is equally important in the understanding of 'classical' multivariate methods. Therefore we shall explore it in some detail.

We saw that if, for a single normally distributed variable, we plotted the probability z as ordinate against the values of the variable x as abscissa we obtained a curve resembling a cross-section through a bell. When we have two variables X_1 and X_2 it is helpful to envisage a solid model whose base is defined by the variables on orthogonal axes x_1 and x_2 and the probability z

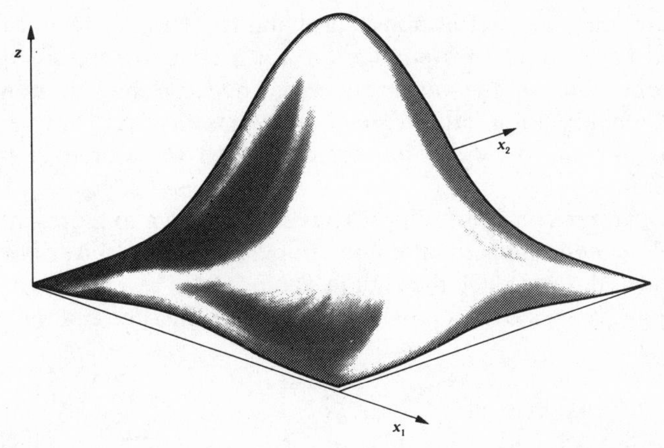

FIG. 5.4 Perspective view of a bivariate normal probability surface

represented vertically. The values of z lie on a curved surface shaped very like a bell (see Fig. 5.4). The equation of the surface is

$$z=\frac{1}{2\pi\sigma_1\sigma_2(1-\rho^2)^{1/2}}\exp\left[-\left\{\frac{(x_1-\mu_1)^2}{\sigma_1^2}-\frac{2\rho(x_1-\mu_1)(x_2-\mu_2)}{\sigma_1\sigma_2}+\frac{(x_2-\mu_2)^2}{\sigma_2^2}\right\}\Big/2(1-\rho^2)\right]. \qquad (5.6)$$

Its five parameters are

μ_1 and μ_2, the means of X_1 and X_2 respectively,

σ_1 and σ_2 the corresponding standard deviations,

and

ρ, the correlation coefficient between X_1 and X_2.

The distribution has several interesting properties.

1 Its peak lies directly above the means μ_1 and μ_2.
2 The two variables X_1 and X_2 are themselves normally distributed, so that projections of the surface parallel to the x_2 and x_1 axes are normal curves with variances σ_1^2 and σ_2^2 on the planes x_1z and x_2z respectively.
3 Any vertical section through the surface has a normal distribution.
4 Any horizontal section is an ellipse, and is a 'contour' line of equal probability.
5 In the univariate case the expression $(x-\mu)^2/\sigma^2$ in the exponent was distributed as χ^2. So here the expression in square brackets is also distributed as χ^2, this time with two degrees of freedom. As we shall

see, this enables us to choose probability contours to enclose any desired proportions of a population. Such contours are analogous to confidence limits in the univariate case. They also show the connection between correlation and regression geometrically. This has applications in many fields of investigation, and is covered in the next chapter.

In Fig. 5.5 three concentric ellipses have been drawn to represent the 50, 90, and 99 per cent limits of an infinite population with a bivariate normal distribution and from which the data in Table 5.1 are regarded as a sample. The axes have been moved so that the origin is at the centre of the ellipses.

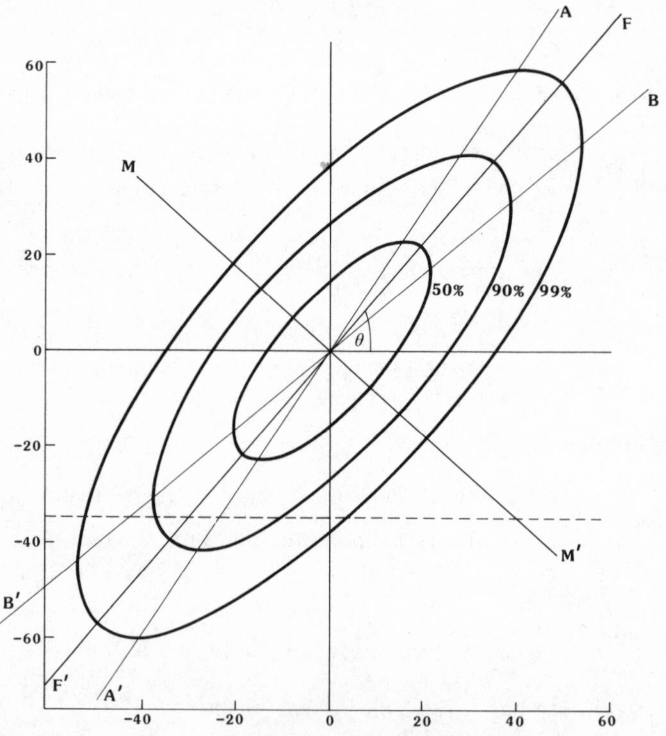

FIG. 5.5 Confidence ellipses for 50, 90, and 99% probability for the population represented in Fig. 5.1

The major and minor axes of the ellipses, the principal axes, lie on the lines FF′ and MM′ respectively. The line AA′ bisects all horizontal chords to the ellipses, hence it joins the mean values of x_1 for each value of x_2, $x_{1.2}$. It cuts each ellipse where the tangent to the ellipse is horizontal. It is known as the line of regression of X_1 on X_2. Its equation, assuming the coordinates of the centre are (0, 0), is

$$x_1 = b_1 x_2 \tag{5.7}$$

where

$$b_1 = \rho \frac{\sigma_1}{\sigma_2}.$$

Similarly, the line BB′ is the regression line linking the mean values of x_2 for every value of x_1, $x_{2.1}$. It cuts the ellipses where their tangents are vertical, and its equation is

$$x_2 = b_2 x_1 \tag{5.8}$$

where

$$b_2 = \rho \frac{\sigma_2}{\sigma_1}.$$

The quantities b_1 and b_2 are known as regression coefficients.

Constructing confidence ellipses

Confidence ellipses aid our understanding of a bivariate population, and therefore we should be able to draw them. We can do so if we first determine the orientation and lengths of their principal axes. Principal axes are considered algebraically in Chapter 8. Here we shall anticipate a little, drawing especially on the geometric representation given there.

Orientation. Let the angle between the principal axes and the original axes be θ. In Chapter 8 (p. 129) we see that any point with coordinates x_1, x_2 on the original axes can be referred to a new pair of perpendicular axes rotated through an angle θ by two new coordinates y_1, y_2, which are

$$\begin{cases} y_1 = x_1 \cos\theta + x_2 \sin\theta \\ y_2 = x_2 \cos\theta - x_1 \sin\theta \end{cases} \quad . \tag{5.9}$$

As mentioned earlier, when the principal axes of a configuration of points coincide with the reference axes there is no correlation between the variates. Thus, if we rotate the original axes to principal axes the new variates that we obtain are uncorrelated. The sum of their products is therefore zero. So, by multiplying equations (5.9) for each point, summing, and dividing the sum by the number of degrees of freedom, we obtain

$$0=(s_2^2-s_1^2)\sin\theta\cos\theta+c(\cos^2\theta-\sin^2\theta)$$

$$=(s_2^2-s_1^2)\frac{\sin 2\theta}{2}+rs_1s_2\cos 2\theta, \qquad (5.10)$$

where s_1^2, s_2^2, c, and r are the variances of X_1 and X_2, and their covariance and their correlation coefficient respectively. The angle θ is given by

$$\tan 2\theta=\frac{2rs_1s_2}{s_1^2-s_2^2}. \qquad (5.11)$$

To find the orientation of the principal axes for the configuration shown in Fig. 5.1, we insert the values $s_1^2=322.63$, $s_2^2=386.05$, and $r=0.752$ into (5.11) to give

$$\tan 2\theta=-8.362,$$

and, taking due care with signs,

$$\theta=48°24'.$$

Standard deviations. The lengths of the major and minor axes of the confidence ellipses of a bivariate normal distribution are proportional to the standard deviations along them. Let the estimates of the standard deviations from a set of points be u_1 and u_2. Then by squaring equations (5.9) for each point, summing, dividing by the degrees of freedom, and finally adding, we obtain

$$u_1^2+u_2^2=s_1^2+s_2^2. \qquad (5.12)$$

Further, from (5.6), the value of z for the central ordinate, i.e. for $x_1=\bar{x}_1$ and $x_2=\bar{x}_2$, is

$$z=\frac{1}{2\pi s_1s_2\sqrt{(1-r^2)}} \qquad (5.13)$$

For the new variates defined by the principal axes, r is zero and

$$z=\frac{1}{2\pi u_1u_2}. \qquad (5.14)$$

Since the value of z must be the same whether it is calculated from the new variates or the original ones,

$$u_1 u_2 = s_1 s_2 \sqrt{(1-r^2)}. \tag{5.15}$$

Thus we have two equations, (5.12) and (5.15), in two unknowns that can be solved to find u_1 and u_2.

Inserting the values of s_1, s_2, and r gives

$$u_1^2 + u_2^2 = 322.62 + 386.05 = 708.68$$

$$u_1 u_2 = 17.95 \times 19.65 \times \sqrt{(1-0.752^2)} = 232.50;$$

therefore

$$2u_1 u_2 = 465.0.$$

Adding and subtracting these gives

$$u_1^2 + 2u_1 u_2 + u_2^2 = 1173.68$$

and

$$u_1^2 - 2u_1 u_2 + u_2^2 = 243.68.$$

On taking square roots we obtain

$$u_1 + u_2 = 34.26$$

$$u_1 - u_2 = 15.60,$$

so that

$$u_1 = 24.93 \quad \text{and} \quad u_2 = 9.33.$$

We now have all the information needed to construct ellipses for any level of confidence. The probability density referred to the principal axes is

$$z = \frac{1}{2\pi u_1 u_2} \exp\left[-0.5\left\{\frac{y_1^2}{u_1^2} + \frac{y_2^2}{u_2^2}\right\}\right]. \tag{5.16}$$

All terms involving the correlation coefficient disappear since it is zero. The term in the square brackets is distributed as χ^2; so by setting first $y_2 = 0$ and then $y_1 = 0$ we can find the values of y_1 and y_2 where the ellipse cuts the major and minor axes. To construct the ellipse containing 90 per cent of the population represented in Fig. 5.1, and hence excluding 10 per cent, χ^2 with two degrees of freedom is 4.605, $\sqrt{\chi^2}=2.15$; the ends of the major axis are $y_1=\pm 53.60$ and those of the minor axis $y_2=\pm 20.06$.

Confidence ellipses so constructed apply strictly only for distributions that are bivariate–normal. Many pairs of soil variable have roughly such a distribution, and for them the ellipses are likely to be good enough approximations. However, since principal axes are defined as that pair of orthogonal axes through the data points for which the coordinates are uncorrelated, the angle θ that they make with the original axes of measurement, and the standard deviations along them, u_1 and u_2, hold, whatever the distribution. Unlike most univariate statistics, multivariate methods are often exploratory and descriptive in situations where the underlying distributions are quite unknown or immaterial. Transformation to principal axes can then be a valuable means of elucidating structure in the data, as we shall see in Chapter 8.

Normal correlation theory is restricted to quantitative variables. There are other techniques for measuring relations between qualitative characters. They include *rank correlation* for ranked characters and measures of *association* or *contingency* for binary and unordered multi-state characters.

Matrix representation

The information contained in the variances and covariance can be expressed as a matrix, **A**, with elements a_{ij}:

$$\mathbf{A}=\begin{bmatrix} a_{11} & a_{12} \\ a_{21} & a_{22} \end{bmatrix}.$$

The diagonal elements a_{11}, a_{22} are the variances of X_1 and X_2 respectively. The off-diagonal elements a_{12}, a_{21}, which are equal, contain the covariance. If each element is divided by the square root of the product of the diagonal elements in the same row and column the result is the *correlation matrix* **R**. Its diagonal elements will clearly all equal 1, and the off-diagonal elements contain the correlation coefficient:

$$\mathbf{R}=\begin{bmatrix} 1 & r \\ r & 1 \end{bmatrix}.$$

The original observations will have consisted of n pairs of measurements, and these too can be arranged in a matrix **X**, conventionally with n rows and 2 columns. If the elements, x_{ij}, of this matrix are adjusted so that they are deviations from the variate means, x_j, then we can form the matrix **S** by

$$\mathbf{S} = \mathbf{X}^T\mathbf{X}. \tag{5.17}$$

This is the matrix of sums of squares and products (of deviations from the means) and is usually found first. Dividing by $n-1$ gives matrix **A**:

$$\frac{1}{n-1}\mathbf{S} = \mathbf{A}. \tag{5.18}$$

S is sometimes known as the *dispersion matrix*; but so is **A**. Therefore, it is safer to avoid ambiguity and refer to **S** as the *sums of squares and products* (abbreviated to SSP) *matrix* and **A** as the *variance–covariance matrix*, respectively.

Later we shall see how, by operating on matrix **A** or **R**, we can derive the principal axes of ellipses.

Extension to more than two variates

The principles of correlation extend to more than two variables, and in a geometric representation we simply add one new dimension to the character space for each additional variable. For three variables the scatter of individuals occupies a three-dimensional space, and it can be displayed by constructing a model or a stereogram. It requires a considerable stretch of the imagination to envisage the associated probability distribution, since this extends into a fourth dimension. However, the probability envelopes, corresponding to the contours of the bivariate case, occupy only three dimensions. For normal distributions they are ellipsoidal, and their size and shape are determined as in the two-dimensional case by the variances on their principal axes. Their orientation is determined by the covariances, of which there are now three, namely c_{12}, c_{13}, and c_{23}. We can again represent the information as a variance–covariance matrix, this time with three rows and three columns:

$$\mathbf{A} = \begin{bmatrix} a_{11} & a_{12} & a_{13} \\ a_{21} & a_{22} & a_{23} \\ a_{31} & a_{32} & a_{33} \end{bmatrix}.$$

As before, the variances lie on the principal diagonal and the three covariances off the diagonal: and the matrix is symmetric; i.e., $a_{ij} = a_{ji}$ for all i and j.

For more than three variables it requires an even further stretch of the imagination, since the scatter of points must be envisaged in a space of at least four dimensions. Yet all the principles that we have demonstrated for two and three dimensions hold for as many dimensions as we have measured variables. The SSP, variance–covariance, and correlation matrices all take one more row and one more column for each additional variate. Thus, for p variates these matrices are of size $p \times p$. In particular, the probability density for any multivariate-normal distribution can be written in general in matrix notation as

$$z = \frac{1}{(2\pi)^{p/2}|\mathbf{A}|^{1/2}} \exp\{-\tfrac{1}{2}(\mathbf{x}-\boldsymbol{\mu})\mathbf{A}^{-1}(\mathbf{x}-\boldsymbol{\mu})^T\}, \tag{5.19}$$

where $\mathbf{A}$ is now the population variance–covariance matrix, $\mathbf{x}$ is a vector of values, and $\boldsymbol{\mu}$ is the mean vector.

Matrices of this kind which represent the relationships among variates are sometimes known as *R-matrices*, as distinct from *Q-matrices*, which represent relations among individuals and which we shall consider in Chapter 7.

6

REGRESSION

> Madness is hereditary:
> you get it from your children.
>
> Anon.

Before pursuing the principle of correlation into multivariate analysis, there is another closely related matter to deal with. It is the topic of regression. The central idea of regression is that of *dependence*: of one variable's depending on one or more others in some sense. Examples of such dependence are numerous in soil science. For instance, cation exchange capacity depends on organic matter content, and pH on the degree of base saturation. In Fig. 5.1 plastic limit can be thought of as depending on the content of clay + silt. Behind these statements is the notion of causation or control. In some instances a physical relationship is obvious. Thus, organic matter has many exchange sites and its presence in soil contributes to the exchange capacity of the soil. Regression alone, however, cannot show whether there is a direct causal relationship between variables. Furthermore, it does not matter from the point of view of regression analysis which variable is thought of as cause and which as effect.

The ideas of regression, however, are not restricted to situations in which there is obvious cause and effect. For example, the cation exchange capacity of the soil does not determine its plastic limit, although they are likely to be related. Nevertheless, we may still wish to consider the dependence of one on the other. This is the basis of prediction which is central to regression. For instance, soil surveyors cannot usually make as many measurements as they would like on all the properties of interest. In particular, they make only a few measurements of properties that are expensive to determine. If, however, they could establish a close relation between such properties and those that are cheap to measure then they would be able to predict the former from the latter. In these circumstances the variable to be predicted is treated as dependent, and the predictor as the independent variable. There can be more than one predictor. For example, de la Rosa (1979) attempted to predict engineering properties of the soil from organic matter, clay content, and cation exchange capacity recorded in soil survey.

Regression analysis is a valuable technique. Apart from prediction, it can be used for interpolation, calibration, and in certain circumstances for description. It is also very easy to perform, and indeed to misuse out of context. Users must be aware of the restricted circumstances in which regression is valid. The subject is well covered in standard texts, notably by Williams (1959), Sprent (1969), and Draper and Smith (1981). Gower's (1983) review is also instructive.

Bivariate regression

Consider first the situation illustrated in Figs. 5.4 and 5.5 of the previous chapter in which two random variables, X_1 and X_2, are related in such a way as to have a bivariate-normal distribution and are defined by equation (5.6). Suppose that the value of X_1, for any sample of soil, is known accurately, having been measured, and that we want to predict X_2. Clearly, for each value x_1 there is a spread of values of x_2. The best prediction, therefore, will be the expected value of X_2 given that $X_1 = x_1$. This is the conditional mean of X_2, which is written as

$$\mathrm{E}[X_2|x_1] = \rho \frac{\sigma_2}{\sigma_1} x_1, \tag{6.1}$$

where σ_1 is the standard deviation of X_1, σ_2 is the standard deviation of X_2, and ρ is the product–moment correlation coefficient as defined in Chapter 5. Usually the scales of measurement will not be centred, and account must be taken of the means, μ_1 and μ_2; so (6.1) becomes

$$\mathrm{E}[X_2|x_1] = \mu_2 + \rho \frac{\sigma_2}{\sigma_1}(x_1 - \mu_1). \tag{6.2}$$

By varying x_1 we obtain the corresponding values for x_2 as

$$x_2 = \mu_2 + \rho \frac{\sigma_2}{\sigma_1}(x_1 - \mu_1). \tag{6.3}$$

This is the equation of a straight line passing through the centroid (μ_1, μ_2) of the distribution with a gradient $\rho(\sigma_2/\sigma_1)$: it is the regression line of X_2 on X_1, and it is shown as BB′ in Fig. 5.5. It cuts the ellipses of equal probability where their tangents are vertical. If we denote the gradient by

$\beta = \rho(\sigma_2/\sigma_1)$ we can write the equation in the usual form for a straight line:

$$x_2 = \alpha + \beta x_1, \tag{6.4}$$

where $\alpha = \mu_2 - \beta\mu_1$ is the intercept on the ordinate. The gradient β is the regression coefficient.

In the above X_1 is treated as the independent variable or predictor and X_2 as depending on it and being predicted. Equally, their roles can be reversed to obtain the equation

$$x_1 = \mu_1 + \rho\frac{\sigma_1}{\sigma_2}(x_2 - \mu_2), \tag{6.5}$$

which defines the locus of $E[X_1|x_2]$. This line, with gradient $\beta_2 = \rho(\sigma_1/\sigma_2)$ and passing through $\mu_1 = \mu_2 = 0$, is shown as AA′ in Fig. 5.5. It cuts the ellipses of equal probability where their tangents are horizontal. It is the line we should use if we wished to predict values of X_1 from X_2. In regression analysis it is usual to designate the independent variable as X and the dependent variable as Y. This follows the convention in mathematics in which a variable y is treated as a function of another variable x, though herein lies a trap, as we shall see.

Estimation

Estimating the regression equation is straightforward provided that the data represent the population without bias. Combining the above definition of β with equations (5.1)–(5.3) and replacing X_1 by X and X_2 by Y, the regression coefficient is simply the covariance of X and Y divided by the variance of X. If there are n pairs of observations, $(x_1, y_1), (x_2, y_2), \ldots, (x_n, y_n)$, then we can estimate it by

$$\hat{\beta}_{Y.X} = b_{Y.X} = \frac{\sum_{i=1}^{n}(x_i - \bar{x})(y_i - \bar{y})}{\sum_{i=1}^{n}(x_i - \bar{x})^2}, \tag{6.6}$$

where $\bar{x}$ and $\bar{y}$ are the means of the observed X and Y. The regression line passes through $\bar{x}, \bar{y}$, and its intercept on the ordinate

$$a = \bar{y} - \beta\bar{x} \tag{6.7}$$

estimates α, the true intercept.

The data in Table 5.1 can be used to illustrate this. To predict the plastic limit, Y, of the soil from the content of silt + clay, X, we compute

$$b_{Y.X}=\frac{265.63}{322.45}=0.824,$$

and $a_{Y.X}=35.45-0.824\times 61.26=-15.0$, and the regression equation is

$$y=-15.0+0.824x.$$

Figure 6.1 shows this line *fitted* to the data.

We could attempt equally well to predict the content of silt + clay from the plastic limit of this case This would give

$$b_{X.Y}=\frac{265.63}{388.26}=0.684,$$

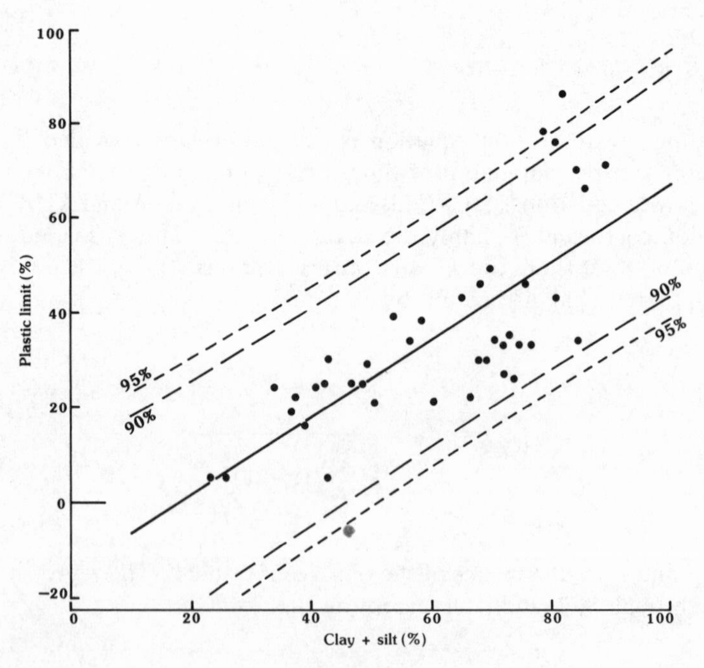

FIG. 6.1 Scatter diagram of data from Table 5.1 with regression of plastic limit on clay + silt fitted. The curved lines join the confidence limits at the 90 and 95% levels for individual predictions

and

$$a_{X.Y} = 61.26 - 0.684 \times 35.45 = 37.01,$$

and the regression equation would be

$$x = 37.01 + 0.684y.$$

Gauss linear regression

Another situation in which we often wish to use regression is where only one of the variables is random. Examples abound in agricultural field experiments in which a treatment of, say, nitrogen fertilizer is applied to the soil at several rates and the yield of crop that it produces is measured. Similarly, different amounts of gypsum or lime might be added to the soil and the resulting changes in the soil's strength, degree of aggregation, or acidity measured. In each case, the amounts of fertilizer or ameliorant are chosen deliberately, but the responses to a given dressing fluctuate in a random manner. This situation is depicted in Fig. 6.2, where the response variable (yield, strength, or acidity) is represented on the ordinate and the amount of treatment applied on the abscissa. For each predetermined amount of treatment there is a distribution in the response, shown here as a normal curve that should be thought of as standing perpendicular to the plane of the paper.

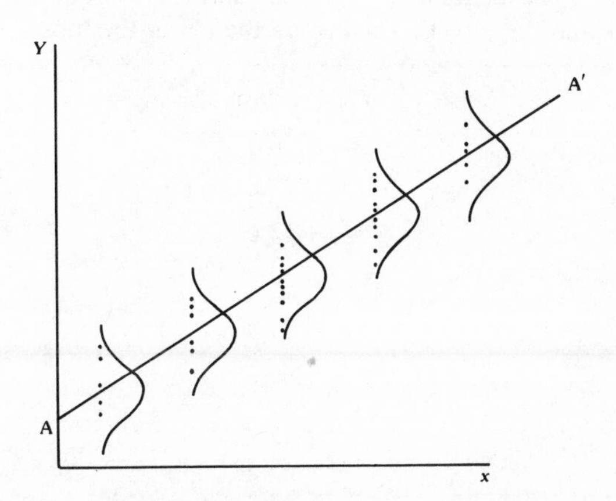

Fig. 6.2 Diagram illustrating the Gauss linear model of regression. The curves of the normal distribution should be thought of as projecting perpendicular to the page

Assuming that the relation between the variables is linear, our regression model is now

$$y_i = \alpha + \beta x_i + \varepsilon_i, \qquad (6.8)$$

where y_i is the value of the random variable Y for a given value of the mathematical variable x_i, the treatment in this case, and ε_i is a random variable with a mean of zero. In the absence of any information to the contrary, the residuals are usually assumed to be normally distributed with a common variance σ^2 and independent of one another. The regression line is

$$y = \alpha + \beta x$$

and is that shown in Fig. 6.2. It is estimated by the method of least squares, usually attributed to Gauss. The computational procedure for fitting the regression line is exactly the same as that for the bivariate-normal distribution, and this fact often obscures the essential difference between the two types of variation.

A recent study by Goulding *et al.* (1989) illustrates this situation The soil of the Park Grass Experiment at Rothamsted is naturally acid, and it has been made more so by the addition of nitrogen fertilizer (Johnston *et al.* 1986). In 1975 lime was added at several rates and the change in pH, ΔpH, nine years later measured. The results are given in Table 6.1. Fig. 6.3 shows the data plotted as a scatter diagram and also the regression of ΔpH against the amount of added lime, as the straight line.

TABLE 6.1 The amount of lime added to Park Grass experimental plots and the resulting change in the pH of the topsoil, ΔpH

	Lime applied		ΔpH
	1.9		−0.5
	3.8		−0.5
	12.6		+0.5
	13.8		+1.1
	16.3		+1.1
	19.5		+1.3
	20.7		+2.3
Means	12.7		0.757
Variances	53.38		1.0229
Standard deviation	7.31		1.01
Covariance		7.073	
Correlation coefficient		0.96	

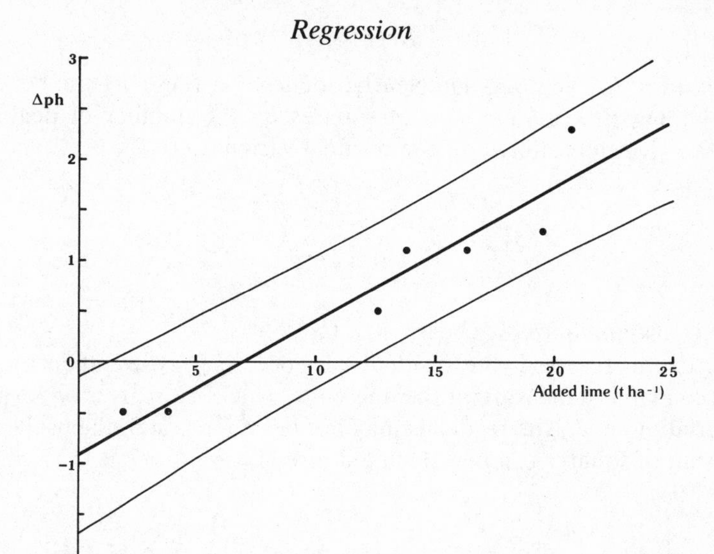

Fig. 6.3 Scatter diagram of data from Table 6.1 with the regression of ΔpH on added lime. The curved lines pass through the confidence limits for individual predictions at the 90% level. The data were kindly provided by Dr K. W. T. Goulding

The values of x need not be chosen in advance for the Gauss linear model to apply. If, instead of performing an experiment to determine the effects of gypsum or lime, an investigator were to conduct a survey of such treatments he would obtain numerous different values that depended on the amounts the farmers had applied. These may still be regarded as mathematical variables.

For the Gaussian model there is no regression of x on Y corresponding to to the regression of X on Y in the bivariate case, though see the discussion on calibration below.

Residual variances, prediction, and confidence intervals

In both the bivariate and Gaussian regression the procedure for finding the regression equation minimizes the sum of squares of the differences between the observed values of Y and those given by the equation. That is, it minimizes

$$\sum_{i=1}^{n} (y_i - \hat{y}_i)^2,$$

the residual sum of squares (RSS). The $y_i - \hat{y}_i$, $i = 1, 2, \ldots, n$, known as the *residuals*, are the vertical distances between the plotted points and the lines in Figs. 6.1 and 6.3, and clearly, the smaller these are the better the fit. Dividing the residual sum of squares by the number of degrees of freedom gives an estimate of the residual variance,

$$\hat{\sigma}^2_{Y.X} = s^2_{Y.X} = \frac{1}{n-2} \sum_{i=1}^{n} (y_i - \hat{y}_i)^2, \tag{6.9}$$

which is also minimized.

The extent to which the residual variance, $s^2_{Y.X}$, is less than the total variance in Y is a measure of the degree to which the regression 'explains' the variation in Y. The residuals may not be of interest in themselves, but their sum of squares can be calculated directly as:

$$\text{RSS} = \sum_{i=1}^{n} (y_i - \bar{y})^2 - \left\{ \sum_{i=1}^{n} (x_i - \bar{x})(y_i - \bar{y}) \right\}^2 \Bigg/ \sum_{i=1}^{n} (x_i - \bar{x})^2. \tag{6.10}$$

From the equations (5.1)–(5.4) we can express the numerator in the second term on the right-hand side of (6.6) as

$$r^2 \sum_{i=1}^{n} (x_i - \bar{x})^2 \sum_{i=1}^{n} (y_i - \bar{y})^2.$$

Substituting this in (6.10) gives

$$\text{RSS} = (1 - r^2) \sum_{i=1}^{n} (y_i - \bar{y})^2. \tag{6.11}$$

In words, $1 - r^2$ is the proportion of the total variance in Y that remains after allowing for the variation in X. Its complement, r^2, is the proportion accounted for by the regression. Thus, these quantities are analogous to what we called the relative variance of classification and the intraclass correlation in Chapter 4.

Applying (6.10) and (6.11) to the example of the regression of plastic limit on silt + clay content gave the following results. The total sum of squares (of deviations from the mean) of Y is 15 055.85. When the regression line is fitted RSS = 6550.24, giving a residual variance of $s^2_{Y.X} =$ 172.37. The ratio of the two sums of squares is 6550.24/15 055.85 = 0.455, which equals $1 - r^2$. Thus, somewhat less than half the variance in Y remains in the residuals from the regression. In like manner, the total sum of squares for change of pH in our second example is 6.1374 and the RSS after fitting the regression is 0.5138. In this case no more than 8.4 per cent of the variance remains after fitting the regression: the regression equation explains 91.6 per cent of the variation in ΔpH.

Before using a regression equation we need to satisfy ourselves that our estimate of the slope of the line, b, is not such that we could have obtained it easily by chance. It must be significantly larger (or smaller) than zero. The variance of b about the true value β is estimated by

$$s_b^2 = s_{Y.X}^2 \bigg/ \sum_{i=1}^{n} (x_i - \bar{x})^2. \tag{6.12}$$

The appropriate significance test is then based on

$$t = b/s_b \tag{6.13}$$

with $n-2$ degrees of freedom. This test is equivalent to that for the correlation coefficient, equation (5.5).

In our example $s_{Y.X}^2$ is 172.37 and $\Sigma_{i=1}^{n}(x_i - \bar{x})^2$ is 12 575.32, giving the variance of the regression coefficient $s_b^2 = 0.01371$ and $t = b/s_b = 7.02$. This far exceeds the tabulated value of t for 38 degrees of freedom at $P = 0.001$, and we therefore judge it significant.

Having satisfied ourselves that b is significant, we may wish to place confidence limits on β. This is straightforward, since $(b-\beta)/s_b$ is distributed as t. If we wanted the 90 per cent symmetrical confidence interval then the limits would be $b - t_{0.05}s_b$ and $b + t_{0.05}s_b$. In the example β would lie in the range $0.824 - 0.236$ to $0.824 + 0.236$.

We can now consider the errors of prediction. To do so, it will be helpful to think of the value of Y of an individual as having contributions from three sources: the mean of Y, the regression of Y on X or x, and the residual from the regression, ε. If we use the subscript 0 to refer to values of this new individual then we can represent the three sources by the equation

$$y_0 = \mu + \beta(x_0 - \bar{x}) + \varepsilon_0. \tag{6.14}$$

We have already encountered the variance of the residuals, $s_{Y.X}^2$, in (6.9). If the regression equation were known exactly then this would be the only source of uncertainty. Since we usually have only estimates of μ and β, we have to recognize that these are also subject to error and swell the error in any prediction that we make. The difference between a true value, y_0, and its estimate, $\hat{y}_0$, is therefore composed of three error terms, as follows:

$$y_0 - \hat{y}_0 = (\mu - \bar{y}) + (\beta - b)(x_0 - \bar{x}_0) + \varepsilon_0. \tag{6.15}$$

In addition to the residual variance, we must estimate the variances associated with μ and β. The variance of $\bar{y}$ is $s^2_{Y.X}/n$. The variance of the second term in (6.15) is

$$s^2_b(x_0-\bar{x})^2 = s^2_{Y.X} \Big/ \sum_{i=1}^{n} (x_i-\bar{x})^2. \tag{6.16}$$

Assuming that these three sources of variation are independent, the three components may be summed to provide the variance of any regression estimate:

$$s^2_{Y0} = \frac{s^2_{Y.X}}{n} + \frac{s^2_{Y.X}(x_0-\bar{x})^2}{\sum_{i=1}^{n} (x_i-\bar{x})^2} + s^2_{Y.X}, \tag{6.17}$$

which is often expressed in the form

$$s^2_{Y0} = s^2_{Y.X}\left\{1 + \frac{1}{n} + \frac{(x_0-\bar{x})^2}{\sum_{i=1}^{n} (x_i-\bar{x})^2}\right\}. \tag{6.18}$$

Confidence limits for y are found readily by using the t distribution as $\hat{y} \pm ts_{Y0}$. For example, the limits for predicting plastic limit from clay + silt content are shown as continuous lines in Fig. 6.1. The closer pair span the 90 per cent interval, the wider pair, that for 95 per cent probability. Notice that the lines are curved. This is because the confidence limits depend on x_0. The second term on the right-hand side of (6.17) shows this. From (6.15) it can be seen that any error in the estimate of β is magnified more strongly in the regression estimate the further we depart from the mean of X. The confidence interval is least at $\bar{x}$, and widens as $|x_0-\bar{x}|$ increases. If we attempt to extrapolate by taking x beyond the range of our data we might find the confidence interval too wide for worthwhile prediction. Fig. 6.3 shows the 90 per cent confidence limits for predicting change in pH from adding lime.

Sometimes an investigator wants to place a confidence zone about the regression line itself. In this case he needs the confidence limits for the average value of Y for a given X. The variance of this average is obtained readily from the model, equation (6.12), and equation (6.16) without the final error term, thus:

$$s^2_{Y0} = \frac{s^2_{Y.X}}{n} + \frac{s^2_{Y.X}(x_0 - \bar{x})^2}{\sum_{i=1}^{n} (x_i - \bar{x})^2}. \quad (6.19)$$

The predicted value is $\hat{y}$, as before, but the confidence limits, given by $\hat{y} \pm ts_{Y0}$, are narrower.

Finally in this section, readers must be aware of another and usually more serious hazard of extrapolation. The regression equation was derived to fit a particular and finite set of data. It expresses the relation between two variates within the range observed. There can be no guarantee that the same relation holds beyond that range. It is clear from Fig. 6.1 that the equation for linear regression cannot hold when the clay + silt content is less than about 20 per cent. And in Fig. 6.3 we cannot increase the pH indefinitely by adding lime. In many instances the population regression is curved, but the curvature is not detected because the sample spans only a small part of the range.

Calibration

It often happens that a property of the soil that we want to know could be determined substantially without error if we were prepared to take the necessary time and trouble. Nevertheless, if the determination is time-consuming or expensive then we might prefer to predict it from some other property that is quicker and cheaper to measure, but which gives only approximate results. This is a very common situation in soil laboratories. The concentrations of soil constituents are often determined spectrographically or by some photoelectric means. The analyst, therefore, makes up a series of standards of known concentration and then measures the response on the instrument. He then draws a calibration line from the results from which he can estimate the concentration of other constituents in the soil.

Thus the analyst has a set of fixed values, x_i, $i = 1, 2, \ldots, n$, of a mathematical variable for each of which he has an observation y_i which is subject to error. In these circumstances the Gauss linear model applies; there is only the one regression, namely that of Y on x; and the calibration line is found by fitting this regression equation and thereby minimizing the deviations in Y. The aim of calibration, however, is to estimate x from Y, sometimes called the *inverse estimation*, and the analyst wants to use the equation in the form

$$x = (y - a)/b_{Y.x}. \quad (6.20)$$

This calculation is straightforward. Finding confidence limits, or *tolerance limits* as they are more usually called when dealing with calibration, is more difficult. Nevertheless, a formula based on normal distributions of the deviations in Y and of $b_{Y.x}$ is available. To simplify the expression, let s and b stand for $s_{Y.x}$ and $b_{Y.x}$ respectively, and define g as

$$g = t^2 s^2 \Big/ \left\{ b^2 \sum_{i=1}^{n} (x_i - \bar{x})^2 \right\}, \tag{6.21}$$

where t is the Student's t at the chosen level of probability for the appropriate degrees of freedom. Then the confidence limits are given by

$$l = \hat{x}_0 + \left[(\hat{x}_0 - \bar{x})g \pm \frac{ts}{b} \sqrt{\left\{ \frac{(\hat{x}_0 - \bar{x})^2}{\sum_{i=1}^{n} (x_i - \bar{x})^2} + \frac{n+1}{n}(1-g) \right\}} \right] \Big/ (1-g). \tag{6.22}$$

In the situation that we have considered the errors in Y are likely to be small: we should regard the analytical technique as poor otherwise. So s^2 is likely to be small in relation to $\Sigma_{i=1}^{n}(x_i - \bar{x})^2$ and g negligible. In these circumstances (6.22) reduces to

$$l = \hat{x}_0 \pm \frac{ts}{b} \sqrt{\left\{ 1 + \frac{1}{n} + \frac{(\hat{x}_0 - x)^2}{\sum_{i=1}^{n} (x_i - \bar{x})^2} \right\}}. \tag{6.23}$$

Hodgson *et al.* (1976) described another interesting application of regression to calibration. They wanted to know the proportions of silt and clay in the soil. Determining these routinely in the laboratory was too expensive, and so they wished to use hand-texturing in the field to estimate or predict the contents of silt and clay that they would have measured in the laboratory. They took some 200 samples of soil spanning a wide range of textures, estimated the proportions of silt and clay both by hand texturing and laboratory measurement, and then performed a regression analysis. Error was assumed to be present only in the field estimates, and so the regression equation they used was that of the field estimate, a random variable Y, on the laboratory determination, a mathematical variable x. They could then use the inverse relation of equation (6.20) to predict the laboratory values from the field estimates.

This application raises a further consideration. The x values for these particular samples of soil were not chosen deliberately; i.e., they were not fixed. And if the set of samples could be regarded as random then the authors could have computed the regression of x on Y. This would have minimized the deviations in x, and would give somewhat narrower confidence intervals.

Some writers have pointed out that the regression of x on Y would give increased confidence in any event. This choice of regression is a matter of controversy, however, and was aired a few years ago in a series of papers in *Technometrics* (Krutchkoff 1967; Williams 1969; Halperin 1970; Shukla 1972). Readers who have calibration problems in which there are large errors in the predictor should be aware of the different views.

Description

We come now to an aspect of regression analysis that puzzles many scientists, or one that they overlook and as a result perform inappropriate analysis. It is how to express law-like relations. In many instances soil scientists think of the relation between two soil properties or between one measurement and another as having the form

$$y = \alpha + \beta x, \tag{6.24}$$

where y and x are mathematical variables. There is no error, the relation is exact, and y is strictly a function of x and vice versa. The equation describes a *functional relation*. Darcy's law, in which the flow of water through the soil is related to the hydraulic head, is an example.

If we could measure x and y precisely then it would be a simple matter to determine α and β in the equation from just two pairs of observations. In practice, there are usually errors in one or both variables arising from sampling fluctuations and errors in the measurements themselves, so that what we actually measure are

$$\begin{aligned} X &= x + \xi \\ Y &= y + \eta, \end{aligned} \tag{6.25}$$

where ξ and η are the errors incurred in observing x and y respectively. In other words, our observed variables are the random variables X and Y. If we substitute these values into (6.24) we obtain

$$Y = \alpha + \beta X + (\eta - \beta\xi), \tag{6.26}$$

and what seemed a simple matter of regression is clearly not, since X is a random variable and is correlated with the residual $\eta - \beta\xi$. Equation (6.26) describes a *structural relation* between observed random variables that results from an underlying functional relation, equation (6.24), between mathematical variables. The problem is to estimate the parameters α and β, and the procedure for doing this is called *structural analysis*. The subject is involved, and only a brief account is possible here. Readers who wish to pursue it in depth should consult Kendall and Stuart (1967) or Sprent (1969).

Consider a sample on which we have paired observations, which for the ith pair are X_i and Y_i. The true values are x_i and y_i, but because of sampling fluctuation and measurement error the observed values differ from the expected ones by

$$\xi_i = X_i - x_i \quad \text{and} \quad \eta_i = Y_i - y_i.$$

The average values of ξ^2 and η^2, σ_ξ^2 and σ_η^2 respectively, are the error variances of x and y. The error variances can be estimated by taking replicate observations of x_i and y_i in turn for different values of i. Let their ratio be $\sigma_\eta^2/\sigma_\xi^2 = \lambda$. For a bivariate-normal population the slope of the functional relation can then be estimated by

$$\hat{\beta} = [s_Y^2 - \lambda s_X^2 + \sqrt{\{(s_Y^2 - \lambda s_X^2)^2 + 4\lambda c^2\}}]/2c, \tag{6.27}$$

where s_X^2 and s_Y^2 are the variances in X and Y, and c is the covariance of X and Y—following the conventions of (5.1)–(5.3). The line representing the functional relation passes through the mean μ_X, μ_Y, estimated by $\bar{X}$ and $\bar{Y}$, and so α is estimated by

$$\hat{\alpha} = \bar{Y} - \hat{\beta}\bar{X}. \tag{6.28}$$

Clearly, to find β the investigator must know σ_ξ^2 and σ_η^2 or at least their ratio. If he begins with the aim of determining the functional relation then he can plan his sampling with sufficient replication to estimate the error variances. Alternatively, he might be fortunate in knowing them from previous experience.

If the x values are known exactly then $\sigma_\xi^2 = 0$, $\lambda \to \infty$, and the slope of the line, β in (6.27), reduces to

$$\hat{\beta} = c/s_X^2 = b_{Y.X} \tag{6.29}$$

In other words, the functional relation is estimated by the familiar regression of Y on X. It is easier to see what happens when $\sigma_\eta^2 = 0$, i.e. when the y values are known exactly. Here $\lambda = 0$, and (6.27) becomes

$$\hat{\beta} = s_Y^2/c. \tag{6.30}$$

This is simply the reciprocal of the regression of X on Y, i.e.

$$\hat{\beta} = 1/b_{X.Y}. \tag{6.31}$$

These two regression equations represent the extremes in structural analysis. If our knowledge of both x and y is subject to error then β will lie between $b_{Y.X}$ and $1/b_{X.Y}$. Neither regression equation is then appropriate. This important result is all too often overlooked by soil scientists seeking to express functional relationships between properties. Mark and Church (1977) make the same criticism for earth sciences generally. The soil scientist often finds himself analysing data for which no estimates of sampling and measurement errors are available. Therefore, unless he can be sure that the observations of one of the variables are free of error, he has a difficult choice that can be based only on a general understanding of the soil, the field situation, and the techniques used in sampling and measurement. There are three reasonable choices.

Equal errors. The errors in the two variables may be taken to be the same, so that $\lambda = 1$. The equation for the relation is then that of the major principal axis. This minimizes the sum of squares of the deviations perpendicular to the fitted line. The fitted values lie at the intersections of the fitted line with the perpendiculars passing through the observed values. The gradient $\hat{\beta}$ can be çalculated from equation (6.27) or by a principal components analysis, Chapter 8. The choice is reasonable for variables that are measured on the same scales using the same measurement techniques and on the same batches of soil. Thus, the relation between exchangeable sodium and cation exchange capacity might be treated in this way.

If we imagine that the data in Table 5.1 derive from a functional relation between the plastic limit and the silt + clay content of the soil then we can illustrate these options. Fig. 6.4 shows the major principal axis for the data. Its gradient is the tangent of θ in equations (5.9), i.e. $\tan 48°24' = 1.126$. Since it goes through the mean, its equation is

$$y = -33.53 + 1.126x.$$

Proportional errors. The errors in the two variables may be taken to be proportional to their total variances. This gives

$$\lambda = s_Y^2 / s_X^2, \tag{6.32}$$

and the gradient of the equation is simply

$$\hat{\beta} = \pm\sqrt{\lambda}, \tag{6.33}$$

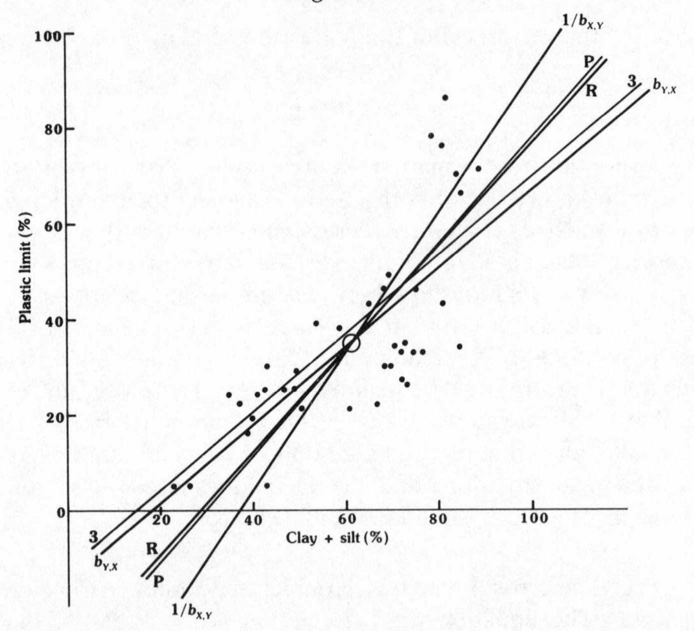

FIG. 6.4 Scatter diagram of data from Table 5.1 with lines representing the options for a functional relation as follows: regression of Y on X, labelled with slope $b_{Y.X}(=0.824)$; regression of X on Y, labelled with slope $1/b_{Y.X}(=1.462)$; principal axis, P, with slope 1.126; reduced major axis, labelled R with slope 1.097; the line joining the centroids of two groups into which the data are divided is indistinguishable from first regression line; the line joining centroids of the two extreme groups, when the data are divided into three, is labelled 3

with the correct sign the same as that of $b_{Y.X}$ or r. This is equivalent to

$$\hat{\beta} = b_{Y.X}/|r|. \tag{6.34}$$

The line with this gradient passing through the centroid of the data is known as *the reduced major axis*. The fitted values lie at the intersections of the fitted line with lines of gradient $-\hat{\beta}$ passing through the observed values, and the sum of the squares of the deviations along these lines is minimized. It also has the interesting property that it bisects the angle between the two regression lines.

With this approach, the gradient for the data in Table (5.1) is $\lambda = 388.26/322.45 = 1.204$. The gradient of the reduced major axis is $\sqrt{1.204} = 1.097$, and the equation of the functional relation is

$$y = -31.77 + 1.097x. \tag{6.35}$$

Unassumed errors. If neither of the above seems reasonable or if too little is known for them to be assumed safely then there is an obvious solution. The pairs of observations can be divided into two groups, preferably equal, roughly about the mid-point of the range of either X or Y. The means of the X and Y for each group are determined, and the line joining their centroids is taken to express the functional relation. Thus, if the centroids of the two groups are $\bar{X}_1$, $\bar{Y}_2$ and $\bar{X}_2\,\bar{Y}_2$ then the gradient of the line is estimated by

$$\hat{\beta}=\frac{\bar{Y}_2-\bar{Y}_1}{\bar{X}_2-\bar{X}_1}. \tag{6.36}$$

This method can be made more efficient by dividing the sample into three groups and choosing the line joining the centroids of the two extreme groups to represent the relationship. Kendall and Stuart (1967) describe the statistical aspects of this approach in detail.

The results of applying this simple procedure to our example are interesting. By dividing the data into two groups we obtain the line

$$y=-15.68+0.835x.$$

This passes through the mean $\bar{X}$, $\bar{Y}$ with gradient $\hat{\beta}=0.835$. It is very close to the regression of Y on X. Dividing the data into three groups and joining the centroids of the two extreme groups of 13 points gives

$$y=-13.58+0.841x.$$

This has a gradient only very slightly more than that calculated from the two groups of 20, but it passes above the centroid. The reason for this discrepancy is that the data are not scattered symmetrically about any straight line. The six points in the upper right of the figure lie somewhat above the general trend through the remainder. In attempting to fit a straight line through the whole set of data, we assumed that such a line could reasonably represent the relation in the population as a whole, and that any departure from that relation in our data was entirely random. After an investigation of this kind we might wish to review this assumption and to fit a curve rather than a straight line.

The replacement of a functional relation by a structural relation and a subsequent structural analysis are designed to deal with those situations where we believe there to be some underlying law that is obscured by the imprecision of our observations. In soil science we have also to recognize that there are many situations where variation is inherent: it is not error. The examples quoted at the beginning of this chapter are of this kind. In such situations we are dealing with a structural relation from the start, and a full description of the relation must recognize its bivariate nature, as in the bivariate-normal distribution, equation (5.6). Sometimes we may wish to think of some line that represents the average relation between X and Y and perform a structural analysis similar to the one described above. This extension is discussed by Kendall and Stuart (1967) and Sprent (1969).

Generalization and matrix representation

The linear regression that we have studied above is the simplest of a large class of regression models. Often, however, we wish to study the dependence of one property of the soil on two or more others. Cation exchange capacity, for example, almost certainly depends on both clay and organic matter content, and in some types of soil it depends on the pH also. Thus we might express the dependence in the form

$$y = \alpha + \beta_1 x_1 + \beta_2 x_2, \tag{6.37}$$

where x_1 and x_2 are values of clay and organic matter content and y is the exchange capacity. A further term, $\beta_3 x_3$, could be added to include the effect of pH. Alternatively, x_1 and x_2 might be geographic coordinates. We should then be expressing y as a function of position on the ground, and the procedure is then generally known as *trend surface analysis*. In this instance the surface is a plane which, unless both β_1 and β_2 are zero, is inclined. More complex expressions can be constructed to take account of evident curvature, for example a quadratic in a single independent variate,

$$y = \alpha + \beta_1 x + \beta_2 x^2, \tag{6.38}$$

or a quadratic trend surface,

$$y = \alpha + \beta_1 x_1 + \beta_2 x_2 + \beta_3 x_1^2 + \beta_4 x_2^2 + \beta_5 x_1 x_2. \tag{6.39}$$

In all the instances mentioned so far the X variables were continuous over some range. However, we might wish to introduce a factor that has only two or a few distinct states. The effects arising from two or more different treatments of the soil or several distinct soil types can be accommodated in regression analysis by defining *dummy variables*. If we have two states of a factor they can be represented by one dummy variable, which takes the value 0 for one state and 1 for the other. If there are three states then two dummy variables are required, taking values 0 and 0 or 0 and 1, or 1 and 0 respectively for the three states.

All the above are examples of a general linear model of regression. They are linear in the parameters $\alpha, \beta_1, \beta_2, \ldots$. It is possible, therefore, to provide a general method for estimating the parameters by matrix algebra.

We shall do this for simple regression, but to make its generality clear we shall change the notation somewhat.

Let us define the regression model:

$$y = \beta_0 + \beta_1 x + \varepsilon. \tag{6.40}$$

Essentially, we have replaced α by β_0 and β by β_1. The coefficients β_0 and β_1 are estimated by b_0 and b_1 from our data. Let S be the sum of the squares of the residuals as before; then

$$S = \sum_{i=1}^{n} (y_i - \hat{y}_i)^2 = \sum_{i=1}^{n} (y_i - b_0 - b_1 x_i)^2. \tag{6.41}$$

We wish to choose the values b_0 and b_1 to minimize S, so we differentiate S first with respect to b_0 and then b_1:

$$\begin{cases} \dfrac{dS}{db_0} = -2 \sum_{i=1}^{n} (y_i - b_0 - b_1 x_i) \\ \dfrac{dS}{db_1} = -2 \sum_{i=1}^{n} x_i (y_i - b_0 - b_1 x_i). \end{cases} \tag{6.42}$$

Setting these to zero gives

$$\begin{cases} \sum_{i=1}^{n} (y_i - b_0 - b_1 x_i) = 0 \\ \sum_{i=1}^{n} x_i (y_i - b_0 - b_1 x_i) = 0. \end{cases} \tag{6.43}$$

By multiplying through and rearranging, we obtain

$$\begin{cases} b_0 n + b_1 \sum_{i=1}^{n} x_i = \sum_{i=1}^{n} y_i \\ b_0 \sum_{i=1}^{n} x_i + b_1 \sum_{i=1}^{n} x_i^2 = \sum_{i=1}^{n} x_i y_i. \end{cases} \tag{6.44}$$

These equations are known as the *normal equations*, and it is clear that their solution leads to the results obtained in (6.6) and (6.7).

Equations (6.44) can be expressed equally well in matrix form. To do so a dummy variate of 1s is added to X so that the data of the predictor form a matrix $\mathbf{X}$, thus:

$$\mathbf{X} = \begin{bmatrix} 1 & x_1 \\ 1 & x_2 \\ \vdots & \vdots \\ 1 & x_n \end{bmatrix}.$$

If $\mathbf{X}$ is pre-multiplied by its transpose, $\mathbf{X}^T$, we obtain

$$\mathbf{X}^T\mathbf{X} = \begin{bmatrix} n & \sum_{i=1}^{n} x_i \\ \sum_{i=1}^{n} x_i & \sum_{i=1}^{n} x_i^2 \end{bmatrix}. \qquad (6.45)$$

Post-multiplying this by the vector of regression coefficients forms $\mathbf{X}^T\mathbf{X}\mathbf{b}$, which is the left-hand side of the normal equations, (6.44). The right-hand side of (6.44) is obtained as $\mathbf{X}^T\mathbf{y}$, where $\mathbf{y}$ is a vector containing the observed values of Y, i.e.,

$$\mathbf{y} = \begin{bmatrix} y_1 \\ y_2 \\ \vdots \\ y_n \end{bmatrix}.$$

The normal equations are thus

$$\mathbf{X}^T\mathbf{X}\mathbf{b} = \mathbf{X}^T\mathbf{y}. \qquad (6.46)$$

They are solved for $\mathbf{b}$ by pre-multiplying both sides by the inverse of $\mathbf{X}^T\mathbf{X}$, so that

$$\mathbf{b} = (\mathbf{X}^T\mathbf{X})^{-1}\mathbf{X}^T\mathbf{y}. \qquad (6.47)$$

The result is quite general, and it can be applied for any number of predictor variates, X, provided that there are enough observations, i.e. provided $\mathbf{X}^T\mathbf{X}$ is not singular. The fitted values are obtained by

$$\hat{\mathbf{y}} = \mathbf{X}\mathbf{b}. \qquad (6.48)$$

The residual sum of squares is $(\mathbf{y}-\hat{\mathbf{y}})^T(\mathbf{y}-\hat{\mathbf{y}})$, which, when divided by the degrees of freedom, $n-2$, gives $s_{Y.X}^2$. The predicted value of Y for any new individual, 0, whose values of X we know can be expressed as

$$\hat{y}_0 = \mathbf{x}_0\mathbf{b},$$

where $\mathbf{x}_0$ is the row vector $[1\ x_0]$ for a single predictor or $[1\ x_{01}\ x_{02}\ .\ .\ x_{0p}]$ for p predictors. Its estimation variance, which we shall state without derivation, is

$$s_{Y0}^2 = s_{Y.X}^2 + s_{Y.X}^2\mathbf{x}_0(\mathbf{X}^T\mathbf{X})^{-1}\mathbf{x}_0^T. \tag{6.49}$$

The method can be modified further to force the regression line or surface through the origin if that seems sensible. It is adapted readily to data in which there are varying degrees of confidence. If, for example, each x, y pair of values is a composite of several observations, then a weight should be assigned to each variate proportional to the number of individual observations that comprise it. This leads to a weighted least-squares solution. There are also closely related methods for fitting nonlinear models to data by least-squares analysis. Readers can pursue these in more specialized textbooks; that by Draper and Smith (1981) is especially recommended.

7

RELATIONS BETWEEN INDIVIDUALS: SIMILARITY

> We would never have learned anything if we had never thought 'This object resembles this other, and I expect it to manifest the same properties.'
>
> Bertrand de Jouvenel

Normal correlation theory, which we described in Chapter 5, was developed early in this century. In many instances it provides the most appropriate means of relating two variables, and we can use it with confidence in pedology. Relations between individuals present more problems. There is no one measure of relation that is obviously best in all circumstances. Consider the following; they illustrate the kinds of comparison that are the essence of soil systematics.

(a) Soil *I* is more acid than soil *J*.
(b) The soil at *C* is stronger than that at *D*.
(c) Soil *E* contains more exchangeable bases than soil *F* does.
(d) The soil in the A horizon of profile *G* is similar to that in its B horizon.
(e) Profile *P* is more like profile *Q* than profile *R*.

It is unlikely that comparison (a) would give rise to any dispute. To compare *I* and *J* quantitatively we should simply take the *difference* of their pH values. We might reasonably take a similar view in the second and third instances, and measure the differences in their shear strengths and total exchangeable bases (TEB) content. However, we might equally reasonably wish to compare them as ratios: the shear strength of soil *C* is four times that of soil *D*, the TEB content of soil *E* is twice that of soil *F*. The latter is often the case, since the difference between the values 2 and 4 meq/100 g soil can be quite as important as that between 30 and 60. We should note that we have already made this judgement for soil acidity by using the logarithmic pH scale. Measuring the similarity of two horizons (d) is more difficult because there are likely to be several characters to take into account, and comparing profiles (e) adds the factor of pattern, the spatial arrangement of horizons or layers within the profile. Although we make comparisons of this sort daily, both in ordinary life and in soil survey and research, we need a formal framework if we are to make them appropriately and consistently; and this we must be able to do if we are to use such comparisons in several of the numerical methods of classification that we shall discuss later.

We shall use the geometric model again to formalize our thinking. We shall represent a single variable as a line, and locate individuals on the line by their values of that variable. The distance separating two individuals is then the measure of their relation. The closer they are, the more alike they are and vice versa. If individuals can be compared better by ratios of their values of the variable then we shall transform the scale to logarithms first. Notice that this is a matter of choice that rests on pedological judgement.

When there are two variables to be considered simultaneously we can plot the positions of individuals as in Fig. 5.1, and measure the likeness between any two individuals as the distance between them. Alternatively, the distance can be calculated by Pythagoras' theorem. If the coordinates of two points i and j are x_{i1}, x_{i2} and x_{j1}, x_{j2} then the distance Δ_{ij} between them is given by

$$\Delta_{ij}=\sqrt{\{(x_{i1}-x_{j1})^2+(x_{i2}-x_{j2})^2\}}. \tag{7.1}$$

Substituting the values of the first two individuals from Table 5.1,

$$\begin{aligned}\Delta_{12}&=\sqrt{\{(48.7-49.4)^2+(25-29)^2\}}\\&=4.06.\end{aligned}$$

The principles hold for any number of dimensions. If there are p variables then we can postulate a p-dimensional character space, and (7.1) generalizes to

$$\Delta_{ij}=\sqrt{\left\{\sum_{k=1}^{p}(x_{ik}-x_{jk})^2\right\}}. \tag{7.2}$$

The distance Δ is often known as the *Pythagorean distance*, *Euclidean distance*, or *taxonomic distance* between the individuals. It increases with the number of characters involved in the comparison, so it is often divided by $\sqrt{p}$ to give an 'average' distance,

$$d_{ij}=\sqrt{\left(\frac{\Delta_{ij}^2}{p}\right)}, \tag{7.3}$$

which is actually the square root of the average of the squared distances. Strictly, d_{ij} measures the dissimilarity between the individuals i and j. We can convert d_{ij} to a measure of similarity if we scale it first so that it lies in the range 0 for identity to 1 for maximum dissimilarity and then take its complement:

$$S_{ij}=1-d_{ij}. \tag{7.4}$$

The quantity S_{ij} is known as a *similarity coefficient* or *similarity index*. As defined here, it is one of many possible measures of similarity. We shall consider several other possibilities later.

Standardization

When calculating the relation between two variables we found it convenient to standardize the scales of measurement so that we could interpret the result more readily. Standardization is almost essential when we consider relations among individuals. Suppose, for example, we wish to assess the relations between pairs of sampling sites at which we have measured the pH and clay+silt content of the soil. Clay+silt content ranges from perhaps 20 to 90 per cent, as in the data in the Table 5.1, while pH for the same sample is unlikely to exceed the range 5.5–7.5. If we calculate the Pythagorean distances from the data as they are the results will depend largely on the differences in clay+silt content, and scarcely at all on pH. To avoid this the scales are standardized so that they are comparable.

Standardization is usually achieved by dividing all the values of a variable by the standard deviation of that variable in the sample. Every scale then has a standard deviation (and variance) of 1. It would be equally reasonable to divide values by the sample range. However, if the sample range happened to be a small proportion of the known range for a variable it might be better to divide the values by the known range instead. This is equivalent to saying that, because the range of a variable is small in the sample now being studied, that variable is relatively unimportant. Here again, the investigator has a choice that must be based on his experience and judgement.

Similarity matrix

For every pair of individuals i and j there will be a distance Δ_{ij} separating them in character space and a corresponding S_{ij}. Like the covariances of the last section, these can be arranged in a matrix, sometimes known as a *Q-matrix*. If there are n individuals the matrix will be of size $n \times n$. As an example, we can take the first three points in Table 5.1, standardize their values by dividing by the ranges of silt+clay content and plastic limit respectively, and then calculate the average distances between them. We obtain the dissimilarity matrix:

$$\mathbf{D} = \begin{bmatrix} 0 & 0.2095 & 0.4958 \\ 0.2095 & 0 & 0.2934 \\ 0.4958 & 0.2934 & 0 \end{bmatrix}$$

or alternatively the similarity matrix:

$$\mathbf{S} = \begin{bmatrix} 1.000 & 0.7905 & 0.5042 \\ 0.7905 & 1.0000 & 0.7066 \\ 0.5042 & 0.7066 & 1.0000 \end{bmatrix}.$$

Both are symmetric.

Other measures of likeness

Pythagorean distance has been introduced as a measure of relationship between individuals because it derives directly from the geometry of our model and is intuitively appealing. However, it is by no means the only such measure. Many have been proposed, and readers requiring a reasonably full account of them should consult Sneath and Sokal (1973). Here we shall consider just a few measures that have been used in pedology.

We shall see that Pythagorean distance is not necessarily satisfactory if we consider the following situation. Suppose three properties have been measured on each of five individuals. When standardized their values are:

P_1	0.0	0.0	0.0
P_2	0.0	0.0	1.0
P_3	0.0	1.0	1.0
P_4	1.0	1.0	1.0
P_5	0.0	0.0	0.0

In the geometric representation they occupy positions in a three-dimensional character space with these values as coordinates. The first four individuals occupy different corners of a unit cube, while the fifth coincides with the first. The Pythagorean distances of the first four points from P_5 are respectively

$$0,\ 1.0,\ 1.41,\ \text{and}\ 1.73.$$

The last three distances are respectively a side, a diagonal of a face, and a diagonal through the cube. It might be thought that a difference in one character has made a disproportionately large contribution to the distance and hence to the calculated dissimilarity. We could avoid that in this instance by using squared distances, but we should encounter similar difficulties with distances to individuals in other positions.

Mean character distance

A reasonable alternative to Pythagorean distance is the sum of the absolute differences in each dimension,

$$\sum_{k=1}^{p} |x_{ik} - x_{jk}|,$$

which is known in topology as the *Manhattan* or *city-block metric*. The dissimilarities between P_5 and P_1, P_2, P_3, and P_4 are now 0, 1.0, 2.0, and 3.0. When divided by the number of dimensions the sum of the differences becomes

$$d_{ij} = \frac{1}{p} \sum_{k=1}^{p} |x_{ik} - x_{jk}|. \qquad (7.5)$$

In this form it is usually known as *mean character distance* (Cain and Harrison 1958), though it had been used much earlier in taxonomic work.

Gower's similarity coefficient

The above two measures assume that the characters on which the comparison is based are continuous variables, though binary characters can be included. In some instances we shall want to include binary and multistate characters. For example, we might wish to compare soil profiles on their clay content, pH, presence of coatings (binary), and type of structure (unordered multistate) simultaneously. Gower (1971) devised a general measure of similarity to accommodate these, given by

$$S_{ij} = \frac{\sum_{k=1}^{p} z_{ijk} w_{ijk}}{\sum_{k=1}^{p} w_{ijk}}, \qquad (7.6)$$

where z_{ijk} is a value for the comparison for the kth character, and w_{ijk} is the weight assigned to it. For continuous variables

$$z_{ijk} = 1 - \frac{|x_{ik} - x_{jk}|}{r_k},$$

where r_k is the range of the character. As mentioned above, r_k can be either the range within the sample being studied or the range known to exist in the population at large. In either case division by r_k standardizes the characters. For qualitative characters, $z_{ijk}=1$ if $x_{ik}=x_{jk}$, and 0 otherwise. The weight w_{ijk} is set conveniently to 1 when a valid comparison can be made between i and j for the character k, and to 0 if x_{ik} or x_{jk} or both are unknown or inapplicable.

A point of some subtlety arises with binary characters of the present-or-absent kind for which presence is regarded as important and absence as unimportant. For example, the presence of erratic stones in the soil of a region showing no clear evidence of glaciation and the presence of montmorillonite where the soil clay is dominantly kaolinite might be thought significant, whereas their absence would be of little interest. In these circumstances two soil specimens or profiles may be regarded as alike when both possess the character, and unlike when one possesses the character and the other does not; but no comparison is made when both lack the character. They are accommodated in formula (7.6) by setting $w_{ijk=0}$ when $x_{ik}=x_{jk}=0$. Rayner (1966), who first used Gower's coefficient for matching soil horizons, chose to treat the presence of earthworms, manganese concretions, and iron concretions in this way. He called them *dichotomies*. He treated porosity likewise, scoring it as 1 when mentioned in the survey record, and 0 otherwise. In doing so, he illustrates another feature of survey data. For a surveyor often records only what is present in the soil at a sampling site, little realizing that by ignoring absent characters he could be implying, perhaps justifiably, that they are relatively unimportant in the context of the survey. This leads to uncertainty in the interpretation of the records. Binary characters of this kind should therefore be distinguished from other qualitative characters, and should be firmly recorded as absent when they are absent.

When all data are quantitative Gower's coefficient is the mean character distance. When all are binary and matches between absent characters are ignored, it is the same as that of Jaccard (1908) and Sneath (1957). Further, it can be converted readily to a distance (Gower 1966) by

$$\Delta_{ij}=\sqrt{\{2(1-S_{ij})\}}. \tag{7.7}$$

Canberra metric

A measure that has no geometric equivalent, but which has been used in several soil studies (Moore and Russell 1967; Campbell *et al.* 1970; Webster and Burrough 1972a, b), is the *Canberra metric* developed by Lance and Williams (1967a). It is a dissimilarity defined as

$$d_{Cij}=\sum_{k=1}^{p} |x_{ik}-x_{jk}| \Big/ (x_{ik}+x_{jk}). \tag{7.8}$$

It is attractive because it is a measure whose values depend solely on the individuals being compared, and the data need not be standardized beforehand. Differences are measured as proportions of the data, and hence this metric is appropriate for ratio scales. Qualitative characters can be included, as in Gower's general similarity coefficient. However, prior standardization is swift by computer, and proportional differences can be included in any of the coefficients mentioned earlier by transforming the scales to logarithms. It can be used only when all values of characters are positive. Its major weakness is that its validity is in doubt for scales with an arbitrary zero, e.g. colour hue, pH. Its advantages are therefore less than might appear at first.

Generalized distance

In choosing characters for multivariate comparisons and analysis, one will omit any character that is necessarily correlated with another already included. For example, the proportion of stones in soil would not be included if the percentage of fine earth had already been selected: total soil=fine earth+stones; the two are perfectly correlated, and to include both would simply duplicate data.

However, it often happens that many characters that an investigator wishes to include are correlated to some extent. If the correlation coefficients are small (in absolute value) they can be disregarded. If they are large then the investigator should consider whether to take them into account when measuring likeness. The situation is illustrated in Fig. 7.1a, which shows the scatter of the 40 individuals of Fig. 5.1 but on standardized scales. Consider the relationships between the points X, Y, and Z. X is nearer to Y than it is to Z in the ordinary sense. However, the distance XZ is measured approximately parallel to the long axis of the configuration, whereas XY is approximately at right angles, and so much of the difference

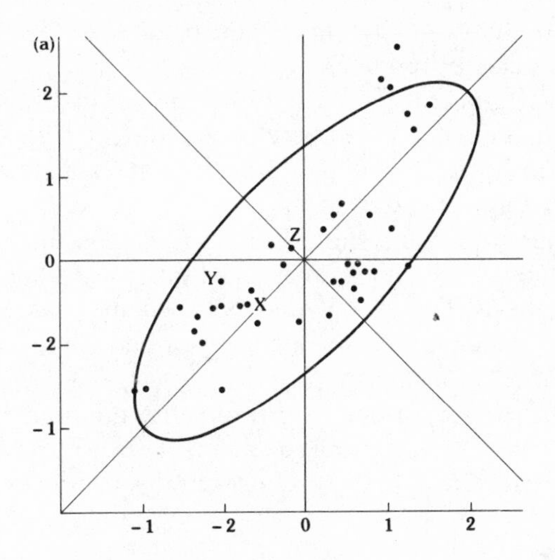

FIG. 7.1(a) Scatter of points shown in Fig. 5.1 replotted on standardized scales. The correlation coefficient is 0.752

between the two distances could be attributed to the correlation. If we were to add a third soil property, such as cation exchange capacity, which like plastic limit depends to a large extent on the amount of clay in the soil, it is likely that we should still have an obliquely orientated elongated configuration. This elongation could be attributed to the control of two of the

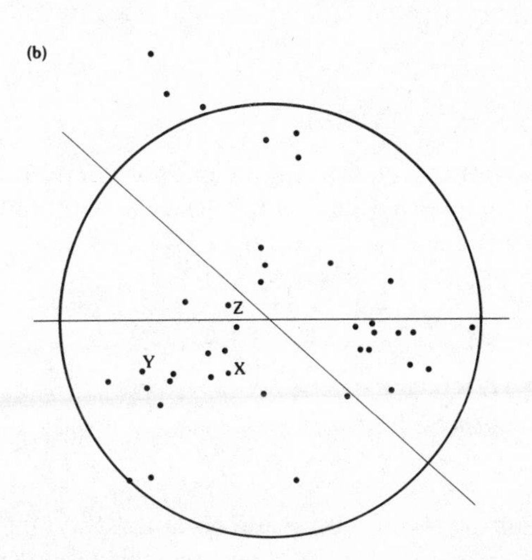

FIG. 7.1(b) The same points plotted on oblique axes cutting at the angle $\cos^{-1}$ 0.752, i.e. 41°12′

properties, plastic limit and cation exchange capacity, by a large component of the third, clay+silt content.

The dominating influence of clay+silt content in this example, and correlation effects in general, can be eliminated if we rotate the axes with respect to one another until each pair is inclined at an angle whose cosine is the correlation coefficient between the two variates. When this is done, and provided the scales are standardized as here, the distribution will no longer be elongated. The ellipse of Fig. 7.1a is transformed into a circle, Fig. 7.1b, and the distances XY and XZ become more nearly equal. Distances between individuals in this transformed space can be used as measures of their relationships.

It would be far too laborious to measure the relations between many pairs of points in this way, and impossible if there were more than three variates. But we can calculate the distances in this transformed space from the quadratic form

$$D_{ij}^2 = (\mathbf{x}_i - \mathbf{x}_j)\mathbf{A}^{-1}(\mathbf{x}_i - \mathbf{x}_j)^T, \tag{7.9}$$

where $\mathbf{x}_i$, $\mathbf{x}_j$ are the row vectors of values for the individuals i and j, and $\mathbf{A}$ is the variance–covariance matrix. The quantity D_{ij} is the *generalized distance* between the two individuals, and it takes into account any correlation that there might be between the characters. It is in principle the same as Mahalanobis's generalized distance between groups of individuals that we shall meet later (p. 151). If there is no correlation then $\mathbf{A}$ contains the variances of characters in the leading diagonal and zeros elsewhere, so that formula (7.9) simply gives the squared standardized Pythagorean distance.

Correlation coefficient

The product–moment correlation coefficient has been much used in some fields of study for comparing individuals. (See Sneath and Sokal 1973 for references.) It is given by

$$r_{ij} = \frac{\sum_{k=1}^{p} (x_{ik} - \bar{x}_i)(x_{jk} - \bar{x}_j)}{\sqrt{\left\{\sum_{k=1}^{p} (x_{ik} - \bar{x}_i)^2 \sum_{k=1}^{p} (x_{jk} - \bar{x}_j)^2\right\}}}. \tag{7.10}$$

The characters must be standardized and be centred so that their means are zero if results are to be sensible. When this is done r_{ij} approximates to the cosine of the angle subtended by i and j at the origin. The smaller this angle is, the more alike are the individuals judged to be.

Correlation coefficients calculated for pairs of individuals need to be viewed with caution. The set of points x_{1k}, x_{2k}, $k=1, 2, \ldots, p$, is unlikely to be even approximately bivariate-normal. Thus, statistical theory has no place in interpreting a value of r_{ij}. The size of r_{ij} does not depend primarily on the sizes of the original measurements, only on their angular relations. Any two individuals, however far apart, would be judged very similar if they happened to lie in much the same direction from the origin. A user should therefore be sure that he wants this angular measure before he chooses it. When many characters are included in a comparison r_{ij} is often found to give results comparable to those obtained using a distance measure. But this is no justification for its use. (See Sneath and Sokal 1973, and Eades 1965 for further discussion of this point.)

The correlation coefficient was used in exploratory studies on the comparison of soil profiles by Moore and Russell (1967) and Cuanalo and Webster (1970). However, it is now generally regarded as inappropriate and is not recommended.

Missing values

A tiresome problem in studies of relations between individuals arises when some values of variables are missing for some of the individuals. In fact, it occurs in most kinds of multivariate analysis. Some analyses cannot proceed unless the data matrix is complete. How the problem is solved will vary according to the method.

Missing values occur where one or more properties could not be or were not recorded for some reason. This is different from examining a sample of soil and finding that the property of interest, say mottling, is absent. In this latter situation the correct entry in the record would be zero. On the other hand if a sample could not be obtained, for instance where a subsoil sample could not be taken where the soil was very shallow, then no observation can be made, and this must be reflected by an appropriate code in the records. Zero is inappropriate here.

One way of dealing with missing values is to replace them by suitable values. In the absence of better information, the best value to insert for a variate is its mean. If other information is available then a better estimate may be inserted. Provided that there are only few missing values, estimating them in either way is unlikely to affect the outcome appreciably or to mislead.

Many of the formulae for calculating similarities between individuals can be adapted to deal with missing values without inserting estimates. Similarities are brought to a common scale by dividing the sum of similarities, distances, or whatever for single characters by the number of characters involved in each comparison of individuals. If we take mean character distance between individuals i and j as an example then we have

$$d_{ij} = \frac{1}{p_{ij}} \sum_{i=1}^{h} |x_{ik} - x_{jk}| \Big/ r_k, \tag{7.11}$$

where p_{ij} is the number of attributes for which $|x_{ik} - x_{jk}|$ can be calculated. If either x_{ik} or x_{jk} is missing or inapplicable then there is one less element in the sum $\Sigma|x_{ik} - x_{jk}|$, and p_{ij} is diminished by 1.

Anderson (1971) points out that, since this method represents similarity for the missing variates by the similarity for those that are present, it is unlikely to be satisfactory when characters are not strongly related. It is sounder to estimate the similarity for a missing variate by the average of all similarities for that variate; and this is equivalent to replacing missing values by the mean for the variates concerned, as before.

Character weighting

A further problem that arises in multivariate comparisons that does not occur in univariate studies is how to choose characters and their relative weights. So far we have assumed that we know the values of all characters of the individuals we wish to compare, and though we introduced the idea of standardization to bring characters to comparable scales, we avoided the matter of the weights to be assigned to the different characters. With one exception, we have considered only formulae that assign weight to characters equally. The principle, usually attributed to the eighteenth-century French botanist Michel Adanson, seems inappropriate to some pedologists; others point out that the inclusion of some characters and the omission of others in an analysis is in effect differential weighting.

Characters need not be given equal weight for mathematical reasons. Gower's formula, equation (7.6), for calculating similarities leaves its user free to choose whatever weights, w_{ijk}, he considers reasonable, and similar adjustments can be made to other formulae. But differential weighting does raise serious logical difficulties. How should such weights be chosen? If the aim is simply to express relations in existing data, whether as similarities between individuals or in summary form as a classification or ordination (see later), then there is no obvious case for assigning more weight to some characters than to others. If we are to collect data by which to assess relationships then the logical starting point is to record those characters that are obvious and those that are known to be important. Characters known to be unimportant should be excluded, in effect given zero weight; otherwise they blur the picture. However, the relative importance of characters can be judged only in relation to purpose, and it is only when we know what purpose we aim to serve that we can choose differential weights rationally. If we lack this information then there is no reasonable alternative, *a priori* at least, to giving equal weight to the characters for which we have or can obtain data easily.

Concomitant variables

A more subtle problem arises when the soil properties in which we are interested are influenced by one or more others that are themselves of no interest. For example, the water content or suction of the soil at the time of sampling is transient, and we should be unlikely to want to use it when comparing two profiles for general purposes. Yet if consistence, strength, and colour are used for comparing profiles then any variation in water content or suction will be embodied in our comparison. In this situation water content is a *concomitant variable*. Provided this is recognized, its effects can be eliminated by standard regression methods as in the analysis of covariance. Marriott (1974) discusses the matter at some length and shows how to adjust the main data accordingly.

Special problems of comparing soil profiles

The above measures take no account of any relations between different parts of the individuals being compared. They assume further that if an individual has several parts the properties of any one part can be matched to those of corresponding parts of individuals. This correspondence between parts of two or more individuals is known in biology as *homology*. In pedology it is a question of recognizing corresponding horizons or levels in different soil profiles.

The measures that we have discussed so far can be applied to soil profiles for which we have information from several layers provided we can recognize corresponding layers in all the profiles being studied. If we have recognized, say, four horizons then we compare each property of the first horizon in one profile with the same property of the first horizon of the second profile, and repeat for each horizon in turn. Recognition of horizons involves the subjective judgement of the pedologist, of course. To avoid this profiles can be described at several fixed depths, though at the cost of losing information on homology. In both cases each property in each layer is treated as an independent variate. Despite the comments of Lance and Williams (1967c), this is theoretically sound, and we can use generalized distance as the measure of dissimilarity to remove effects of correlation between layers if we wish.

Nevertheless, the treatment depends on the layers' being comparable, and it may be necessary to decide which layers are homologous before comparing profiles. To avoid such judgement Rayner (1966) devised a means for comparing profiles as follows. Initially each horizon in each profile is considered as an individual, and similarity coefficients are computed between all pairs of horizons in all profiles. The similarity between two profiles, say P_1 and P_2, is then computed from the similarity coefficients between their horizons. The first horizon of P_1 is compared with each horizon of P_2 in turn. The most similar horizon in P_2 is chosen as the best match. The second horizon in P_1 is then compared with each horizon in P_2, starting with the best match found for horizon 1 and proceeding downwards. The third horizon of P_1 is compared with the horizons in P_2, starting with the best match for horizon 2, to find the best match for it, and so on until all horizons in P_1 have been matched with ones of P_2. The roles of P_1 and P_2 are then reversed and the procedure is repeated. Matching need not be symmetrical; the best match between a horizon of P_1 among those in P_2 is not necessarily the best match when the roles are reversed, as Rayner shows. Having found which pairs of horizons match, the similarity between P_1 and P_2 is finally computed as the average of the similarities between the matched horizons.

Rayner's zig-zag method is attractive. Not only does it allow soil at different depths or in different horizons to be compared, it also allows similarities between profiles with different numbers of horizons to be calculated automatically. It also takes account of the order in which horizons occur down the profile; so if, for example, one profile were turned upside down then substantially different similarities with other profiles would result. Whether the advantages of the method over the simpler techniques justify the increased computing is uncertain.

Williams and Rayner (1977) compared a similarity matrix calculated in this fashion with matrices derived by the simpler techniques described earlier. They found little difference in the patterns of relations revealed. Further comparative study is needed to assess the actual advantages of Rayner's method over the simpler ones, and to identify situations that justify the substantially increased computing that it entails.

Two other methods for measuring the similarities between soil profiles have been tried. The first (Moore *et al.* 1972) computes for each profile the relation between depth and each soil property as a polynomial function. The coefficients of the function are then treated as variates from which the similarities are calculated. The second, due to Dale *et al.* (1970), is based on transition matrices. It first classifies the horizons or levels according to their similarities and then considers the order in which the horizons occur in the profile. The more closely the order of horizons in one profile matches that in another, the more similar are the profiles judged to be. The method was explored by Moore *et al.* (1972) and by Norris and Dale (1971), but seems to have few advantages in practice.

Although we have condemned prior weighting of characters when matching soil specimens or profiles, a reasonable case can be made for assigning more weight to characters of the topsoil than to those in the deep subsoil, at least in a biological context, and especially when the soil has been described at equal intervals down the profile. Russell and Moore (1968) considered this situation, and they proposed that the weight w given to each level should decline exponentially with increasing depth according to the formula

$$w=c\ \exp(-cx), \tag{7.12}$$

where x is the depth and c is a parameter set by the user. When a profile has been described by layers bounded by upper and lower limits x_1 and x_2, the expression on the right-hand side of (7.11) can be integrated between x_1 and x_2 to give the weight for each layer as

$$w=\int_{x_1}^{x_2} c\ \exp(-cx)dx. \tag{7.13}$$

Russell and Moore judged from their experience that a value of c=0.02 (and depth expressed in centimetres) fairly reflected the change in biological activity down a profile, but suggested that other values of c in the range 0.01–0.04 might be tried. The way in which their comparisons are biased in favour of the upper parts of profiles is shown in Fig. 7.2. Notice that this is no different in principle from comparing profiles horizon by horizon and treating thin horizons near the surface on an equal footing with thicker ones beneath.

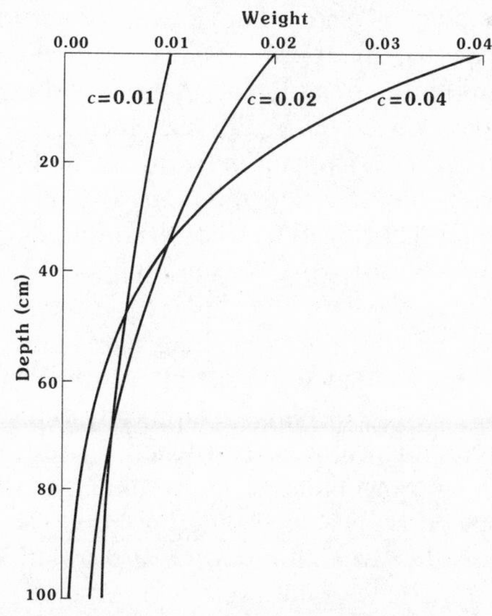

FIG. 7.2 Graphs showing how the weight changes down a profile when using Russell and Moore's weighting function

8

ORDINATION

'What is the use of a book,' thought Alice, 'without pictures or conversations?'

Lewis Carroll, *Alice in Wonderland*

When just one soil property has been measured on a set of individual specimens, profiles, or sampling sites we can represent the measured values by their positions on a single scale or line. The relation between any pair of individuals can be represented by the distance between them, and relations among several individuals can be appreciated simultaneously from their relative positions on the line. If we wish we can classify the individuals by simple dissection of the line (Chapter 4). Relations are slightly more difficult to assess if we have two properties, but a scatter diagram shows the similarities between individuals and among groups of individuals. When there are many properties of interest we can measure the relation between any pair of individuals by calculating their similarity or dissimilarity (Chapter 7). And although we may postulate the scatter of individuals in a space of many dimensions, it is almost impossible to envisage their positions in it and the relations among more than two individuals simultaneously. The traditional means of overcoming this difficulty has been to classify the individuals, and in Chapter 7 we introduced the idea of multivariate relations as a prelude to numerical classification. However, classification is not the only way of expressing such relations, and for continuously variable material like soil it is by no means always the most satisfactory way.

An alternative is to arrange the individuals along one or a few new axes chosen so as to preserve as much as possible of the original information. The arrangement can then be displayed graphically as histograms or scatter diagrams, or in three dimensions by constructing models or stereograms. This reduction of an arrangement in many dimensions to one in a few dimensions has become known in ecology as *ordination*. The term was apparently introduced by Goodall (1954) as a translation of the German *Ordnung*, and it is the one we shall use.

Several ordination methods have been proposed, some specially suited to qualitative data (see later, p. 141). Hole and Hironaka (1960) first attempted ordination of soil using a method devised by Bray and Curtis (1957) for ecological work. Since then, however, most ordination studies, both in pedology and in other branches of science, have used vector methods. It is to these that we now turn our attention, and to principal component analysis in particular, which has proved to be one of the most valuable means of exploring relations among soil profiles.

Principal components

Consider again Fig. 5.1. Forty points are scattered in two dimensions that represent the two soil characters silt+clay content and plastic limit. The variances on these axes are 322.6 and 386.0 respectively. The two characters are correlated to the extent that r=0.752, and the configuration of points is elongated, rising from left to right. In Chapter 5 we found the principal axes of this configuration, and calculated the variances on these axes as 621.5 for the long one and 87.1 for the short. Thus the proportion of the total variance (88 per cent) accounted for by the longer principal axis is considerably larger than that represented by either of the original axis. If we were prepared to forgo the remaining 12 per cent then we could represent our data in just one dimension by projecting the points orthogonally on to the longer principal axis. This is the basis of ordination by principal components.

When only two characters have been measured we need not lose information in this way in order to see all the relations. But the principle of the method extends to many dimensions, and it is then of considerable practical value. In general, the analysis finds the principal axes of a multidimensional configuration and determines the coordinates of each individual relative to these. This is equivalent to rotating the configuration to new axes such that (1) the sum of the squares of the perpendicular distances from the first axis to the points representing the individuals is the least; (2) a second axis is chosen at right angles to the first to minimize the sum of the squares of the perpendicular distances from the points to it; (3) the third and subsequent axes, all perpendicular to one another, are chosen similarly. We can then display the relations among individuals in the plane defined by the first two principal axes, or a model of the first three, and these displays will be more informative than those on any other set of two or three axes. It is sometimes helpful also to project the scatter on to other pairs of low-order axes.

The good sense of component analysis for ordination depends on the variances of the original variates being approximately equal. Otherwise the orientation of the principal axes will be controlled largely by those characters with the largest variances. It is worth amplifying the earlier remarks on measurement scales at this stage. The results of a principal component analysis depend on the scales on which the original observations are recorded. Using again the illustration of clay+silt content and pH, components calculated from such data are likely to be determined largely by clay+silt content, since this is likely to have a much larger variance than pH. As in the calculation of similarities between individuals, we can avoid this effect by standardizing the scales so that all have a standard deviation of 1. This should be done as a matter of course when the variates are measured on different scales. The variance–covariance matrix thereby becomes the correlation matrix, and in ordination studies involving properties measured on different scales components are calculated on this matrix.

Rotation and projection are important both in ordination and as the means of understanding several other multivariate methods. Therefore, we shall consider them in detail.

Method

Fig. 8.1 illustrates the transformation of the data that we require. The origin has been moved to the centre O of the configuration to make the transformation as clear as possible. One of the points P of Fig. 7.1 is located at (x_1, x_2) on axes OX_1 and OX_2. The principal axes make an angle θ, which we shall ultimately wish to determine, with the original axes, and having found θ we shall require the values of y_1 and y_2. By simple geometry

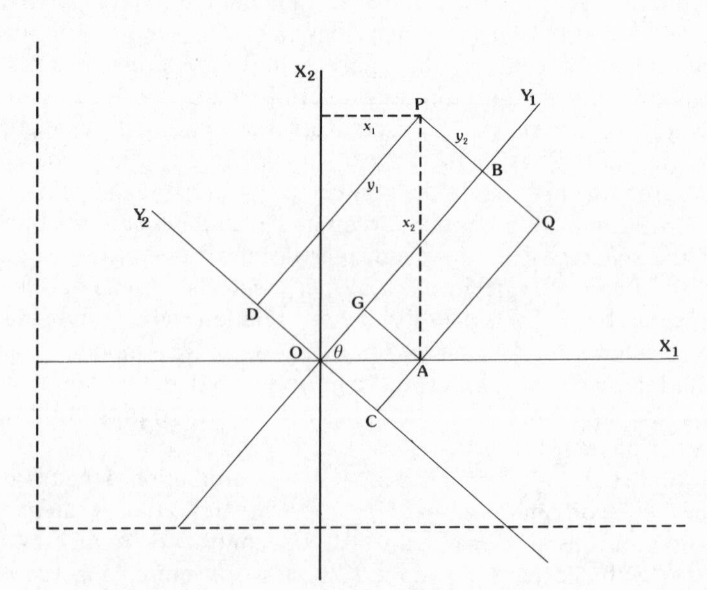

FIG. 8.1 The geometry of axis rotation

we can see that the line OB is made up of two parts:
(1) OG, which is the projection of OA on OY_1 of length $x_1 \cos\theta$, and
(2) GB, the projection of AP on OY_1, of length $x_2 \sin\theta$, or alternatively $x_2 \cos(90°-\theta)$.

Similarly, the line OD consists of

(1) CD=PQ, the projection of PA on the perpendicular from P on to OY_2, of length $x_2 \cos\theta$

less

(2) OC, the projection of OA on OY_2 and of length $x_1 \sin\theta$.

Thus, the coordinates of P relative to the principal axes are

$$\begin{cases} y_1 = x_1 \cos\theta + x_2 \sin\theta \\ y_2 = -x_1 \sin\theta + x_2 \cos\theta. \end{cases} \tag{8.1}$$

Equations (8.1) constitute a linear transformation, which we can write in matrix form:

$$\mathbf{y} = \mathbf{x}\mathbf{C}^T,$$

where **x** and **y** are row vectors $\mathbf{x} = [x_1 x_2]$, and $\mathbf{y} = [y_1 y_2]$ for the individual, and

$$\mathbf{C} = \begin{bmatrix} c_{11} & c_{12} \\ c_{21} & c_{22} \end{bmatrix} = \begin{bmatrix} \cos\theta & \sin\theta \\ -\sin\theta & \cos\theta \end{bmatrix}. \tag{8.2}$$

Since $\sin\theta = \cos(90°-\theta)$, the elements of **C** are the cosines of the angles between the new axes and the old.

We should notice that

$$\mathbf{C}\mathbf{C}^T = \begin{bmatrix} \cos\theta & \sin\theta \\ -\sin\theta & \cos\theta \end{bmatrix} \begin{bmatrix} \cos\theta & -\sin\theta \\ \sin\theta & \cos\theta \end{bmatrix}$$

$$= \begin{bmatrix} \cos^2\theta + \sin^2\theta & -\cos\theta\sin\theta + \sin\theta\cos\theta \\ -\sin\theta\cos\theta + \cos\theta\sin\theta & \cos^2\theta + \sin^2\theta \end{bmatrix}. \tag{8.3}$$

So

$$\mathbf{C}\mathbf{C}^T = \begin{bmatrix} 1 & 0 \\ 0 & 1 \end{bmatrix} = \mathbf{I}. \tag{8.4}$$

We obtain the identity matrix, **I**, by post-multiplying the matrix of direction cosines by its transpose. Notice also that if we pre-multiply both sides of (8.4) by $\mathbf{C}^{-1}$ it becomes

$$\mathbf{C}^T=\mathbf{C}^{-1}.$$

Thus, the transpose of **C** equals its inverse.

If we could determine the angle θ we should be able to find the coordinates y_1,y_2, of any point x_1,x_2. We know that the new variables will be uncorrelated, so that their variance–covariance matrix will be a diagonal matrix with zeros off the diagonal. By convention we represent the variances on the principal axes by λ_1 and λ_2, and the new variance–covariance matrix can be written

$$\frac{1}{n-1}\mathbf{Y}^T\mathbf{Y}=\begin{bmatrix}\lambda_1 & 0\\ 0 & \lambda_2\end{bmatrix}=\mathbf{\Lambda}. \tag{8.5}$$

Since $\mathbf{Y}=\mathbf{X}\mathbf{C}^T$, we have

$$\frac{1}{n-1}\mathbf{Y}^T\mathbf{Y}=\frac{1}{n-1}\mathbf{C}\mathbf{X}^T\mathbf{X}\mathbf{C}^T,$$

and since $\{1/(n-1)\}\ \mathbf{X}^T\mathbf{X}=\mathbf{A}$, the variance–covariance matrix (Chapter 7) is

$$\mathbf{\Lambda}=\mathbf{C}\mathbf{A}\mathbf{C}^T. \tag{8.6}$$

Post-multiplying by **C** and remembering that $\mathbf{C}\mathbf{C}^T=\mathbf{I}$, we have

$$\mathbf{\Lambda}\mathbf{C}=\mathbf{C}\mathbf{A},$$

which on multiplying out gives

$$\begin{bmatrix}c_{11}\lambda_1 & c_{12}\lambda_1\\ c_{21}\lambda_2 & c_{22}\lambda_2\end{bmatrix}=\begin{bmatrix}c_{11}a_{11}+c_{12}a_{21} & c_{11}a_{12}+c_{12}a_{22}\\ c_{21}a_{11}+c_{22}a_{12} & c_{21}a_{12}+c_{22}a_{22}\end{bmatrix}$$

and on rearrangement

$$\begin{bmatrix}c_{11}(a_{11}-\lambda_1)+c_{12}a_{21} & c_{11}a_{12}+c_{12}(a_{22}-\lambda_1)\\ c_{21}(a_{11}-\lambda_2)+c_{22}a_{21} & c_{21}a_{12}+c_{22}(a_{22}-\lambda_2)\end{bmatrix}=\mathbf{0}, \tag{8.7}$$

where **0** is the 2×2 matrix with all elements zero. If we assume that the c_{ij} are unknowns then we can find them by equating the elements of the first rows of (8.7) to give two simultaneous equations:

$$\left.\begin{aligned} c_{11}(a_{11}-\lambda_1)+c_{12}a_{21}=0 \\ c_{11}a_{12}+c_{12}(a_{22}-\lambda_1)=0 \end{aligned}\right\}. \tag{8.8}$$

Similarly for the elements of the second rows:

$$\left.\begin{aligned} c_{21}(a_{11}-\lambda_2)+c_{22}a_{21}=0 \\ c_{21}a_{12}+c_{22}(a_{22}-\lambda_2)=0 \end{aligned}\right\}. \tag{8.9}$$

One solution of these would be $c_{11}=0$ and $c_{12}=0$, but in that event all values of y_1 would be zero also, and clearly this is not the solution we require. If a system of homogeneous equations such as that above is to have other than this trivial solution then the determinant of the coefficients of the equations must be zero. Thus,

$$\begin{vmatrix} a_{11}-\lambda_1 & a_{21} \\ a_{12} & a_{22}-\lambda_1 \end{vmatrix}=0. \tag{8.10}$$

Similarly,

$$\begin{vmatrix} a_{11}-\lambda_2 & a_{21} \\ a_{12} & a_{22}-\lambda_2 \end{vmatrix}=0, \tag{8.11}$$

and in general

$$\begin{vmatrix} a_{11}-\lambda & a_{21} \\ a_{12} & a_{22}-\lambda \end{vmatrix}=0. \tag{8.12}$$

On multiplying the determinant out we obtain a quadratic in λ,

$$\lambda^2-(a_{11}+a_{22})\lambda+a_{11}a_{22}-a_{12}a_{21}=0, \tag{8.13}$$

which, since it is a quadratic, has two roots λ_1 and λ_2. These are known as *latent roots*, *characteristic roots*, or *eigenvalues* of the variance–covariance matrix $\mathbf{A}$. They are found by solving the determinantal equation, generally written

$$|\mathbf{A}-\lambda\mathbf{I}|=0. \tag{8.14}$$

The associated vectors $\mathbf{c}_1$ and $\mathbf{c}_2$ are known as the *latent vectors* or *eigenvectors*. In a quadratic of the form $x^2 - bx + q = 0$, b is the sum of the roots and q the product. Equation (8.13) gives

$$\lambda_1 + \lambda_2 = a_{11} + a_{22} \tag{8.15}$$

and

$$\lambda_1 \lambda_2 = a_{11} a_{22} - a_{12} a_{21}. \tag{8.16}$$

In words, the sum of the roots equals the sum of the variances of the original variates, and their product equals the product of the variances minus the square of the covariance, i.e. the determinant of **A**. This is the same result as we found by a different route in Chapter 5.

If now we recall (8.7) and the simultaneous equations (8.8) and (8.9) derived from it, we have, for example, on rearranging,

$$c_{12} = \frac{a_{11} - \lambda_1}{a_{21}} c_{11}. \tag{8.17}$$

Since $c_{11}^2 + c_{12}^2 = 1$, we can solve for c_{11} and c_{12}.

Let us illustrate the procedure using again the data on particle size and plastic limit in Table 5.1. The variance–covariance matrix is

$$\begin{bmatrix} 322.63 & 265.31 \\ 265.31 & 386.05 \end{bmatrix},$$

and hence we need to solve

$$\begin{vmatrix} 322.63 - \lambda & 265.31 \\ 265.31 & 386.05 - \lambda \end{vmatrix} = 0.$$

Therefore

$$\lambda^2 - (322.63 + 386.05)\lambda + 322.63 \times 386.05 - 265.31^2 = 0,$$

and thus

$$\lambda_1 = 621.53 \text{ and } \lambda_2 = 87.14.$$

Substituting these values in (8.17) gives us

$$c_{12} = -\frac{322.63 - 621.53}{265.31} c_{11}$$

$$= 1.127c_{11}.$$

Since $c_{11}^2 + c_{12}^2 = 1$, we have

$$c_{11}^2 + (1.127c_{11})^2 = 1,$$

hence

$$2.270c_{11}^2 = 1.$$

Therefore $c_{11} = 0.6638$ and $c_{12} = 0.7479$, and since $c_{11} = \cos\theta$ we have $\theta = 48°24'$ as before.

Having determined the elements of **C** in this way, we can calculate the coordinates for any point given its values on the original axes of measurement using (8.2).

We have pursued the analysis for two variates in detail. The principles are quite general, however, and we can always proceed by solving the determinantal equation.

$$|\mathbf{A} - \lambda\mathbf{I}| = 0.$$

If there are p variates there will be p roots. It is also worth remembering the following properties that are associated with rigid rotation to principal axes:

(a) $$\sum_{i=1}^{p} \lambda_i = \sum_{i=1}^{p} s_i^2, \tag{8.18}$$

(b) $$\prod_{i=1}^{p} \lambda_i = |\mathbf{A}|, \tag{8.19}$$

(c) $$\mathbf{c}_i\mathbf{A} = \lambda_i\mathbf{c}_i, \tag{8.20}$$

(d) $$\mathbf{c}_i\mathbf{c}_i^T = 1, \tag{8.21}$$

where $\mathbf{c}_i$ is a row vector $[c_{i1}\ c_{i2}\ \ldots\ c_{ip}]$ of **C** and $i = 1, 2, \ldots, p$.

Two further points need mention. If any variate depends linearly on one or more others then the matrix **A** is positive semi-definite (p.s.d.) and at least one root will be zero; the individuals are then distributed in fewer than p dimensions. Secondly, if any two roots are equal then the distribution in the corresponding plane is circular, and though we may place orthogonal axes in that plane their orientation is quite arbitrary. There are thus many vectors that can be associated with these roots.

In practice, principal component analysis is usually performed where there are many variates, and hence large matrices. It would be very laborious indeed to find their latent roots except by computer, and few analyses would be attempted without one. Most scientific computer centres now have library programs for finding latent roots and vectors.

For ordination it is usual to rank the roots from the largest to the smallest and to accumulate the sum of the roots as each is considered in turn. Since the sum of all the roots equals the sum of the p original variances, we can then see readily what percentage of the total variance the first k roots ($k<p$) account for. It is hoped that the first few roots will represent a large proportion of the total variance. In most soil studies to date this hope has been realized: four or five roots out of a total of between 20 and 40 have commonly accounted for about 70 per cent of the variance of standardized variates.

The coordinates of the individuals, often known as their *component scores*, are computed for the first few dimensions. Using these, the individuals can be displayed to advantage by plotting their positions in planes defined by pairs of these axes in turn. The other dimensions can usually be ignored. A display on axes 1 and 2 is the most informative, of course.

Interpretation of latent vectors

Although a component analysis is often carried out to display relations between individuals, those relations can often be understood better if some meaning can be attached to the component axes. The matrix of latent vectors contains the information for this interpretation. An element c_{jk} with a value near 1 means that the axis representing the jth original variate is closely aligned to the kth component axis, and hence makes a large contribution to that component. Conversely, if c_{jk} is near 0 the two axes are nearly at right angles, and the contribution of variate j to component k is small. By examining the latent vectors for those variates that load heavily it may be possible to give the principal axes a physical interpretation. Such interpretation is by no means assured. The principal components are no more than mathematical constructs, and they have no direct physical meaning. A projection of the vectors on to a plane may be helpful (see for example Fig. 8.2), and its interpretation can sometimes be improved by a rotation that emphasizes those coefficients, c, that are most significant. We describe this later.

Examples

Two examples will illustrate how component analysis can be used to display and interpret relations in multivariate soil populations.

The first is from a study reported by Cuanalo and Webster (1970) on data recorded in a survey of soil in west Oxfordshire by Beckett and Webster (1965b). Sites had been chosen by sampling randomly 17 physiographically distinct types of land. Several morphological and physical

TABLE 8.1 Latent vectors of correlation matrix from west Oxfordshire soil study

Variates		Vectors for topsoil (13 cm)				Vectors for subsoil (38 cm)		
		1	2	3		1	2	3
Colour hue (Munsell)	1	−0.1573	0.0197	0.0310	16	−0.2024	−0.0430	0.0970
Colour value (Munsell)	2	0.0003	0.1968	0.2295	17	−0.1643	0.0377	0.2001
Colour chroma (Munsell)	3	0.2371	0.0225	0.0018	18	0.1726	0.0975	−0.0699
Colour of mottles (Munsell)	4	0.0173	−0.0580	0.1785	19	0.0236	−0.0316	0.1006
Mottle abundance	5	−0.1062	−0.0718	−0.0417	20	−0.1280	0.0423	0.3094
Clay + silt	6	−0.2347	0.2509	0.0532	21	−0.2536	0.1437	0.1496
Fine sand	7	0.1605	−0.3557	0.0448	22	0.1352	−0.3701	0.0164
Stone content	8	0.1703	0.1750	−0.1070	23	0.2066	0.1597	−0.1009
Plastic limit	9	−0.2171	0.2475	−0.0886	24	−0.2118	0.1753	−0.0550
Liquid limit	10	−0.2008	0.2725	−0.0881	25	−0.2359	0.2038	−0.0224
pH	11	0.0049	0.0728	−0.5226	26	0.0117	0.1074	−0.5261
Soil strength in summer	12	0.2248	0.2113	0.2365	27	0.2430	0.2345	−0.0032
Soil strength in winter	13	0.2434	0.1538	0.1351	28	0.2312	0.2277	−0.0604
Matric suction in summer (pF_1)*	14	0.1746	0.2676	0.1793	29	0.1815	0.2248	0.1243
Matric suction in winter (pF_1)	15	0.1508	0.0857	−0.0630	30	0.2405	0.0679	0.0716

* $pF_1 = \log_{10}$ (cm water + 50).

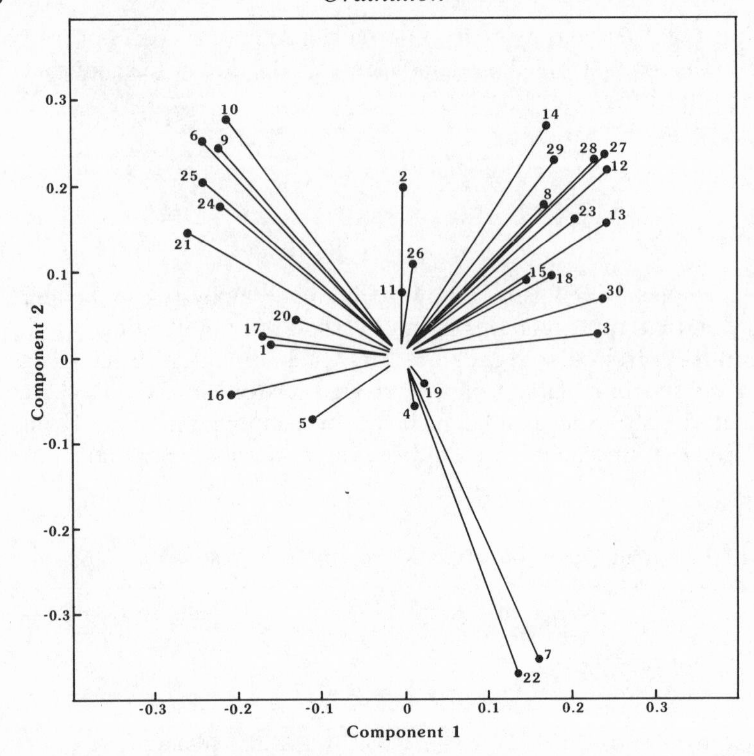

FIG. 8.2 Graph of vectors showing the contribution the variates listed in Table 8.1 make to the first two principal components

properties of the soil had been measured at 13 cm and 38 cm depths at each site. Here 15 properties at each of the two depths for 85 sites have been analysed. The variates are listed in Table 8.1. They were standardized to unit variance and centred at the origin. The six largest latent roots of their variance–covariance matrix, which on standardized variates is also the correlation matrix, are given in Table 8.2. The first is much the largest, and the first two, out of 30, account for nearly half of the variance in the sample. Projection of the population scatter on to the plane defined by the first two principal axes gives much the most informative single display of relations in the whole space, and Fig. 8.3 illustrates the relations among the 85 sampling sites in this way.

In the original interpretation the first principal axis, the horizontal one, was thought to represent gleying, while the second, shown vertically, was thought to represent texture. A more careful examination of the latent vectors shows that interpretation is not so simple (Fig. 8.2). Bright colour (large chroma) and large matric suction undoubtedly contribute strongly to component 1, but so does clay+silt content (in the opposite sense). Fine sand content contributes strongly to component 2. Atterberg limits and

TABLE 8.2 Latent roots of correlation matrix for 30 variates in west Oxfordshire soil study

Order	Root	Percentage of variance	Cumulative %
1	9.695	32.32	32.32
2	4.833	16.11	48.43
3	2.342	7.81	56.24
4	1.776	5.92	62.16
5	1.668	5.56	67.72
6	1.304	4.35	72.06

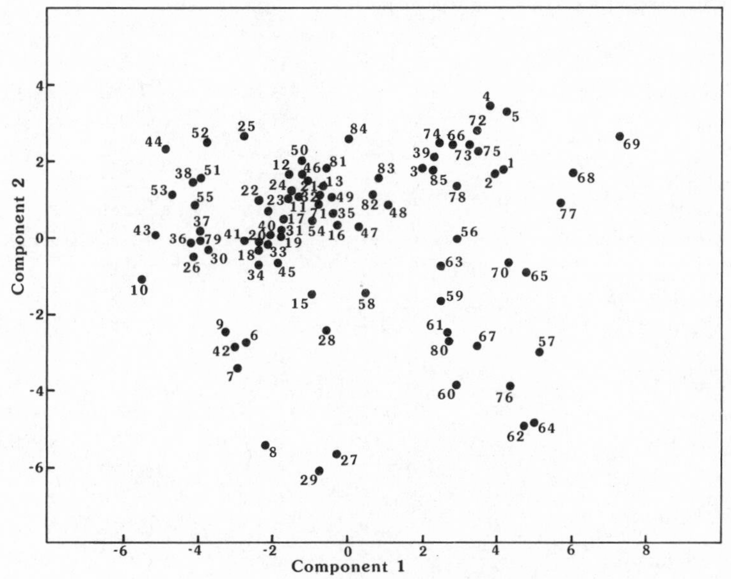

FIG. 8.3 Scatter of 85 sampling sites in west Oxfordshire projected on to the plane of the first two principal components

several other properties contribute strongly to both components. The vector diagram (Fig. 8.2) suggests that if an interpretation is required then further rotation would help, and this is discussed later (p. 139).

A second example is taken from research by Kyuma and Kawaguchi (1973) in which they ranked sites for paddy rice according to the 'chemical potential' of the soil to produce the crop. They analysed laboratory measurements of 23 mainly chemical properties of soil from 40 rice-growing sites in South-East Asia. The latent roots and vectors are given in Table 8.3. The first component accounts for 36 per cent of the variance, and the first four for over three-quarters. Large contributions to the first

TABLE 8.3 First four latent roots and vectors for correlation matrix for 23 chemical properties of 41 paddy soils

	Roots			
	1	2	3	4
Root	8.252	3.903	3.069	2.129
Percentage of variance	35.9	16.9	13.4	9.2
Cumulative %	35.9	52.8	66.2	75.4
	Vectors			
Variates	1	2	3	4
1 pH	0.126	−0.223	−0.324	−0.079
2 Electrical conductivity	0.285	−0.188	−0.080	0.105
3 Total C	0.144	0.389	0.100	0.039
4 Total N	0.149	0.385	0.140	−0.003
5 NH_3–N	0.097	0.186	0.205	0.153
6 Bray-P	0.101	0.165	−0.231	0.450
7 Exchangeable Ca	0.250	0.038	−0.097	−0.355
8 Exchangeable Mg	0.308	−0.115	−0.112	−0.161
9 Exchangeable Na	0.275	−0.184	−0.085	−0.187
10 Exchangeable K	0.288	−0.114	−0.45	0.001
11 Free Fe	0.045	−0.243	0.371	0.145
12 Free Mn	0.189	−0.236	0.200	0.257
13 Silt	0.044	−0.029	0.477	−0.077
14 Moisture	0.272	0.228	0.105	−0.072
15 Kaolin	−0.237	−0.117	0.080	0.237
16 Illite	−0.057	−0.184	0.254	−0.044
17 Cation exchange capacity	0.303	0.128	−0.014	−0.231
18 Available Si	0.262	−0.157	−0.092	−0.101
19 Total P	0.164	0.334	0.146	0.196
20 0.2 N HCl–P	0.134	0.180	−0.182	0.436
21 0.2 N HCl–K	0.305	−0.116	−0.036	0.067
22 Reducible Mn	0.144	−0.279	0.198	0.287
23 Sand	−0.180	0.012	−0.376	0.184

component derive from exchangeable cations, cation exchange capacity, electrical conductivity, and available silicon. Kyuma and Kawaguchi concentrated on the first component, presumably because it accounts for so much of the variance, and also because it seems to represent chemical characteristics that are desirable in a paddy soil. They found that morphologically similar soils possess similar values of the first component; young marine sediments have the largest values; sandy, kaolinitic ones have the smallest values; soils on clayey marine alluvium and swamp deposits have above average values; brown soils on more weathered materials have values somewhat below average. Kyuma (1973a, b) later refined the concept of chemical potential and combined it with other soil properties to predict rice yield.

Rotation of principal components

The ease with which principal components can be interpreted in any physical sense is largely fortuitous. We saw in the example from Oxfordshire that, although we obtained an informative two-dimensional display of the relations between sampling sites, we could not easily attribute physical meaning to the new axes.

There have been numerous attempts, especially in psychology, to obtain meaningful variates from combinations of others using methods that are known collectively as 'factor analysis'. Some of the methods allow the user to specify which original variates are to be combined to derive a specific factor, and it seems that much of the popularity of factor analysis derives from this: the scientist can embody his experience, or prejudice, in the analysis. The scientist thereby imposes structure on his data, and clearly this is undesirable where the aim of the ordination is to reveal structure. Readers who wish to pursue the topic should consult one of the standard texts, e.g. Harman (1976), and Lawley and Maxwell (1971). They should also read the criticism of factor analysis by Reyment *et al.* (1984) and several of the references given there.

There are methods, however, whereby meaningful factors can be obtained with relatively little guidance from the investigator. These are simply analytic rotations of the principal components. We shall consider only one of these, the Varimax rotation, developed by Kaiser (1958) and probably the most widely used in other sciences. The principle of the method is intuitively simple. The user first selects a few principal components on the assumption that they represent adequately the scatter in the data. This is a subjective decision, of course, but it is the only one the investigator needs to make in this method. The distribution of the points is projected into the space defined by these components, and the configuration is then rotated rigidly to new axes so that, as far as possible, each original axis of measurement is aligned closely to one of the new factor axes and at right angles to all others. In this way each original variate contributes strongly to one of the factors and little to the others. The factor axes remain orthogonal, and in their new positions they are said to have a 'simpler' structure; that is, they are simpler to interpret.

Kaiser defined the simplicity of a factor i as the variance of its squared loadings, say V_i. Thus, if we have p original variates and we select the first k components then we shall have a $k \times p$ matrix **M** of loadings, and

$$V_i = \frac{1}{p} \sum_{j=1}^{p} \left\{ m_{ij}^2 - \frac{1}{p} \sum_{j=1}^{p} m_{ij}^2 \right\}^2. \tag{8.22}$$

Then for the whole matrix

$$V = \sum_{i=1}^{k} V_i. \tag{8.23}$$

Kaiser's original technique was to maximize V.

Formula (8.22) gives equal weight to all variates regardless of the proportion of their variances retained in the principal components. This seemed less than satisfactory, and Kaiser considered that a better measure would be obtained by dividing the squared elements of **M** by their corresponding *communalities*. The communality of a variate, h^2, is a measure of the proportion of its variance that is represented by the k

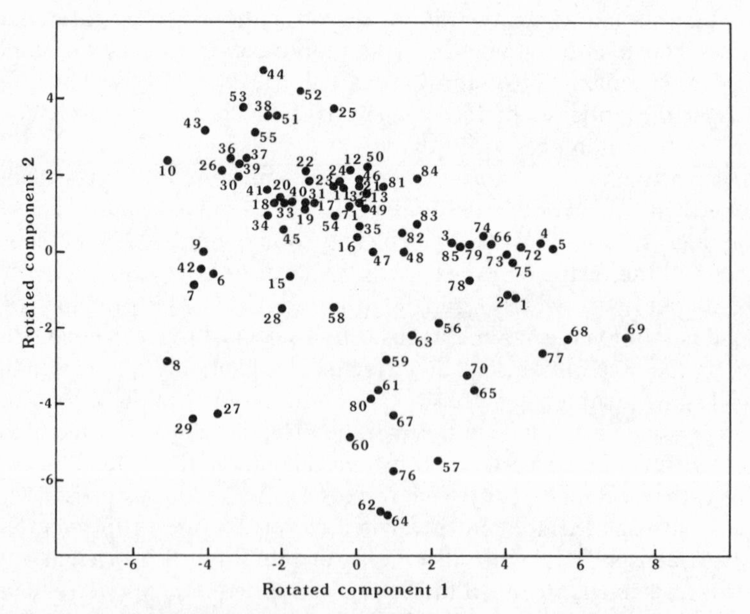

FIG. 8.4 Scatter of 85 sampling sites rotated in the plane of the first two principal components by Varimax criterion

principal components, and for the jth variate is calculated as

$$h_j^2 = \sum_{i=1}^{k} m_{ij}^2. \tag{8.24}$$

The Varimax criterion calculated is thus:

$$V' = \sum_{i=1}^{k} \left[\frac{1}{p} \sum_{j=1}^{p} \left\{ \frac{m_{ij}^2}{h_j^2} - \frac{1}{p} \sum_{j=1}^{p} \frac{m_{ij}^2}{h_j^2} \right\}^2 \right]. \tag{8.25}$$

The values of m_{ij} are modified, subject to the condition that they retain the relations in the original configuration, and V' is increased by iteration until a maximum is found. Readers can find the details in Kaiser (1958).

An application of the technique to the same data from Oxfordshire will serve as an illustration. In this instance the first two components have been rotated, and the results are shown in Figs. 8.4 and 8.5. By confining the rotation to the first two dimensions the relations of the individuals to one

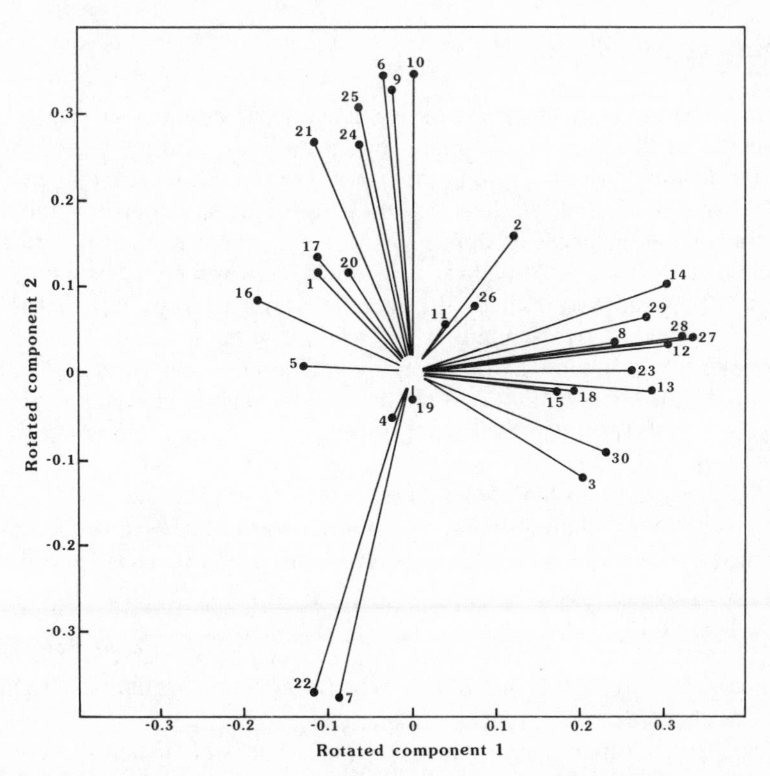

FIG. 8.5 Rotation of vectors by Varimax method. The variates are listed in Table 8.1

another in the plane are retained exactly. Thus Fig. 8.4 is still the most informative two-dimensional display. Similarly, the angular relations among the vectors remain unchanged (Fig. 8.5). However, it is clear that there has been a clockwise rotation through approximately 40°, and this enables us to give some meaning to the new axes. The first axis, the horizontal one, now strongly represents colour chroma, average matric suction, and strength of the soil—all properties associated with the water regime of the soil. The driest, strongest, and brightest-coloured soil profiles are on the right; the wettest, weakest, and dullest are on the left. The second axis has strong contributions from particle size and plastic and liquid limits, properties embodied in the texture of the soil. The profiles of heaviest texture are at the top, the lightest at the bottom. It is worth noting that there are very few profiles falling immediately to the left of centre with negative values of the second factor. In our interpretation there are very few moderately wet, light-textured profiles. If the soil is light then it is likely to be dry or very wet, whereas heavier textured soil shows almost the full range in wetness.

Principal coordinates

Principal component analysis, as we have seen, adheres strictly to the geometry of the Euclidean model. Though we may eventually project the scatter from many dimensions into few, the distances between plotted points, and hence the relations that they represent, are approximations to the Euclidean distances as defined in Chapter 7. We have also seen that Euclidean distance is not always the most appropriate measure of the likeness between two individuals, and that if our data are not metric we might be unable to calculate it anyway. Several methods have been proposed for displaying relations among individuals on metric scales in few dimensions from data that are all or mainly qualitative. These include Kruskal's (1964) *multidimensional scaling* and *correspondence analysis* (Benzécri, 1973), or *reciprocal averaging* as it is termed by Hill (1973, 1974). Gower (1966) has shown however, that provided a suitable measure is used for calculating similarities or dissimilarities between individuals, it is possible to find their coordinates relative to principal axes. He calls the method *principal coordinate analysis*. It has been used to elucidate structure in several soil populations by Rayner (1966, 1969), Campbell *et al.* (1970), and Webster and Butler (1976). Oliver and Webster (1989) have also used principal coordinates as a basis for applying spatial constraint in soil classification.

Principal coordinates are calculated from a matrix of 'distances' between individuals. So the first step in the analysis is to calculate a distance d_{ij} between every pair of individuals i and j. We may use one of the measures described in Chapter 7. Alternatively, if we have measured likeness using one of the similarity indexes S then we can scale it in the range 0 (for maximum possible dissimilarity) to 1 (for identity) and take

$$d_{ij} = \sqrt{\{2(1-S_{ij})\}}. \tag{8.26}$$

From the distances between individuals we form the matrix **Q**, with elements $q_{ij} = -\frac{1}{2}d_{ij}^2$. If there are n individuals then **Q** is of order $n \times n$. The matrix **Q** is now adjusted by subtracting from each element the corresponding row and column means and adding the general mean. Thus we form **F** with elements

$$f_{ij} = q_{ij} - \bar{q}_i - \bar{q}_j + \bar{\bar{q}}. \tag{8.27}$$

The latent roots and vectors of **F** are found, and the vectors are arranged as columns in a $n \times n$ matrix **C**. The rows then represent coordinates of the points, thus:

$$\mathbf{C} = \overbrace{\begin{matrix} 1 & 2 \ldots n \end{matrix}}^{\text{vectors}} \begin{bmatrix} c_{11} & c_{12} \ldots c_{1n} \\ c_{21} & c_{22} \ldots c_{2n} \\ \cdot & \cdot\cdot\cdot\cdot\;\cdot \\ \cdot & \cdot\cdot\cdot\cdot\;\cdot \\ \cdot & \cdot\cdot\cdot\cdot\;\cdot \\ c_{n1} & c_{n2} \ldots c_{nn} \end{bmatrix} \left.\begin{matrix} 1 \\ 2 \\ \cdot \\ \cdot \\ \cdot \\ n \end{matrix}\right\} \text{points.} \tag{8.28}$$

The vectors are now normalized so that the sums of squares of their elements equal their corresponding latent roots. This transforms the matrix **C** into a new matrix, say **G**, with elements

$$g_{ik} = \sqrt{\left\{ c_{ik}^2 \lambda_k \Big/ \sum_{i=1}^{n} c_{ik}^2 \right\}}. \tag{8.29}$$

That is,

$$\mathbf{G}^T\mathbf{G} = \mathbf{\Lambda}, \text{ and } \mathbf{G}\mathbf{G}^T = \mathbf{F}.$$

Gower shows that when this transformation is made, and starting from the matrix **Q** as defined above, the square of the distance between any two points i and j, whose coordinates are the ith and jth rows of **G**, equals d_{ij}^2. The latent vectors scaled in this way represent exactly the distances between individuals and define their positions relative to principal axes. The matrix **F** must be p.s.d.; i.e., it must have no negative roots, otherwise the individuals cannot be represented in Euclidean space. Many kinds of similarity matrix are p.s.d. and can therefore be analysed in this way (Gower 1966). Note that the positions of n points can always be represented in $n-1$ dimensions, so that **F** always has at least one zero root.

As in principal component analysis, it is usual to find that a few roots are much larger than the others, so that a good representation can be obtained in a few dimensions. Since we transformed the vectors so that $\Sigma_{i=1}^{n} g_{ik}^{2} = \lambda_k$, contributions to the distances between points from vectors corresponding to small roots must be small themselves, and can often be ignored.

When the starting matrix **Q** consists of Pythagorean distances calculated as in equation (7.6) Gower's method gives the same results as a principal component analysis. This is true not only for fully quantitative variates but also for binary characters (scored 0 or 1), for which principal component analysis is often thought to be invalid. However, its main attraction is that it can be used when Pythagorean distance is inappropriate or impossible to calculate because the data are wholly or partly qualitative.

Although principal coordinate analysis is more versatile than classical component analysis, the latter is preferable from two standpoints. First, component values are linearly related to the values of the soil properties: the elements of the latent vectors can be considered as the contributions made by the original variates to the new components, and often enable the new axes to be interpreted in a pedologically meaningful way. Principal coordinate analysis does not allow this because the information is contained in the similarity matrix and is lost during the course of the analysis. Some meaning can be given to the axes, however, by relating them to the original data. One way of doing this is by computing Snedecor's F statistic for each variate and then ranking the variates in the order of these values.

Second, if there are many individuals and few variates then component analysis is much the more efficient ordination method. By suitable use of backing store an almost unlimited number of individuals can be included in a component analysis. In a principal coordinate analysis the $n \times n$ matrix of similarities or distances, or its $n(n+1)/2$ lower triangle at least, has to be held in memory, and the size of this memory therefore imposes a ceiling on the number of individuals that can be handled in a single analysis. However, coordinates relative to principal axes can be found for additional points as described below.

Additional individuals

When we have carried out an ordination we may wish to add new individuals to the scatter diagrams; i.e., we may wish to calculate their coordinates relative to the principal axes already found. After a component analysis there is no difficulty: we simply compute for a new point i_{i+1},

$$\mathbf{y}_{n+1}=\mathbf{x}_{n+1}\mathbf{C}^T. \tag{8.30}$$

When principal axes have been found from the matrix of distances or similarities the task is less easy, but Gower (1968) has again provided a solution.

We proceed by first calculating the distances between the new $(n+1)$th individual, and all the initial individuals, $i=1, 2, \ldots, n$. Let us denote these distances by $d_{n+1,i}$. We also require the distances between the original individuals and the centroid of the configuration, d_i. Their squares lie in the diagonal of the matrix $\mathbf{F}$. This follows from $\mathbf{G}\mathbf{G}^T=\mathbf{F}$ and the fact that, in forming $\mathbf{F}$ from $\mathbf{Q}$, we transferred the centroid of the n points to the origin. So

$$d_i^2=f_{ii}=\bar{\bar{q}}-2q_{i\cdot}. \tag{8.31}$$

We now compute a row vector $\mathbf{u}$ with elements

$$u_i=d_i^2-d_{n+1,i}^2. \tag{8.32}$$

The coordinates of point i_{n+1} relative to the principal axes are then given by

$$\mathbf{g}_{n+1}^T=\tfrac{1}{2}\mathbf{\Lambda}^{-1}\mathbf{G}^T\mathbf{u}^T. \tag{8.33}$$

9

ANALYSIS OF DISPERSION AND DISCRIMINATION

> The whole character and fortune of the individual are affected . . . by the perception of differences.
>
> R. Waldo Emerson, *Nature Addresses Discipline*

In the previous chapter we saw how the joint dispersion of several variables could be expressed by a matrix of their sums of squares and products (the SSP matrix) or of their variances and covariances. The population was envisaged as scattered in a multidimensional character space with a variance–covariance matrix defining an ellipsoid within which some predictable proportion of the population is likely to occur. When there are two or more groups of individuals on which we have measured several characters we can calculate a separate variance–covariance matrix for each group. These groups may represent separate populations, and we may envisage their dispersions as a set of separate ellipsoids. The ellipsoids will overlap more or less, depending on the differences between the groups and on the proportion of the population in each that we wish to encompass. Fig. 9.1 illustrates a likely situation for two variates. The three ellipses represent equal probability 'contours' for three groups, whose centroids are at A, B, and C.

We shall now want to ask questions about such a situation:

1 How different are the groups from one another?
2 How do we describe the variation within them on average?
3 Can we compare this variation within groups with the total spread, and thereby obtain a measure of the distinctness of the classes and hence the effect of classification?

These are essentially the same questions as we asked in Chapter 4 when studying the analysis of variance. There we saw that, if we had classified a population, we could divide the total variance of that population or of a sample from it into that within the classes and that between them. The within-class variance measured the average dispersion within the classes, and the general utility of the classification. The between-class variance measured the differences between or among the classes, and when expressed as a ratio of the total variance showed how effective the

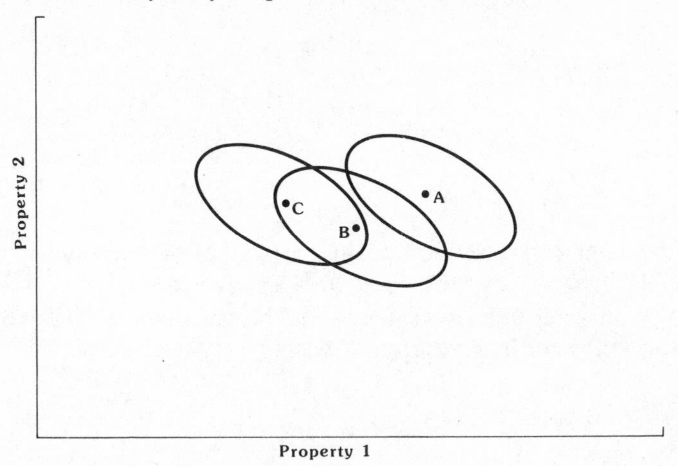

FIG. 9.1 Ellipses representing probability contours of three groups of equal dispersion with centres at A, B, and C

classification was. If we generalize these measures to the multivariate situation we get the multivariate analysis of variance, or, more succinctly, the *analysis of dispersion*, and this will enable us to answer the above questions.

In the multivariate case we have another question, which was trivial with only one variate:

4 If we have one or more new individuals, to which class or classes should we assign them, or to which class or classes do they belong?

This is the question of allocation or identification. It is often loosely termed 'classification', but this is unfortunate. The population must have been classified previously; the classification must already exist. The answer involves an extension of the analysis of dispersion, and for that reason we deal with it in this chapter also.

The measures embodied in the analysis of dispersion—the average dispersion within groups, the differences between the classes, and a comparison of the average dispersion within groups with that of the total—enable us to quantify the concept of the 'goodness of classification'. The same ideas also underlie the methods of optimal non-hierarchical classification which we shall describe in Chapter 11. These methods aim to optimize one of several possible test criteria, which derive from combinations of the above measures that comprise the analysis of dispersion.

A preliminary answer to the first question is simple. Whereas for a single variate we could compare means, in the multivariate case we represent each group by a vector of means, say $\mathbf{m}_k$ for the kth group. If there are j variates, then

$$\mathbf{m}_k = \begin{bmatrix} m_{1k} \\ m_{2k} \\ \vdots \\ m_{jk} \end{bmatrix}.$$

This vector defines the position of the centroid of the ellipsoid.

A simple measure of difference between two groups is the distance between their centroids, which can be represented by the vector of differences between their mean vectors, say

$$\mathbf{d} = \mathbf{m}_1 - \mathbf{m}_2. \tag{9.1}$$

This takes no account of the degree to which ellipsoids overlap, and we shall return to this later.

The second question involves the idea of average dispersion. Any two of the ellipses as drawn in Fig. 9.1 can be slid, without rotation, and superimposed on the third. Thus, apart from their positions they are identical, and clearly any one of them could represent the dispersion within any of the groups. In a real situation they will not fit exactly on top of one another, but, provided they are reasonably similar, each can be regarded as an estimate of within-group dispersion in general, and by combining them we can get a better estimate. We calculate, therefore, a pooled within-group variance–covariance matrix from the sums of squares and products of the deviations of the individual values from their respective group means.

If a group k contains n_k individuals on which we have measured p variates we can represent the data as a $n_k \times p$ matrix, $\mathbf{X}_k$. Its SSP matrix $\mathbf{W}_k$ will be of order p with elements

$$w_{kij} = \sum_{a=1}^{n_k} (x_{ai} - \bar{x}_i)(x_{aj} - \bar{x}_j) \tag{9.2}$$

where $\bar{x}_i$ and $\bar{x}_j$ are the means for variates i and j respectively for that group; i.e., $\mathbf{W}_k$ contains the sums of squares and products of deviations about the group means. The SSP matrix $\mathbf{W}$ for all groups, of which there are say g, is simply the sum of the SSP matrices for the individual groups, i.e.

$$\mathbf{W} = \sum_{k=1}^{g} \mathbf{W}_k. \tag{9.3}$$

It has $\Sigma_{k=1}^{g}(n_k - 1)$ degrees of freedom, so by dividing **W** by this value, we obtain the within-group variance–covariance matrix, **V**:

$$\mathbf{V} = \frac{1}{\sum_{k=1}^{g}(n_k - 1)} \mathbf{W}. \tag{9.4}$$

This answers our second question, and we are ready to tackle the third.

The dispersion in the whole set of data can be represented by the matrix of sums of squares and products of deviations about the general mean. Let this be **T**. We can obtain the total variance–covariance matrix **A** from it by

$$\mathbf{A} = \frac{1}{n-1} \mathbf{T}, \tag{9.5}$$

where $n = \Sigma_{k=1}^{g} n_k$, i.e. the total number of individuals. We can also calculate a between-groups SSP matrix **B** analogous to the between-groups sum of squares in analysis of variance. Like **W** and **T**, it is of order p, and its elements are

$$b_{ij} = \sum_{k=1}^{g} n_k(\bar{x}_{ki} - \bar{\bar{x}}_i)(\bar{x}_{kj} - \bar{\bar{x}}_j), \tag{9.6}$$

where $\bar{\bar{x}}_i$ and $\bar{\bar{x}}_j$ are the general means of the ith and jth variates. **B** and **W** sum to **T**.

In the analysis of variance we calculated mean squares and compared them, both to assess the effectiveness of classification and to provide a test of significance. As it happens, in multivariate analysis comparisons are made from SSP matrices. We compute the ratio of the determinant of the within-groups SSP matrix to the determinant of the total SSP matrix:

$$L = \frac{|\mathbf{W}|}{|\mathbf{T}|}. \tag{9.7}$$

This ratio is Wilks's Criterion (Wilks 1932). It can vary between 1 and 0, and is like the relative variance that we introduced for the univariate case, except that the relative variance was the ratio of two mean squares, not the sums of squares. If L is 1 then **W** and **T** are the same: there are no differences between the groups; i.e., $|\mathbf{B}| = 0$, and classification achieves nothing. For a given set of data **T** is constant, and L depends only on **W**.

The less the dispersion within the groups, i.e. the more compact they are, the smaller is $|\mathbf{W}|$ and hence so is L. If the population has been classified in several ways L can be used to compare them and to identify the best in the sense of the one in which the classes are most compact. It is worth noting that L does not depend on the measurement scale. In fact, any linear transformation of the data will give the same result.

When the individuals for which we have data constitute a sample from a larger population L is an estimate of a population parameter Λ. The sampling distribution of L was worked out by Wilks, and therefore we can test L for significance. We can calculate $n \ln L$, which is distributed approximately as χ^2, with $p(g-1)$ degrees of freedom. A somewhat more accurate test can be performed if n is replaced by $n-1-(p+g)/2$ in the above formula.

In using Wilks's Criterion we must observe the following. The data matrix must not be over-defined; there must be more individuals than variates ($n>p$), otherwise $\mathbf{T}$ and $\mathbf{W}$ will be singular and hence will have zero determinants. In fact, n must be greater than $p+g$ for $\mathbf{W}$ to be non-singular and the outcome sensible. Equally, any property that is invariant within classes will give $|\mathbf{W}|=0$ and should be avoided. Finally, $\mathbf{T}$ and $\mathbf{W}$ will be singular if there is any exact linear relation between any pair of variates or if any variate is an exact linear function of several others.

Homogeneity of dispersions

In the last section our analysis was based on the understanding that the variance–covariance matrices for the groups were equal or nearly so. We introduced the idea of sliding the ellipses of Fig. 9.1 so that they were superimposed. It is quite possible that ellipses representing the groups in a population cannot be superimposed by sliding them to new positions. Fig. 9.2 illustrates the possibilities. Take ellipse A as reference. Ellipse B is much larger than A; there is much more dispersion in group B than in group A. Ellipse C is much narrower than A; the correlation in this group is much stronger. Ellipse D is the same size and shape as A, but has a different orientation; it represents a correlation of different sign. Other ellipses could be drawn that differ from A in more than one of these respects. If any of these situations occurs in our data then we should want to know and to consider whether a general comparison among the group mean vectors can be meaningful.

Inequality among the group dispersions can be tested by the formula

$$c = -\ln \prod_{k=1}^{g} \left\{ \frac{|\mathbf{V}_k|}{|\mathbf{V}|} \right\}^{n_k - 1}, \tag{9.8}$$

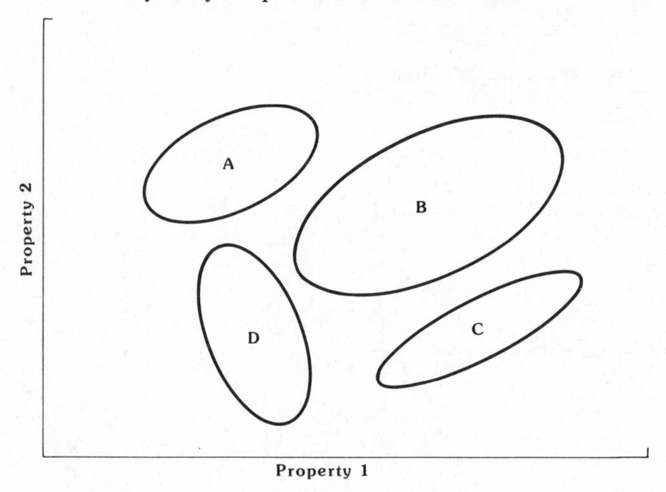

FIG. 9.2 Ellipses representing probability contours of four groups with different dispersions

where $\mathbf{V}_k$ is the variance–covariance matrix for the kth group and $\mathbf{V}$ the pooled within-group variance–covariance matrix as before. The quantity c is distributed as χ^2 with $p(p+1)(g-1)/2$ degrees of freedom. To simplify computing the formula is converted to

$$c = -\sum_{k=1}^{g} (n_k - 1) \ln \frac{|\mathbf{V}_k|}{|\mathbf{V}|}. \tag{9.9}$$

It might seem desirable to make this test before comparing group means, but it does not necessarily give practical guidance. If the sample were large and the result non-significant then we might proceed to further analysis with confidence. If not, we should be in a dilemma. Are the dispersions so different that further analysis would be invalid? And can we be sure that departure from normality is not the cause of the result? Like Bartlett's test (see Chapter 4), this test is more sensitive to such departure than tests based on Wilks's Criterion. Experience suggests that analysis of dispersion and discriminatory analysis are fairly robust and that it is reasonable to proceed unless differences among the sample dispersions are very large.

Mahalanobis distance

We return now to the matter of differences between groups that we left earlier. Our preliminary measure of the difference between two groups was the difference between their mean vectors. This took no account of the

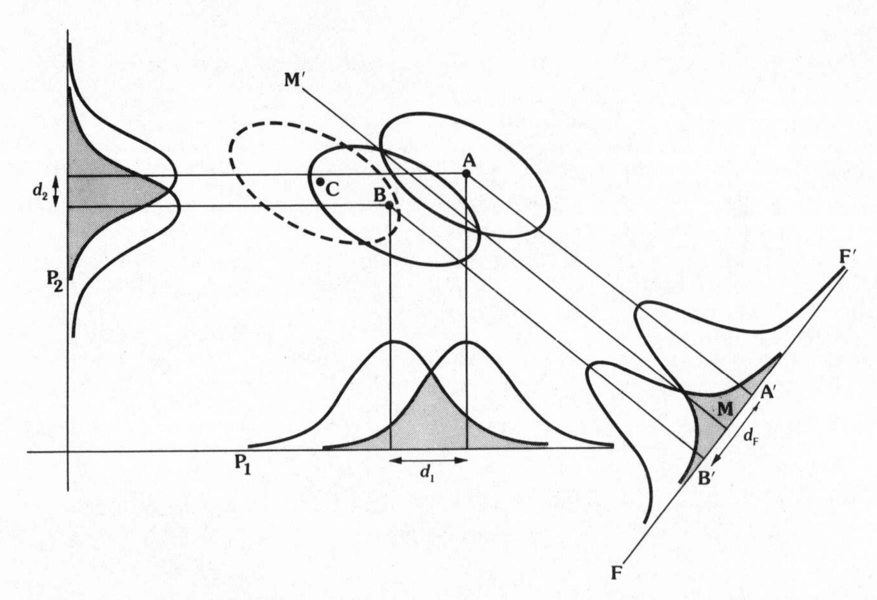

FIG. 9.3 Elliptical probability contours for three groups. Graphs of normal distribution of original variates for groups A and B are represented on axes P_1 and P_2 and of the discriminant function on axis FF′

variation within the groups. Yet we tend to attach less importance to a given difference when variation within groups is large than we do when it is small. This was so, for example, in the case of a single variate. If d is the difference between means and σ the standard deviation then we tend to judge a difference between two groups in terms of the ratio d/σ. The idea extends to the multivariate situation.

Figure 9.3 represents the same three groups as Fig. 9.1, though this time graphs of the normal distribution have been added for the variates on their axes P_1 and P_2. If the difference between the groups A and B on axis 1 is d_1, and the within-group standard deviation for that variate is σ_1, assumed to be the same for both groups, then we can measure the difference between the two groups as d_1/σ_1 for that variate. Similarly, the difference for variate 2 is d_2/σ_2. The separation between the groups on axis 2 is small, and the overlap in their distributions (shown shaded) is large; d_2/σ_2 is consequently small. On axis 1 the separation is larger and the overlap of their distributions less; so d_1/σ_1 is larger than d_2/σ_2. However, on neither axis is the measure of separation d/σ as large as it could be. It has a maximum value that can be illustrated as follows. A line MM′ is drawn through the points of intersection of the ellipses. A second line FF′ is drawn perpendicular to MM′. The centroids of A and B are projected parallel to MM′ on to FF′ and graphs of their distributions drawn also on FF′. A line such as FF′ is often known as a discriminant axis. The separation d_F on FF′ divided by the standard deviation of these distributions σ_F is a unique measure of the differences between the groups. It can be represented in matrix notation very simply as the quadratic form:

$$D^2 = \mathbf{d}^T\mathbf{V}^{-1}\mathbf{d}, \tag{9.10}$$

where **d** is the vector of differences between the means of the two groups and **V** is the pooled within-group variance–covariance matrix as before. This measure of difference between two multivariate groups or populations is due to Mahalanobis (1927), and values of its square root D are known as *Mahalanobis distances*. The meaning of 'distance' in this sense will be made clear later.

Addition of a third property extends the geometry into three dimensions, so that the ellipses become ellipsoids which intersect on a plane. We can add further dimensions if we wish, and the principles of the method hold. In general, MM′ is a hyperplane with a discriminant axis FF′ normal to it. There is always a single value of D, representing the difference between two groups and equivalent to the maximum value of d/σ, which can be found analytically.

The correlation between properties within groups is important. If there were no such correlation then the ellipses would be circles, and the standard deviations on all projections would be equal (provided that the original variates had been standardized first). In that event the maximum separation between groups could be measured along the lines joining their centroids, and the Mahalanobis distance would be equal to the Pythagorean distance divided by the within-group standard deviation. When there is a correlation this is not so, as can be seen clearly in Fig. 9.3.

A second effect of correlation can be seen if we consider a third group in Fig. 9.3 with its centroid at C. The distance separating C from B is about the same as that between B and A: B is equally far from A and C. However, when measured in terms of d/σ the difference between A and B is much larger than that between B and C, and on these grounds we consider B to be more like C than like A. Intuitively, this makes sense.

Two groups can be compared using either Mahalanobis D^2 or Wilks's Criterion, L. The two statistics, though not exactly equivalent, are closely related. In general, the larger D^2 is, the smaller is L, and vice versa. When D^2 is used for a comparison based on sample data it estimates Δ^2_M, the squared Mahalanobis distance between two populations. Its significance can be tested by calculating

$$U = \frac{n_1 n_2 (n_1 + n_2 - p - 1)}{p(n_1 + n_2)(n_1 + n_2 - 2)} D^2, \tag{9.11}$$

which is distributed as an F ratio with p and (n_1+n_2-p-1) degrees of freedom, and where n_1 is the number of individuals in the first group, n_2 is the number of individuals in the second group, and p is the number of variates.

The quantity $\{n_1 n_2/(n_1+n_2)\}D^2$ is Hotelling's T^2, developed independently by Hotelling (1931). Finally, if there is only one variate then U reduces to the square of Student's t.

Geometric representation of Mahalanobis distance: canonical variates

In Chapter 7 we saw how we could represent the generalized distance between points in a population in which two variates were correlated by standardizing the variates and rotating the axes so that the elliptical distribution became circular. The generalized distance could then be displayed in the transformed space and measured if necessary. Provided the dispersions within the groups are approximately equal, we can carry out a similar transformation to represent Mahalanobis distances between groups. In this case we standardize the within-group variances and rotate the axes so that the ellipses representing the distributions within the groups become circles. The distances we seek are then the distances between the group centroids in this space. The axes of the space are known as *canonical axes*. They have their origin at the centroid of the whole distribution. If the groups contain equal numbers of individuals then the directions of the axes are such that the sum of the squares of the distances of the group centroids to the first axis is a minimum; the second lies at right angles to the first and is chosen to minimize the sum of the squares of the distances of the group centroids to it; and so on. The procedure is analogous to that of finding the axes in a principal component analysis. If the groups are not of equal size, then the directions will be approximately as above, though they will still be at right angles to one another. If there are only two groups then one canonical axis exactly represents the distance between their centroids and is equivalent to the line FF′ in Fig. 9.3. If there are three groups then, unless the three centroids are collinear, there will be two canonical axes. In general, if there are k groups and p variates the transformed space will have either $k-1$ or p dimensions, whichever is the smaller.

The values of canonical variates for group centroids (mean canonical points) and for individuals are determined in a way similar to that for finding principal components. In this case we have to find the latent roots and vectors of the matrix $\mathbf{W}^{-1}\mathbf{B}$.

The latent roots $\lambda_1, \lambda_2, \ldots,$ are defined by the determinantal equation

$$|\mathbf{W}^{-1}\mathbf{B}-\lambda\mathbf{I}|=0. \tag{9.12}$$

The equation

$$(\mathbf{W}^{-1}\mathbf{B} - \lambda_i\mathbf{I})\mathbf{c}_i = 0 \tag{9.13}$$

gives the ith canonical vector. It is equivalent to

$$(\mathbf{B} - \lambda_i\mathbf{W})\mathbf{c}_i = 0, \tag{9.14}$$

from which form it is usually solved. The equations are similar to those in principal component analysis. The obvious difference is that $\mathbf{W}^{-1}\mathbf{B}$ has replaced the total variance–covariance matrix. The canonical vectors define the relations between the original variates and the canonical axes. By multiplying the two we obtain the canonical variates needed for plotting. Thus, if the centroid of a group is represented by its mean vector $\mathbf{m}$ then its position on the ith canonical axis is

$$z_i = \mathbf{m}^T\mathbf{c}_i. \tag{9.15}$$

Canonical variate values of individuals can be found likewise.

When there are two or three groups the relations between group centroids can be represented exactly on a plane graph. When there are more than three groups a display in two dimensions is only an approximation to these. Nevertheless, projection on to the plane of the first two canonical axes is the best display and will sometimes account for a large proportion of the variation among groups.

As pointed out earlier, the probability contours for the groups on the plane of canonical axes are circular. The variances are standardized to unity. Therefore, it is a simple matter to add contours for any desired probability. These have radius $\surd\chi^2$ with two degrees of freedom. In some instances it is helpful to display confidence regions for group centroids. For the kth group this has a radius of $\surd(\chi^2/n_k)$, where n_k is the size of the group. It is interesting to note that when the groups are of equal size the same projection of group centroids can be found by a principal coordinate analysis of the matrix of Mahalanobis D between the groups (Gower 1966).

Numerical example

We illustrate the procedures by extending the analysis originally performed by Webster and Burrough (1974) using data recorded to see how increasing field effort or scale of mapping affected the quality of classification. One of the regions studied was near Kelmscot (the home of William Morris, the artist and craftsman) on a gravel terrace of the River Thames in Oxfordshire. The soil varies in thickness from about 30 cm to more than

1 m. It is mainly brown calcareous and medium textured, but it is heavier and somewhat gleyed in parts. A rectangular region 1400 m × 600 m was chosen, and three soil maps were made of it, notionally for publication at 1 : 63 360, 1 : 40 000, and 1 : 25 000. Each divided the soil of the region into three classes, namely

1 Badsey series: shallow, medium textured, brown calcareous soil over limestone gravel;
2 Carswell series: deep, more or less grey, mottled soil of heavy texture;
3 Kelmscot series: soil intermediate in character between Badsey and Carswell series.

After the maps had been made, the area was sampled at the 84 intersections of a 100 m grid and the following 21 soil properties measured at each:

— 1st horizon (plough layer, *c*.20 cm thick): hue, value, chroma, degree of mottling, sand, silt and clay contents, available Mg, P, and K, organic matter, cation exchange capacity, structure (size and degree of development), and consistence;
— 2nd horizon (>20 cm): hue, value, chroma, degree of mottling, sand and clay contents; and for the whole profile, depth to mottling.

The sampling sites were grouped according to the mapping classes in which they lay and the classifications analysed as above. The results for the 1 : 63 360 map classification were as follows:

$$|\mathbf{T}| = 2.161 \times 10^{36}$$

$$|\mathbf{W}| = 1.003 \times 10^{36}$$

$$L = \frac{|\mathbf{W}|}{|\mathbf{T}|} = 0.4642.$$

There were 84 test sites (n), three classes (g), and 21 variates (p). Thus chi-square is given by

$$\chi^2 = -\{n - 1 - (p + g)/2\} \ln L$$
$$= -71 \times \ln 0.4642 = 54.5$$

with $p(g-1) = 42$ degrees of freedom.

The probability of achieving this by chance is slightly more than 0.1, and we should usually regard this evidence as insufficient to establish that the three classes differ with respect to the properties measured in this region. Note that the sampling was quite independent of the boundary drawing;

the situation is probabilistic, therefore, and a significance test is appropriate.

We obtain the values for the other classifications similarly, and the results for all three are compared in Table 9.1. They show how a better soil map, using 'better' in the sense of less variation within the mapping units, is made by increasing effort and attention to detail. For the 1 : 40 000 and 1 : 25 000 scale maps the differences among the groups are highly significant.

Table 9.2 gives the Mahalanobis distances between the groups for each classification. On average, the Mahalanobis distances increase as the map

TABLE 9.1 Sample estimates of $|\mathbf{W}|$, L, χ^2, and probability of null hypothesis being true for three map classifications of the Kelmscot area

	Map scale		
	1 : 63 360	1 : 40 000	1 : 25 000
$\|\mathbf{W}\|$	1.003×10^{36}	6.402×10^{35}	3.793×10^{35}
L	0.4642	0.2963	0.1755
χ^2	54.5	86.4	123.5
approx. P	<0.1	<0.001	<0.001

Note
$|\mathbf{T}| = 2.385 \times 10^{36}$ and $p(g-1) = 42$, the number of degrees of freedom, are constant.

TABLE 9.2 Mahalanobis distances between groups for the three map classifications, F ratios, degrees of freedom, and approximate probabilities on null hypothesis

Map	Group comparison	Mahalanobis distance	F ratio	Degrees of freedom in denominator	Approx. probability
1 : 63 360	1, 2	2.70	1.61	40	0.15
	1, 3	1.27	0.87	54	v. large
	2, 3	2.87	0.66	8	v. large
1 : 40 000	1, 2	3.89	1.49	8	0.2
	1, 3	1.07	0.53	50	v. large
	2, 3	3.81	4.66	44	<0.001
1 : 25 000	1, 2	4.96	3.78	14	0.01
	1, 3	2.65	4.09	52	<0.001
	2, 3	3.40	2.93	36	<0.01

scale increases. None are significant for the 1 : 63 360 map. The Mahalanobis distance between groups 2 and 3 is significant for the 1 : 40 000 map, and though the distance between groups 1 and 2 is larger there are too few degrees of freedom to be reasonably sure that this difference is real. For the 1 : 25 000 map all the differences are significant.

The effects of each classification are perhaps grasped best from Fig. 9.4, in which the group distributions are displayed on canonical axes. Since there are only three groups, the group centroids are represented exactly. Distances between them can be measured and will be found to equal the Mahalanobis distances given in Table 9.2. The points representing the individual sampling sites are projections. The circles have radius 2.15, equal to $\sqrt{\chi^2}$ with two degrees of freedom at the 90 per cent confidence level. Thus they show for each group a circle within which 90 per cent of the individuals will lie on average. In Fig. 9.4a for the 1 : 63 360 map there is considerable overlap of these circles: the groups are not well separated. In Fig. 9.4b for the 1 : 40 000 map, group 2 is well separated from groups 1 and 3, but the latter two are scarcely separated at all. Fig. 9.4c for the 1 : 25 000 map shows less overlap among the groups, and groups 1 and 2 are especially well separated. The analysis used to derive the canonical variates is summarized in Table 9.3. The canonical vectors for pre-standardized variates are shown for the 1 : 25 000 map classification in Fig. 9.7.

Allocation

In the preceeding sections of this chapter we have seen how to measure the effectiveness of a classification and to display the groups in the plane of the leading canonical variates. Before describing the individual approaches and methods of classification, we shall consider the problem of

TABLE 9.3 Latent roots of the matrix $\mathbf{W}^{-1}\mathbf{B}$ for three map classifications

	Order	Root	Percentage of trace
1 : 63 360	1	0.635	67.7
	2	0.308	32.3
1 : 40 000	1	1.837	90.6
	2	0.189	9.4
1 : 25 000	1	2.349	77.0
	2	0.701	23.0

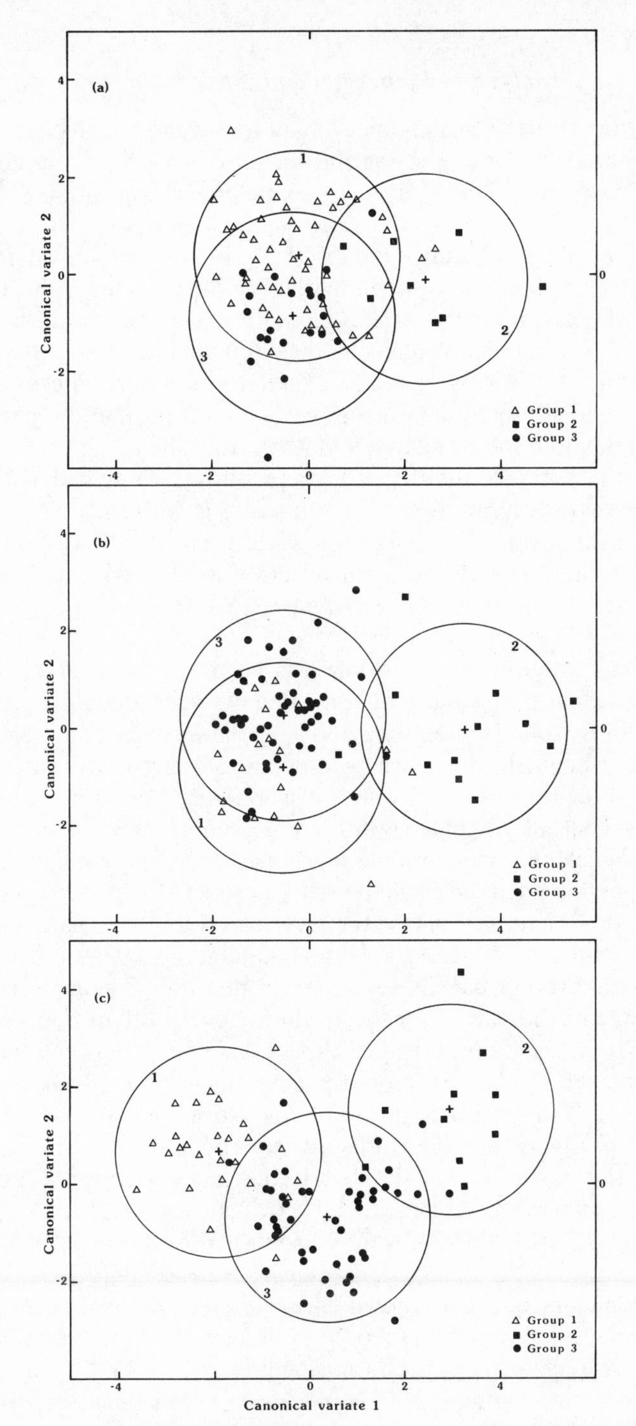

FIG. 9.4 Scatter of Kelmscot sampling sites on planes of canonical axes, (a) for 1 : 63 360 soil map, (b) for 1 : 40 000 soil map, (c) for 1 : 25 000 soil map. Groups are labelled 1, 2, and 3, and 90% confidence circles are drawn around their centroids shown by +

discriminating between classes and, linked to this, how to allocate any new individuals to a class of an existing classification. The general principles are independent of the method used to create the classification in the first instance.

Allocation is the identification of the group to which an individual belongs or the assignment of an individual to the class that it fits best. It is crucial to success in everyday life. Communication would be hopelessly inefficient if we could not allocate the individual and the particular to general groups of objects or action. Similarly, scientists must be able to identify particular individuals or samples with more general classes so that they can talk about their medium and learn from the experience of others. Allocation or identification by the layman, and by many scientists in their specialities, is largely intuitive. Consequently it is not necessarily consistent. Different people can assign things differently, and any one person might judge matters differently on different occasions. One has only to attend a meeting of pedologists to realize that this is just as true in soil science as it is in everyday life.

Taking soil as an example, pedologists have sought criteria that would enable individual occurrences of soil to be assigned unequivocally to the correct or most appropriate group and so avoid inconsistency. The criteria have usually been the presence or absence of simple attributes such as calcium carbonate, or critical values of variables, e.g. 35 per cent of clay and 8 per cent of organic matter. Assignment often involves several decisions, which are made in some prescribed order, as in a key. Using soil characters in this way is equivalent to attempting to identify planes through character space that (a) separate the classes of soil in question, and (b) are orthogonal to the character axes. Unfortunately, the commonly recognized classes of soil are not usually separated in this way. Their properties are interdependent and they overlap in almost every dimension. So planes orthogonal to the soil property axes do not distinguish between the classes. Using simple criteria in these circumstances inevitably leads to some wrong identifications and some inappropriate allocation. Some pedologists have attempted to overcome this difficulty by changing the classification—redefining the classes so that simple criteria can be used for identification. They have not notably succeeded and they have dodged the central issue, which is to allocate individual soil specimens, profiles, or sites to existing classes.

To discriminate between known classes we need divisions through the character space that are oblique to the character axes, and that might need to be curved. One of the most promising means of achieving this is by the classical procedure known as *multiple discriminant analysis*, which was developed by Fisher (1936) and Rao (1948, 1952). Discriminant analysis is being used increasingly for identifying items from automatically recorded data, especially in the analysis of digital imagery from remote sensors.

Studies by Norris and Loveday (1971) and Webster and Burrough (1974) show that discriminant analysis has considerable potential in routine soil survey.

Principle of multiple discriminant analysis

This is an extension of those principles discussed earlier in this chapter. Consider Fig. 9.3 again. The ellipses with centres at A and B represent two groups of individuals on which two properties have been measured. Suppose now that we have another individual on which the same two properties have been measured and we wish to allocate it to one of the two existing groups. Which group is the more appropriate to allocate it to? This can be decided by projecting its position orthogonally on to FF′, the line that shows the difference between the groups to best advantage. The point M on that line is the cutting value. If our new individual, when projected, lies above and to the right of M then it is assigned to the class whose centre is at A; if it lies below and to the left of M then it is assigned to the class whose centre is at B. For this reason FF′ is sometimes known as a *discriminant axis*—it is an axis that allows the best discrimination between individuals of the two groups. Notice that in this example both of the original properties would be relatively poor discriminators if used independently; there is no horizontal or vertical line that separates the groups as well as the line MM′ does. The principle extends to as many properties as we wish; MM′ is in general a hyperplane in one dimension less than the number of properties, but there is always an axis, such as FF′ orthogonal to MM′, that will provide the best discrimination between the groups.

If there are more than two groups, however, this simple solution will not serve, for we need to consider simultaneously the possibility of allocation to every group. Rao (1952), in his solution of this problem, considers the discriminant space to be divided into regions, each of which is associated with one of the groups. Any one region is that part of the space defined by the canonical axes that is nearer to its associated group centroid than it is to any other centroid. Discriminant functions are calculated for an individual and serve to identify the region in which that individual falls. Distances in the discriminant space are Mahalanobis distances, and it seems more natural to express relations in these terms. Thus, if we calculate the Mahalanobis distance between a new individual and each group centroid then we can allocate the individual to that group whose centroid is nearest. The two procedures are equivalent.

To identify by discriminant analysis the class to which a new individual belongs we need first representative individuals, preferably chosen at random, from each defined class and the relevant properties measured on each individual. This information is used to calculate the discriminant functions and the weights to be assigned to the different properties, or the transformation of the character space and the group centroids, depending on the way the technique is viewed. In either event it can be regarded as 'calibrating' information. It is sometimes known as a 'training set', since the computer 'learns' to recognize members of each group from it. The other information needed consists of measurements of the same properties on each new individual to be allocated. Mean vectors $\mathbf{m}_j$ for each group, $j, j = 1, 2, \ldots, k$, and the pooled withing-group variance–covariance matrix $\mathbf{V}$ are computed from the training set. The Mahalanobis distances between an individual i and each group centroid can then be calculated from

$$D_{ij}^2 = (\mathbf{x}_i - \mathbf{m}_j)^T \mathbf{V}^{-1} (\mathbf{x}_i - \mathbf{m}_j), \qquad (9.16)$$

where $\mathbf{x}_i$ is the vector of observed values for the individual. The new individual is allocated to that group for which D_{ij}^2 is least.

Character weights and correlation

The technique raises two matters of some importance. The first concerns the relative weights that become assigned to characters for discrimination. In Chapter 7 we saw that, when calculating the similarities between individuals, whether for their own sake or as a preliminary to ordination or classification, we should treat all the properties in our analysis as of equal importance, and hence give them equal weight. There seemed no logical alternative unless we had some special purpose in mind. Our task in discriminant analysis is quite different, and to allocate an individual to the correct or most appropriate class we shall clearly wish to give greatest weight to those characters that discriminate best between classes. Ideally, information shared between several correlated characters should be used no more than once.

We illustrate the point by considering two groups on which two properties have been measured. Suppose initially that the properties are uncorrelated within the groups and that the within-group variances are equal: thus their distributions may be represented by circles of equal probability, as in Fig. 9.5, with centres at A and B. The line MM′ separates the groups best, and it lies at right angles to the line CC′ through A and B, which is a discriminant axis. It is clear that in choosing MM′ we give more weight to property P_1 than than to property P_2. In an extreme situation A and B might have equal value on the P_2 scale, and MM′ would then be vertical; P_2 would then have no discriminating power, and P_1 could be used alone. If MM′ makes an angle of 45° with the horizontal then P_1 and P_2 have equal weight.

The second feature of this situation is that when we wish to identify the

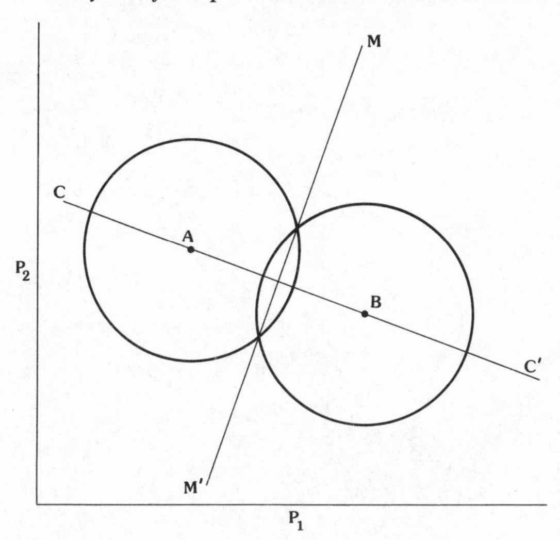

Fig. 9.5 Two classes with centres at A and B with intersecting circular confidence contours

most appropriate group for a new individual we can calculate the Pythagorean distance between it and each group centre, and allocate the individual to the group for which this distance is least. All points above and to the left of MM′ are nearer to A than they are to B; all points below and to the right of MM′ are nearer to B. In this case the Pythagorean and Mahalanobis distances are the same.

When there is correlation between the characters, however, the Mahalanobis distances are not the same as the Pythagorean distances. The situation that arises if we disregard correlation is illustrated in Fig. 9.6. Allocation based on Pythagorean distances is equivalent to choosing CC′ through the group centres as discriminant axis, and the perpendicular QQ′ through the mid-point of AB as separator between the groups. However, the best separator in these circumstances, as we have seen, is the line MM′ through the points of intersection, and the best discriminant axis is some line at right angles to it, say FF′ in Fig. 9.3. Thus, if Pythagorean distances are used for allocation instead of Mahalanobis distances then any new individuals in the hachured area will be assigned to the group centred at B, whereas they belong properly in the other group. The reverse applies to individuals in the stippled zone. The situation will be clear if we recall the transformation to canonical variates: in that transformation the probability contours are transformed into circles. Then a point on a contour at, say, X becomes the same distance from A as any other point, Y, on that contour, and, provided it lies above and to the left of MM′, will be further from B.

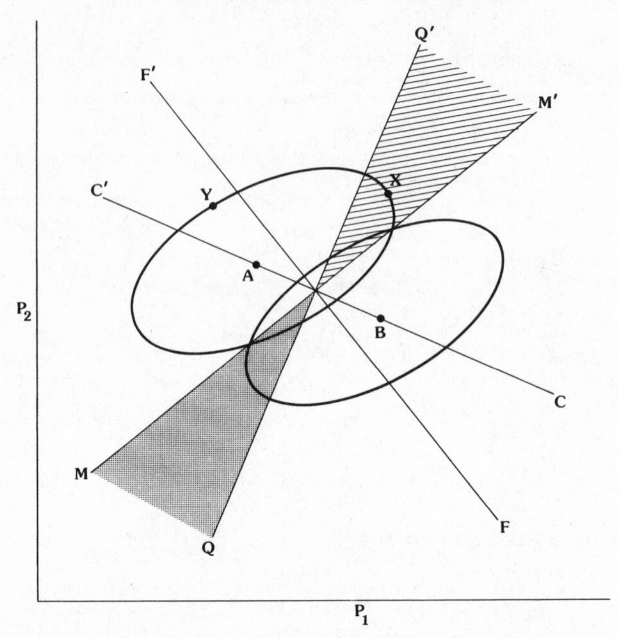

FIG. 9.6 Two classes within which variates P_1 and P_2 are correlated. The ellipses are confidence contours

Finally, we knit together the ideas of canonical variates and discriminant analysis. When there are only two groups there is only one discriminant axis, and this is identical with the one canonical axis. If there are more than two groups then canonical axes will not coincide in general with any of the discriminators between pairs of groups. The first canonical axis, however, is the best single discriminator among all the groups. If there are two or three groups, or only two variates, then we can see from a graph of canonical variates to which group centroid a new individual is nearest in the Mahalanobis sense, and hence to which group it is most likely to belong. Further, the canonical vectors contain the coefficients by which the original data are multiplied to obtain canonical variates. If there are only two groups then there is only one vector, and, provided the original variates are standardized, the coefficients in the vector represent the relative weights (and hence importance) to be attached to each original variate for discrimination. If there are more than two groups then the order of importance is less easy to discern, though the magnitude of the coefficients is still a good guide, and a vector diagram is likely to be helpful. Fig. 9.7 shows the canonical vectors for a soil classification using data from the survey at Kelmscot. The lengths of the vectors indicate the relative importance we should attach to the original variates in deciding the class to which to allocate a new individual. In this example, the colour hue, clay content, degree of mottling, consistence, and organic matter are more important than the other properties.

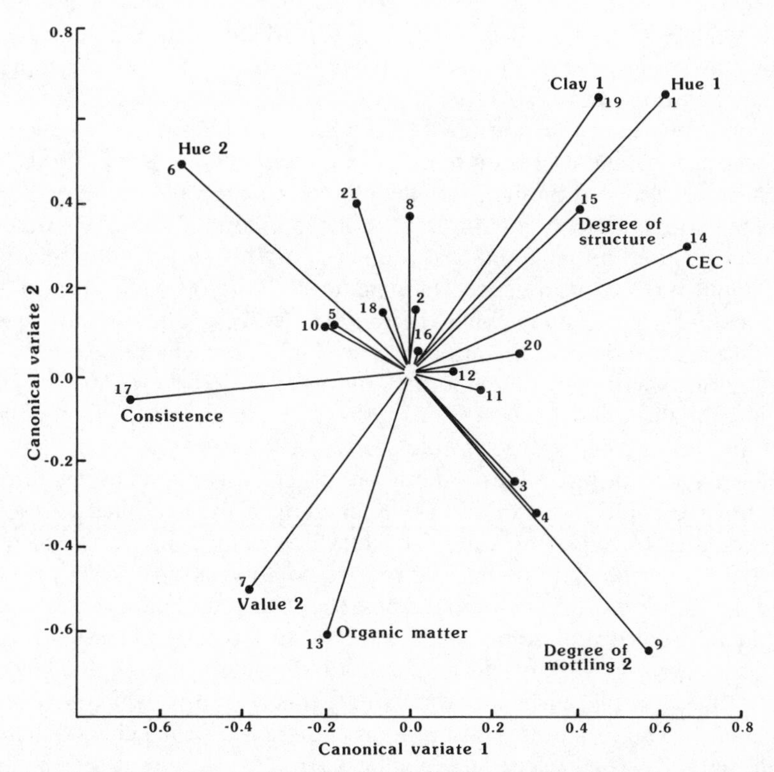

FIG. 9.7 Vector diagram of contributions made by original soil properties to canonical variates for 1 : 25 000 map classification of Fig. 9.4(c). Only the largest contributions are named

Suspending judgement

So far we have assumed that any new individual would be assigned to one or other of the existing groups. This is not always sensible. If the new individual lies a long way from every group then obviously we should want to know, and we might not wish to allocate it to any existing group; we should probably regard it as representing a class that we had not encountered hitherto. The Mahalanobis D_{ij}^2 between an individual i and the centroid of group j is distributed approximately as χ^2 with p degrees of freedom (where p is the number of variates). So for any one potential allocation the confidence with which the individual may be allocated can be determined simply by consulting a table of χ^2. If D_{ij}^2 is large we shall have little confidence. If D_{ij}^2 is small for one value of j and large for all others then we shall be very confident in allocating the individual to group j. If D_{ij}^2 is small for two values of j then we might consider the individual as belonging in some degree to both groups, in a way similar to the joint memberships of a fuzzy classification. Alternatively we could allocate the individual to the nearest in the Mahalanobis sense. Our decision depends on our point of view.

A more subtle problem arises when some of the information needed for the unequivocal identification of an individual is lacking. For example, we might have recorded the morphology of a new soil profile in the field, but not have the laboratory measurements of the kind that are used to create the classification initially and that are necessary for correct identification. We might have only the spectral 'signature' of a crop from an airborne sensor, and want to know what the crop is. Or we might wish to bring a new block of land into cultivation, but are undecided whether it should be used in one way rather than another. In this case we have no measures of its performance, and the best we can do is to match it to the kinds of land that are already being used in the two ways. We are now concerned not only with good or bad allocation but with right or wrong identification. In the first example, we could decide unequivocally to which class the individual belonged by making laboratory determinations. In the second, there is a specific crop whose identity we could verify by going into the field. In the third example there is a best use of the new land even though it would be very troublesome to discover it. Before, we did not need to concern ourselves with borderline cases of allocation. If an individual fell neatly between two groups it was largely immaterial to which group we assigned it. The situation is different now; there is a strong chance that we shall make a wrong decision when we have a borderline case. In these circumstances we might reasonably suspend judgement until we have obtained more information. In the space defined by the available information our new individual falls into a sort of no-man's land. Thus, even if the individual were closer to, say, group A than group B, if the probability of its belonging to group B exceeded, say, 20 per cent, we might decide not to make judgement without further evidence—at least not without some laboratory measurement, further sampling, or field trials respectively in the examples above.

Unequal groups

A further problem of allocation arises if the groups occur with unequal frequencies. In these circumstances we can minimize the risk of wrong identification by considering their relative frequencies or probabilities of occurrence. In Fig. 9.3 our cutting point for discrimination was the mid-point, M, between the projections of A and B. If the group represented by A was more likely to occur than that represented by B then we could reduce the risk of misidentification by moving the cutting point down and to the left in the direction of B. The best position depends on the expected frequencies or probabilities of occurrence. Let these be q_A and q_B. Then we move the cutting point to K such that

$$\mathrm{MK} = \frac{\ln q_A - \ln q_B}{\mathrm{A'B'}}. \tag{9.17}$$

In practice, we allocate an individual i to the group j to which it is nearest in the Mahalanobis sense. If the probabilities of occurrence are known beforehand the Mahalanobis distances can be modified by

$$G_{ij} = D_{ij}^2 - 2 \ln q_j, \tag{9.18}$$

and we can allocate the individual to the group for which G_{ij} is least. It will be evident that, since a probability q_j is always less than 1 and its logarithm is negative, the smaller is q_j the larger is the increment to D_{ij}^2.

A somewhat different situation arises if the dispersions of the groups are unequal. Mahalanobis distances between individuals and group centroids can still be calculated, but using separate variance–covariance matrices for each group, thus:

$$D_{ij}^2 = (\mathbf{x}_i - \mathbf{m}_j)^T \mathbf{V}_j^{-1} (\mathbf{x}_i - \mathbf{m}_j), \tag{9.19}$$

where $\mathbf{V}_j$ is the variance–covariance matrix for the jth group.

Generally speaking, this refinement is justified only when each group is represented by a large sample: with small samples $\mathbf{V}_j$ is subject to large error. Several other points need mention. Whereas with equal $\mathbf{V}_j$ the partitions between groups were planar, with unequal $\mathbf{V}_j$ they are curved. If some groups are more dispersed than others there will be a tendency for individuals to be assigned preferentially to those groups with the larger spreads. If some groups are much more dispersed than others their influence could swamp completely that of the more tightly packed groups, and the technique is not appropriate then.

10

NUMERICAL CLASSIFICATION: HIERARCHICAL SYSTEMS

A fool sees not the same tree
that a wise man sees.
William Blake, *Marriage of Heaven and Hell*

General introduction to classification

Chapter 1 introduced the idea of replacing classification by measurement to describe soil, vegetation, rocks, and so on, precisely and consistently. Quantitative measurement was regarded as superior to qualitative classification. Chapters 7 and 8 described how we could use measurements to relate individual rock or soil samples, soil profiles, and sampling sites to one another, and to see order among them. Nevertheless, it is often helpful to group individuals together even when data are quantitative. More and more measurements are being made now at many sites to describe the soil and other land resources. So many data can be very difficult to interpret and comprehend, and classification often helps to provide a simpler picture. It is also convenient to be able to talk about one or more groups of individuals using names rather than lists of characters and their values. Further, a classification might enable an investigator to economize on expensive measurements. Although qualitative classification may be sufficient to divide land into parcels for planning or management, when quantitative data are available the scientist will usually want to take full advantage of them to classify the land in definable and repeatable ways.

In Chapter 4 we described a simple numerical way of classifying soil from measured values. We called the process *dissection*. The measured range of a property of interest is divided at certain critical or convenient points. If two or three properties are judged to be important then all their scales can be divided to produce a classification still with manageably few groups. When many properties are relevant, however, simultaneous dissection of every scale is not feasible: far too many groups result. To classify complex multivariate populations we need an alternative.

Dissection as described has another potential disadvantage: it takes no account of any natural discontinuities in the population. If the frequency distribution of a variable has several peaks it might be thought that the most sensible or 'natural' places to divide the range would be through the troughs between the peaks. Unless there are critical values at which a scale should be divided, some means is needed for discovering suitable dividing points. When extended to several dimensions it is easier to think of those individuals in the more densely occupied parts of the character space as constituting *clusters* which we hope to isolate in our classification.

The concept of natural classification has a second facet. The term is often used to mean a classification in which the individuals in any one group are generally similar to one another. It applies especially to populations possessing many characters of interest, so that members of a natural group should be similar in many respects. This principle of overall similarity was propounded by the French botanist Adanson in the eighteenth century. Such a grouping should also be useful for many purposes (Gilmour 1937).

The concept of a group in which the individuals share many attributes, but for which no single attribute is either sufficient or necessary to confer class membership, was stated clearly by Beckner (1959). Sokal and Sneath (1963) coined the term *polythetic* to describe such a group. Polythetic classes are not confined to formal taxonomy; they are the norm in everyday life and language. The antithesis of a polythetic group is one that Sokal and Sneath call *monothetic*. It is one in which the possession of one or more attributes is both sufficient and necessary for class membership. Polythetic groups have two disadvantages compared with monothetic ones. First, though they might be generally useful, they cannot be expected to be the most suitable for any particular purpose; indeed, they might not be useful for any desired purpose—see Webster and Butler (1976) for an example. Second, it can be very difficult to create keys for identifying their members and for the allocation of new members. We discussed the principles of allocation in the previous chapter.

The twin threads of division at the gaps and of classes that are generally useful have strongly influenced the search for mathematical procedures for classifying multivariate populations. There have been other influences. Biology has a long tradition of hierarchical classification rooted in Aristotelian logic and a genetic theory to account for hierarchical arrangements of organisms. Thus it has been thought that a classification derived mathematically should also be hierarchical, often in the hope that species, genera, families, etc. would be recognized. Some workers have expressly sought to identify genetic links between members of populations in this way, and they have been more or less successful.

Scientists in other fields, and especially mathematicians, have been more concerned, however, with the nature of the classes that are created. Do the classes have similar dispersions to one another; i.e., are they of about the same categoric level? Are they reasonably homogeneous internally? Must they be disjoint, or should they be allowed to overlap? The latter gives rise to the concept of fuzzy classification, where individuals have a degree of membership to one or more classes simultaneously. Some workers seek specifically to optimize a classification such that the classes are less variable than in any other partition into the same number of classes. Others are concerned that the classes are predictive. And so we have different groups of scientists predisposed to organizing individuals or partitioning populations in different ways for different purposes.

There have also been computational considerations. Elegant algorithms, which run swiftly on computers, have been devised for creating hierarchies. In contrast, creating a classification by minimizing the within-group dispersion is clumsy and usually consumes much computer time. It may not even succeed; a truly optimal classification is generally beyond reach for more than a few tens of individuals, even on the fastest modern computers.

Thus, we find that most attempts to classify individuals mathematically or numerically have sought to create classes within which the members are generally alike and substantially different from the members of other classes. Such methods for creating classifications are often known by the general name of *cluster analysis*. Their aim is to identify clusters in the population. The name is misleading, however, because most methods will create classes whether or not clusters exist in any true sense. Since biologists were the first to attempt mathematical classification, and hierarchies could be created reasonably quickly on the computers of the 1960s, hierarchical methods remained for some time the best developed methods of cluster analysis. Most of the classifications of soil developed by the national soil surveys of the world have also had a hierarchical structure, and it is not surprising that numerical methods to create the same kind of structure became popular with soil scientists. These methods were also favoured in the earth sciences more generally. For this reason we describe them first. However, in spite of the initial flush of enthusiasm in the development of hierarchical methods, little has happened since, and so this chapter remains much as it was in this book's predecessor (Webster 1977). The following chapter describes procedures for non-hierarchical classification, and then reviews the two critically.

Hierarchical classification

A hierarchical classification is one in which individuals belong to small groups, the small groups belong to larger groups, and so on. It may also be thought of as a division of character space in which the whole space is divided into smaller compartments, which in turn are divided into yet smaller compartments. Division or grouping is usually made at a few distinct levels of generalization, known as *categories*. In general, classes in any one category are disjoint, since any appreciable overlap of classes is incompatible with a hierarchy. The methods for creating hierarchies are either agglomerative, putting individuals together into larger and larger groups, or divisive, creating smaller and smaller groups from a single population. However, they do not usually proceed through distinct categories. This chapter describes the general agglomerative strategy and details of particular forms of it, and then discusses it critically. We shall look only briefly at divisive schemes.

Hierarchical agglomerative grouping

The starting point for most agglomerative methods is a set of similarities (or dissimilarities) between individuals. A measure is chosen to represent relations between individuals and its values are calculated for all pairs to form a matrix (see Chapter 7). In practice, the measure is usually expressed as similarity and scaled so that identity is represented by the value 1, and maximum dissimilarity by 0. This matrix contains all the information needed.

To make matters as clear as possible, however, we shall assume that our starting matrix is a dissimilarity matrix and its values are the Euclidean distances between individuals.

The values in the matrix are scanned to find the smallest, i.e. the distance between the closest pair of individuals. These individuals are fused to form a group. From this point there are several possible courses of action. They give rise to the many different methods for agglomeration. The simplest of them is the single-linkage method, and for this reason it is described first.

Single-linkage grouping

This was the first method for creating a hierarchy numerically from a similarity matrix, and was made known largely through the work of Sneath (1957). Rayner (1966) and Moore and Russell (1967) used it in their early work on soil classification. In ecology it is also known as the *nearest-neighbour* strategy (Lance and Williams, 1967b).

After fusion of the closest pair of individuals, grouping proceeds as follows. The matrix of distances is re-scanned, and the second shortest distance is found. If this is between a member of the first group and a third individual then the latter joins the group. If it is between two other individuals then they are fused to form a second group. The process is repeated a third and further times, and fusion is decided as above. If, however, the two individuals in question are in two different groups then the two groups are fused. The process continues until all individuals are contained in a single group.

To illustrate the procedure nine sites have been chosen from a survey of west Oxfordshire for which principal components were calculated in

TABLE 10.1 Values of the first two principal components of nine sites

Site	Components 1	Components 2
6	−2.630	−2.809
14	−2.089	0.726
34	−2.396	−0.661
45	−1.861	−0.729
47	0.367	0.266
57	5.200	−3.023
59	2.481	−1.728
63	2.594	−0.807
70	4.354	−0.732

Chapter 8. Their values are given in Table 10.1. Fig. 10.1 shows the relative positions of the sites in the plane of the first two principal components. In this simple case the hierarchy is constructed from the values of these two components only.

The first fusion is between sites 34 and 45, separated by 0.54 units of distance. The next fusion is of sites 59 and 63 at a distance 0.93, and the third joins site 14 to 34, which already belongs in the group with site 45. Other fusions take place in sequence (the relevant inter-point distances are

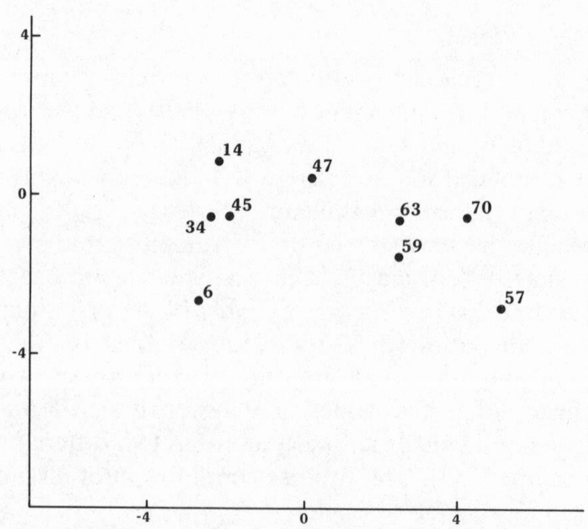

FIG. 10.1 Scatter of nine points in a plane. Their coordinates are given in Table 10.1

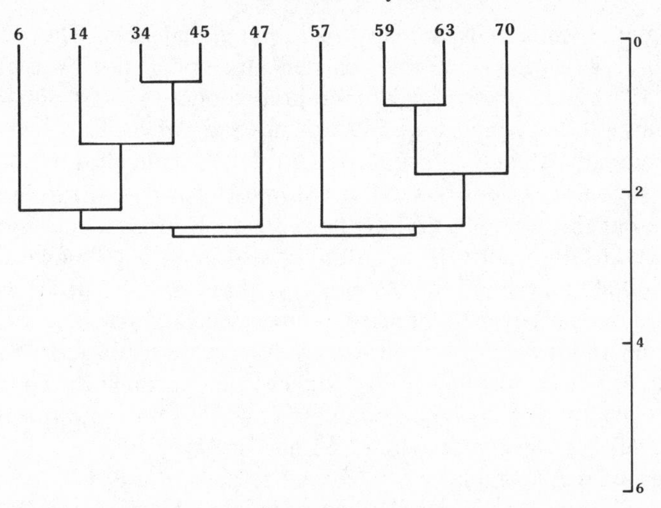

FIG. 10.2 Single-linkage dendrogram for the nine points in Fig. 10.1

shown in Fig. 10.9, discussed on p. 188), and the final fusion is of site 47 to site 63.

The result can be shown as a tree diagram or *dendrogram* (Fig. 10.2). The vertical scale represents distance in character space, and the horizontal lines joining the vertical stems of the tree are placed at positions corresponding to the distances at which fusion occurs. The horizontal dimension has no scale, and the order in which individuals are placed is to some extent arbitrary. Any part of the tree can be rotated about the stem immediately below without changing its structure, as in a mobile.

A classification with some desired number of classes can be obtained readily from the dendrogram by drawing a horizontal line to cut just that number of vertical stems. For example, two classes can be obtained by drawing a line at value 2.45, while a line at 2.0 gives five classes.

A feature of the single-linkage strategy is that each fusion is decided solely on the distance between a pair of individuals. Thus, the distance between a group and another individual is the distance between the individual and the nearest member of the group; similarly, the distance between groups is the distance between their nearest members. The dispositions of the groups as a whole are not taken into account, and this is undoubtedly a weakness of the method.

A closely related feature is that as a group grows it becomes increasingly likely that it has neighbouring individuals that will soon fuse with it. There is a strong tendency, therefore, for individuals to join existing groups in succession rather than form new groups and for existing groups to fuse. The process is known as 'chaining', and its effect is well illustrated in Fig. 11.2. Chaining can result from chain-like structure in the population. With soil it is more often an artifact of the method; at worst it can lead to a grouping that quite fails to recognize obvious clusters. The extreme is shown in Fig. 10.3, where two clusters are connected by a few close individuals. Single-linkage agglomeration applied to this situation will begin among the connecting individuals and proceed by successively adding individuals or small groups from the two main clusters.

Thus, there is a serious risk that a hierarchy formed by single-linkage will misrepresent the configuration of the individuals in Euclidean space, and that a classification derived by cutting it will be poor. To overcome this Wishart (1969a) modified the technique so that fusion occurred only when several individuals were identified as near neighbours. The number of neighbours and the radius within which they are judged to be near can be set to identify true clusters and to avoid chaining through the connections. Wishart calls the method *mode analysis*, and it is an option in the fairly widely available Clustan program (Wishart 1987).

Single-linkage is simple; it will recognize gaps in a distribution; and in the form of mode analysis it will identify clusters. Despite this, the method lost favour to other strategies, some of which we consider now.

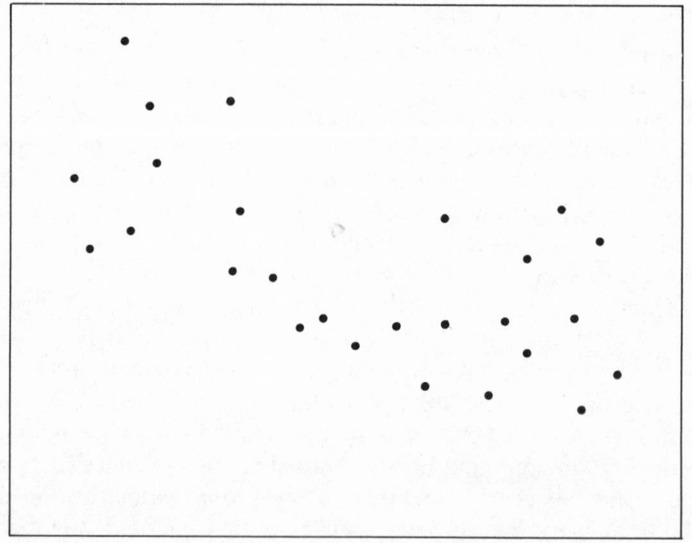

FIG. 10.3 Two clusters of points with chain of connecting points

Centroid method

Perhaps the most attractive fusion strategy from a geometric point of view is Gower's (1967) centroid method. Here a newly formed group becomes, in effect, a synthetic individual whose position in the Euclidean character space is defined by the centroid of the group. Distances between this centroid and other individuals are then calculated and substituted for the

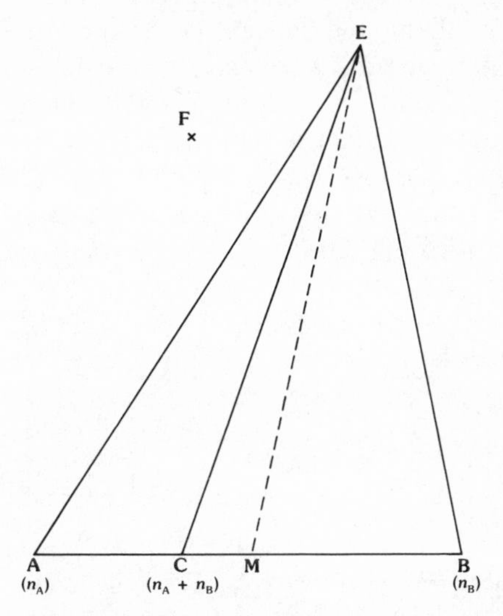

FIG. 10.4 Geometric relationships of centroid strategy. See text for explanation

distances from the individual points included in the pair. Then all distances are re-examined and the procedure is repeated, as described for single-linkage.

The situation at fusion is illustrated in Fig. 10.4. Here a group at A containing n_A individuals is about to fuse with a group at B containing n_B individuals to form a new group that will contain $n_A + n_B$ individuals. This group will be represented by its centroid, C, which lies on the line AB such that the ratio AC:CB = n_B:n_A. Distances to other individuals, the one at E for example, are then calculated as follows:

$$d_{CE} = \sqrt{\left\{\frac{n_A}{n_A + n_B} d^2_{AE} + \frac{n_B}{n_A + n_B} d^2_{BE} - \frac{n_A n_B}{(n_A + n_B)^2} d^2_{AB}\right\}}. \qquad (10.1)$$

Readers might like to verify this for themselves.

When we use the strategy to group the population displayed in Fig. 10.2 the first two fusions are of sites 34 and 45, and of 59 and 63. The third fusion adds site 14 to the group 34 + 45, and as it happens the distance between site 14 and the centroid of 34 + 45 is 1.42. Fusion proceeds until finally the group comprising sites 6, 14, 34, 45, and 47 joins the one consisting of 57, 59, 63, and 70. Fig. 10.5 shows the result. The first three fusions by this method are identical with those achieved using single-linkage. The fourth also joins sites 70 to group 59 + 63. The next fusion of site 47 to the group 14 + 34 + 45 is different, however, and we consider what has happened in this left-hand group. When site 34 joined 45, the synthetic individual representing them took up a position mid-way between them. Site 14 joined them, and the centroid of the three then lay somewhat above this mid-point, where it was nearer to site 47 than to 6. So the group was joined by site 47, and the centroid moved to the right. At each stage it moved further from individual 6. This action is known as *drift*, and can have important consequences. Nevertheless, the centroid method seems to separate the two main groups more clearly than single-linkage does: compare Fig. 10.5 with Fig. 10.2.

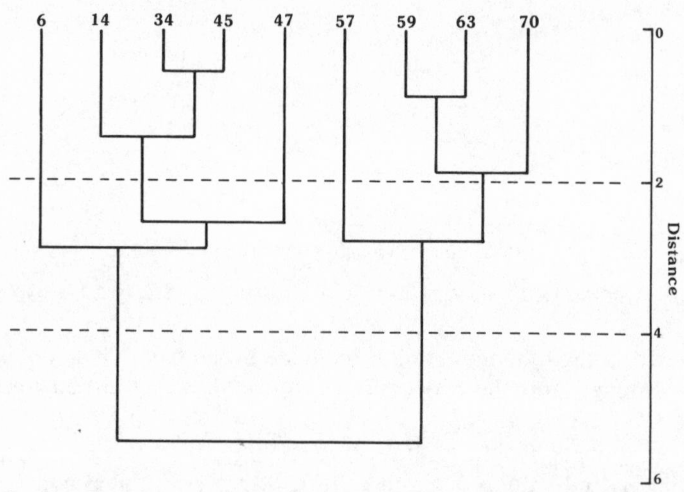

FIG. 10.5 Dendrogram resulting from centroid fusion of nine points in Fig. 10.1

Let us now apply the strategy to a more realistic situation: 40 individuals, on each of which 30 properties have been measured, are to be classified. The individuals have been chosen at random from the 85 analysed earlier (Chapter 8), and their distribution in the plane of the first two principal components is shown in Fig. 10.6. The matrix of distances between individuals is calculated, then examined, and fusion proceeds as before. The result is presented in Fig. 10.7. The general structure of relations evident in the scatter diagram can be seen in the dendrogram. However, it must be remembered that, when the scatter of individuals in a character space of many dimensions is projected, the distances between individuals in the chosen plane represent only approximations to their true relations. Therefore, the precise order of agglomeration cannot be determined from a scatter diagram, and the early fusions often provide information about relations that are obscured in projections on principal axes. It should be noted that the last few fusions in this example are of individuals and small groups. The dendrogram cannot be cut to give a tidy classification with few groups, as we might have wished.

The centroid strategy takes account of the positions of all members of each group in determining fusion. However, despite this and its exact geometric representation, it is still not entirely satisfactory.

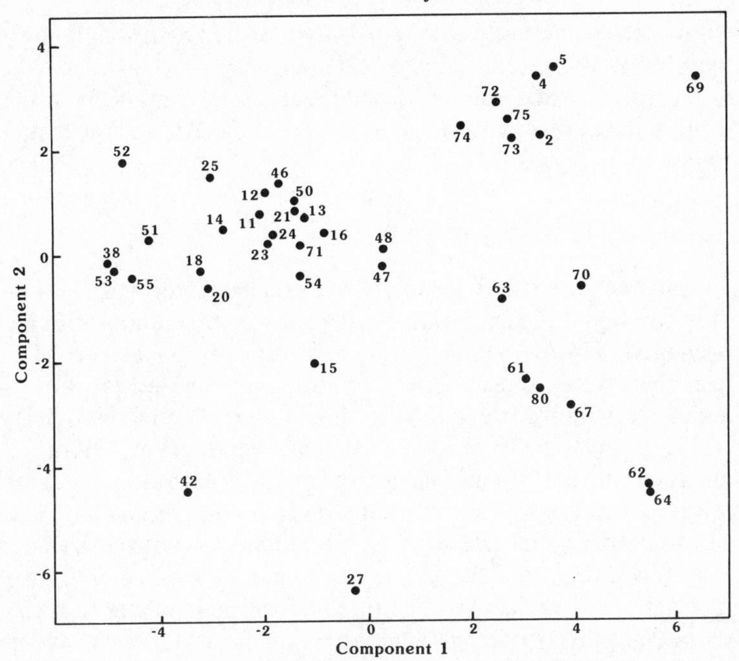

FIG. 10.6 Scatter diagram of 40 soil profiles in west Oxfordshire projected on to a plane of first two principal components

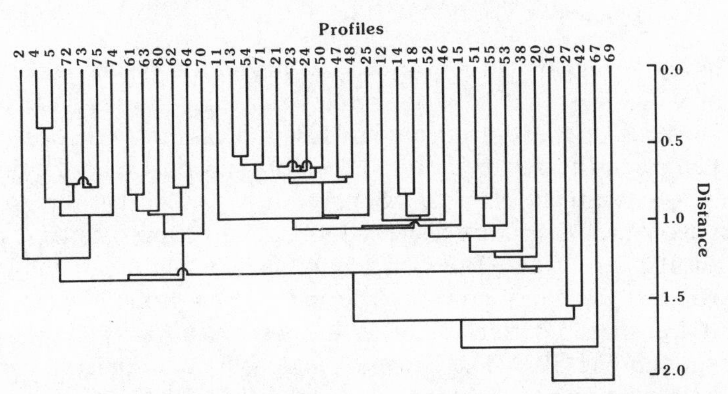

FIG. 10.7 Dendrogram from centroid fusion of 40 west Oxfordshire profiles

Weighted centroid method

In some instances it happens that a pair of groups that are to be fused contain disparate numbers of individuals, so that in equation (10.1) n_A and n_B are very different. If n_A is very much larger than n_B, for example, then the distance from the new group to the group at E is heavily weighted towards the distance AE (Fig. 10.4). If the groups at A and B were well-defined clusters, one of which just happened to be poorly represented in our sample, we should not want the bias that would be present in the group that they form. We should almost certainly prefer groups to have equal weight regardless of the number of individuals they contain. If so then we should place the new group at M, the mid-point on AB, and the distance to E becomes

$$d_{EM} = \sqrt{(\tfrac{1}{2}d^2_{AE} + \tfrac{1}{2}d^2_{BE} - \tfrac{1}{4}d^2_{AB})}. \qquad (10.2)$$

The terms 'weighted' and 'unweighted' need explanation in this context. Normally we should regard formula (10.1) as weighted and formula (10.2) as unweighted. However, the terms were used by Sokal and Michener (1958) in the reverse sense. Their attention was focused on the original individuals, so that by unweighted they meant that all individuals had equal weight. Thus in Fig. 10.4, C is the unweighted centroid of all the individuals constituting the new group comprising the groups at A and B. If the groups at A and B are of different sizes then the original individuals in them do not have equal weight in determining the centroid by equation (10.2), and so in this sense the formula is weighted. This usage has become conventional in numerical classification, and we shall adhere to it to avoid confusion. Weighted centroid agglomeration is also known as *median grouping* (Lance and Williams 1967a).

Group-average method

In general, as agglomeration proceeds successive fusions take place between individuals and between groups that are progressively further apart. However, in the centroid method fusion occasionally creates a group whose centroid is nearer to another group or individual than the distance between its constituent members. Suppose in Fig. 10.4 that the group or individuals centred at E were displaced to a point F, approximately on the perpendicular through C. The distance AB is only a little less than either AF or BF, and when fusion takes place and the new group is centred at F, FC is shorter than AB. Such an event is known as a *reversal*, and several are apparent in Fig. 10.7. The process is no longer monotonic; the order in which fusions take place is not the same as the order of the similarity values at fusion.

Intuitively, reversals seem undesirable, and they can be avoided by using the group-average method developed earlier by Sokal and Michener (1958). In this method, when two individuals or groups join, the distances or similarities between the new group and every other are computed as the *average* distances or similarities between all the members of the new group and those of the others. This destroys the nice geometry of the centroid method, and average distances are usually a little larger than distances between centroids. However, the differences between the two are usually small, and the group-average method ensures that agglomeration proceeds monotonically.

As with the centroid method, either individuals or groups can be given equal weight, as desired.

Complete linkage grouping

This method, also called the *furthest neighbour* strategy (Lance and Williams 1967b), is the exact antithesis of single-linkage grouping. When an individual is considered for admission to an existing group the criterion applied is the distance between it and the furthest member of the group. Similarly, the distance between two groups is regarded as the distance between their most distant members.

Fusion depends again on distances between pairs of individuals. However, as a group grows so its distance from its neighbours, as measured by the above criterion, increases. The result is that groups tend to be hyperspherical and, at some stage in the procedure, to be of roughly equal extent in character space.

The procedure was compared with others by Moore and Russell (1967), but has not otherwise been used in pedology.

Ward's method. A technique due to Ward (1963) enjoys some popularity and has been used in several soil studies. It aims to minimize the 'error sum of squares', i.e. the sum of the squares of the distances between individuals and their group centroids, $\Sigma\, d^2$. Its starting point is a matrix of distances between individuals. At each stage the fusion that takes place is the one that causes the least increase in $\Sigma\, d^2$. As Wishart (1969a) points out, for fusion of two groups A and B with n_A and n_B members respectively, this increase is equivalent to

$$\frac{n_A n_B}{n_A + n_B} d^2_{AB},$$

where d_{AB} is the distance between the group centroids. The method clearly favours the fusion of individuals and small groups.

At first sight Ward's method appears to be the same as optimizing a non-hierarchical classification using the SS_W criterion, described in Chapter 11. This is not so, however. The method is strictly hierarchical; once two individuals or groups are joined they remain so, no matter how inappropriate this might be at coarser levels of classification. The quantity Σd^2 might be near its minimum at some stage in the hierarchy, but it cannot be so at every stage.

Combinatorial representation

Lance and Williams (1966, 1967b) have shown that the above methods for creating hierarchies are variants of a single linear system. To understand this, consider Fig. 10.4 again and its meaning. Two groups at A and B with n_A and n_B members respectively fuse to form a group with n_A+n_B members, notionally at some point, say K, in the vicinity of C and M. We consider the relation between this new group and another group at E. Before fusion the distances or dissimilarities d_{AB}, d_{BE}, and the group sizes n_A and n_B, are all known and can be used to calculate the value d_{EK} from the equation

$$d_{EK}=\alpha_A d_{AE}+\alpha_B d_{BE}+\beta d_{AB}+\gamma|d_{AE}-d_{BE}|. \qquad (10.3)$$

Lance and Williams call this relation *combinatorial*, and they proceed to derive the values of the parameters α_A, α_B, β, and γ for each strategy. They also point out that when $\gamma=0$ agglomeration will proceed monotonically provided that

$$\alpha_A+\alpha_B+\beta\geqslant 1.$$

The parameters for each of the strategies so far considered are as follows.

1. *Unweighted centroid*. Here, provided dissimilarity is defined as the square of the distance between groups, we can readily see from (10.1) that

$$\alpha_A=\frac{n_A}{n_A+n_B}, \quad \alpha_B=\frac{n_B}{n_A+n_B},$$

$$\beta=-\frac{n_A n_B}{(n_A+n_B)^2}=-\alpha_A\alpha_B,$$

and

$$\gamma=0.$$

2. *Weighted centroid*. Dissimilarity is again squared distance, and the parameters are very simply

$$\alpha_A=\tfrac{1}{2}, \quad \alpha_B=\tfrac{1}{2}, \quad \beta=-\tfrac{1}{4}, \quad \text{and } \gamma=0.$$

3. *Unweighted group average*. This is combinatorial for all measures of dissimilarity.

Assume that the group E contains n_E individuals, and that the dissimilarity between the jth individual in it and the ith individual in the group at

A is represented by $1-S_{ij}$. Then the average dissimilarity between the two groups at E and A is

$$d_{AE}=\frac{1}{n_A}\frac{1}{n_E}\sum_{i=1}^{n_A}\sum_{j=1}^{n_E}(1-S_{ij}). \tag{10.4}$$

Likewise

$$d_{BE}=\frac{1}{n_B}\frac{1}{n_E}\sum_{i=1}^{n_B}\sum_{j=1}^{n_E}(1-S_{ij}). \tag{10.5}$$

The average dissimilarity d_{EK} that we require is

$$\begin{aligned} d_{EK} &= \frac{1}{n_E}\,\frac{1}{n_A+n_B}\sum_{i=1}^{n_B+n_A}\sum_{j=1}^{n_E}(1-S_{ij}) \\ &= \frac{n_A}{n_A+n_B}d_{AE}+\frac{n_B}{n_A+n_B}d_{BE}. \end{aligned} \tag{10.6}$$

The parameters for the combinatorial equation are thus

$$\alpha_A=\frac{n_A}{n_A+n_B}, \qquad \alpha_B=\frac{n_B}{n_A+n_B},$$

$$\beta=0, \qquad \gamma=0.$$

It can be seen now that, when dissimilarities are given as distances, the group average distance d_{EK} will be greater than the corresponding centroid distance by

$$\beta d^2_{AB}.$$

4. *Weighted group average*. The parameters for this are even more simple

$$\alpha_A=\alpha_B=\tfrac{1}{2}, \qquad \beta=0, \qquad \gamma=0.$$

5. *Single-linkage*. Here the dissimilarity d_{EK} is equal to the smallest of d_{AE} and d_{BE}. The parameters that give this are

$$\alpha_A=\tfrac{1}{2}, \qquad \alpha_B=\tfrac{1}{2}, \qquad \beta=0, \qquad \gamma=-\tfrac{1}{2}.$$

6. *Complete linkage*. Here we wish to identify d_{EK} with the largest of d_{AE} and d_{BE}, and the parameters are

$$\alpha_A = \alpha_B = \tfrac{1}{2}, \qquad \beta = 0, \qquad \gamma = +\tfrac{1}{2}.$$

7. *Ward's method*. The combinatorial parameters of this method of grouping were found by Wishart (1969b). As in 3 above, we assume that a new group at E contains n_E individuals, and the parameters in equation (9.3) are then as follows:

$$\alpha_B = \frac{n_E + n_A}{n_E + n_A + n_B}; \qquad \alpha_B = \frac{n_E + n_B}{n_E + n_A + n_B};$$

$$\beta = \frac{-n_E}{n_E + n_A + n_B};$$

$$\gamma = 0.$$

The method is strongly space-dilating, see below.

The advantages of a combinatorial strategy are considerable. Once the matrix of similarities or distances between individuals has been calculated the original data are no longer needed, and the space that they occupy in the computer can be released for other purposes. The combinatorial equation can be used to calculate all new similarities as grouping proceeds, though in some instances it is not quite the most efficient way.

Flexible grouping strategy

In a procedure such as complete-linkage agglomeration the distances between groups effectively increase as the groups grow. It appears as though the space around them dilates. The larger a group becomes the less likely it is to fuse with a single individual or small group, and the latter will tend to join together in preference. Therefore, groups tend to be of roughly equal size. The converse is true with single-linkage, and we have noted already its tendency to chain. The centroid methods retain the geometric relationships, and the group-average methods also appear to conserve the space. Nevertheless, chaining can still occur, and when the aim is to produce a classification with few groups this can be a nuisance. Thus the complete linkage method has a potential advantage over the other methods. The tendency to dilate the space is strong, and a strategy with only moderate dilation might be preferred.

To achieve this Lance and Williams (1966) proposed their *flexible strategy*. It is combinatorial, obeying equation (10.3) with the following constraints:

$$\alpha_A + \alpha_B + \beta = 1,$$

$$\alpha_A = \alpha_B,$$

$$\beta < 1,$$

$$\gamma = 0.$$

The strategy is flexible in that the user can adjust β and so vary the degree of space distortion around the growing groups. Values of β near 1 produce very strong space contraction and chaining. As β is decreased this effect is less marked, until at some small negative value (Lance and Williams are unable to define it precisely) space is conserved. Thereafter, decreasing β causes space dilation and increasingly distinct groups in the dendrogram. Lance and Williams (1967b) illustrate these effects by dendrograms in their paper. They recommend $\beta = -0.25$ for general use. This is slightly space-dilating. When $\beta = 0$ the result is the weighted-group-average strategy. Note also that, since $\alpha_A + \alpha_B + \beta = 1$, agglomeration is monotonic.

The flexible strategy is attractive conceptually, but there is the risk that β will be so chosen that it obscures information on the structure of a population that would be revealed in a space-conserving strategy. The strategy has been used with $\beta = -0.25$ by soil scientists in Australia (Moore and Russell 1967; Campbell *et al.* 1970; Moore *et al.* 1972) where this option is standard in several large classification programs. Fig. 10.8 illustrates the use of the strategy on the same 40 sites as in Fig. 10.7 with β set to -0.25, and using the mean character distance as the dissimilarity coefficient. The evident advantages are that (a) all individuals are fused early; (b) chaining, which was to some extent present in Fig. 10.7 no longer occurs; (c) the structure evident in the scatter diagram (Fig. 10.6) still seems well represented; and (d) a classification into a few groups can easily be made by drawing a horizontal line through the lower part of the tree.

We have already given meaning to the axes of the scatter diagram. As it happens, we can give meaningful conventional names to several branches of the tree in this instance. At the two-group level I contains entirely brown earths, while II consists of gleys and gleyed brown earths. At the four-group level A consists of medium textured brown earths, and B of brown sandy soil; C consists of medium and heavy textured gleys and gleyed soil, and D consists of sandy gleys. Further subdivision of C gives C_1, a group of moderately gleyed profiles, and C_2, gleys of heavy texture.

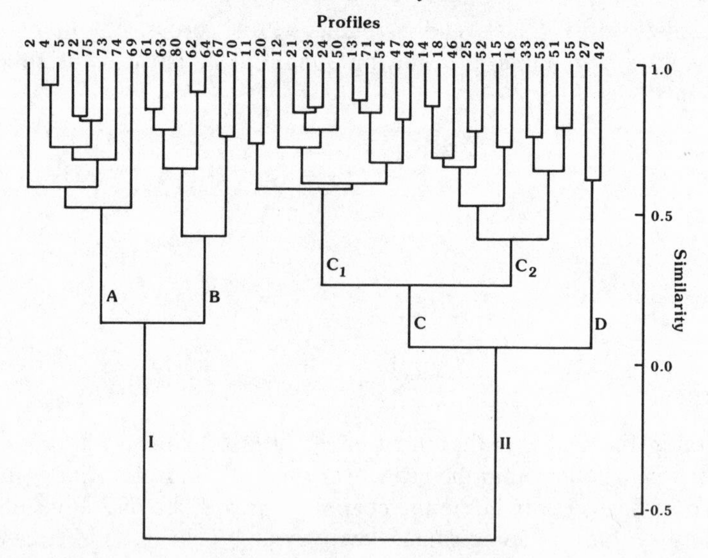

Fig. 10.8 Dendrogram for 40 soil profiles in west Oxfordshire obtained using the flexible strategy with $\beta = -0.25$ and the mean character distance as similarity coefficient

Other hierarchical methods

A number of other agglomerative methods have been devised for hierarchical classification. None have any obvious advantages over those we have discussed in this chapter, but readers who wish to study the subject further should see the much more comprehensive account by Sneath and Sokal (1973) and follow the many references they give.

Divisive grouping

A feature of agglomerative classification that used to be a serious limitation was that the $\binom{n}{2}$ similarity (or dissimilarity) coefficients had to be held in the memory of the computer for n individuals. The process was prohibitively time-consuming otherwise. This imposed a ceiling on the number of individuals that could be classified in a single run, usually between 400 and 500 on large computers. Nowadays, with virtual memory, the similarity matrix can occupy a whole disk, and so the methods can handle some 2000 individuals. Nevertheless, if space for such a matrix is not available a possible alternative is to use a divisive strategy. Divisive methods for creating hierarchical classifications have been used much less than agglomerative ones because to be feasible for a reasonable number of individuals they depend on the states of a single variable and so produce monothetic groups. *Association analysis*, developed by Williams and

Lambert (1959), is the best known of these and has been valuable in ecological classification. The starting point is a set of binary data, the presence or absence of each of p attributes for each of n individuals. In ecology the individuals to be classified are usually quadrats, or 'stands', and the attributes are species, and the result is a classification of vegetation. If the properties of interest are continuous variables, as they are in the main for soil, then each scale is divided into two parts, one of which is assigned the value of 0 and the other 1, and can be considered to represent absence and presence of the attribute. For each attribute the association between it and every other attribute is measured in turn, and the values are combined to give a measure of the degree to which this attribute is associated with others. The measure used is the familiar χ^2. For each comparison between attributes, say i and j, a 2×2 table is formed:

	0	j	Totals
0	n_A	n_B	$n_A + n_B$
i	n_C	n_D	$n_C + n_D$
Totals	$n_A + n_C$	$n_B + n_D$	$n_A + n_B + n_C + n_D$

The values n_A, n_B, n_C, and n_D are the numbers of individuals that lack both i and j, lack i and possess j, lack j and possess i, and possess both i and j respectively. χ^2_{ij} is then computed as

$$\chi^2_{ij} = \frac{(n_A n_D - n_B n_C)^2 (n_A + n_B + n_C + n_D)}{(n_A + n_B)(n_C + n_D)(n_A + n_C)(n_B + n_D)}. \tag{10.7}$$

For each attribute, i, χ^2 is summed over the remaining $p-1$ attributes: $\Sigma_{i=1}^{p} \chi^2_{ij}$, $i \neq j$. The attribute for which $\Sigma\,\chi^2$ is largest is chosen, and the population divided into two groups, one containing those individuals that possess the attribute and the other containing those that do not. These groups are subdivided similarly using the attribute with the second-largest value of $\Sigma\,\chi^2$ and the process is repeated. Division ends either when a predetermined number of groups has been created or a critical value of $\Sigma\,\chi^2$ is reached.

We should notice immediately a disadvantage of the method for fully quantitative data. It provides no means of deciding at which values to divide the measurement scales. It cannot be expected to find gaps in a continuously variable population such as soil unless the scales have been divided first at minima in their frequency distributions.

Edwards and Cavalli-Sforza (1965) proposed a polythetic divisive method of hierarchical classification. Their aim was to create a classification that was optimal in terms of minimizing the sums of squares, SS_W, of the distances between individuals and their class centroids. To find the best subdivision at any one stage they divided the individuals into two groups in every possible way and computed SS_W for each. Unfortunately, the procedure is practicable for only small numbers of individuals since the computing time required increases as a function of $n!$ where n is the total number of individuals. This solution usually takes too long for more than about twenty individuals.

The principle of minimizing SS_W is more general and applies more strongly to non-hierarchical classification, and we discuss this application of the principle in the next chapter.

Minimum spanning tree

In the above account of agglomerative methods we have seen that the methods not only provide means of classification but often give the investigator insight into the structure of a population. In the early stages of fusion they show which pairs of individuals are most similar. When several methods are used in conjunction with an ordination analysis other relations between individuals may become evident. A technique that can help to clarify relations among individuals is the *minimum spanning tree*.

When we have a set of n individuals distributed in character space we can join them together by a network of $n-1$ links such that every individual is connected to every other individual through the network and there are no closed loops. The result is a tree spanning all the individuals. That tree in which the total length of the network is least is the minimum spanning tree (MST).

Topologically, the MST is the exact equivalent of the dendrogram formed by single-link grouping (Gower and Ross 1969). It has two important advantages over the dendrogram, however. The first is that the MST can be derived more efficiently than the corresponding dendrogram. Ross (1969) presents algorithms for its computation and printing, and for deriving a single-linkage dendrogram from it. The second advantage of the MST is that all its nodes occur at positions occupied by individuals. The MST reveals, therefore, not only which pair or pairs of individuals are most alike, but also which pairs of individuals in different branches of the tree are most similar. The tree can be drawn to show these relationships, and one of the most informative ways of doing this is to superimpose the MST on the corresponding scatter diagram. Fig. 10.9 illustrates the MST for the nine points of Fig. 10.1 in this fashion. With only a few points distributed in only two dimensions the MST is especially clear. In this instance the only link that might be in doubt is that between individuals 6 and 34, which is only very slightly shorter than that between 6 and 45.

In Fig. 10.10 the MST for the 40 sites for which dendrograms have been computed is superimposed on their scatter in the plane of their first two

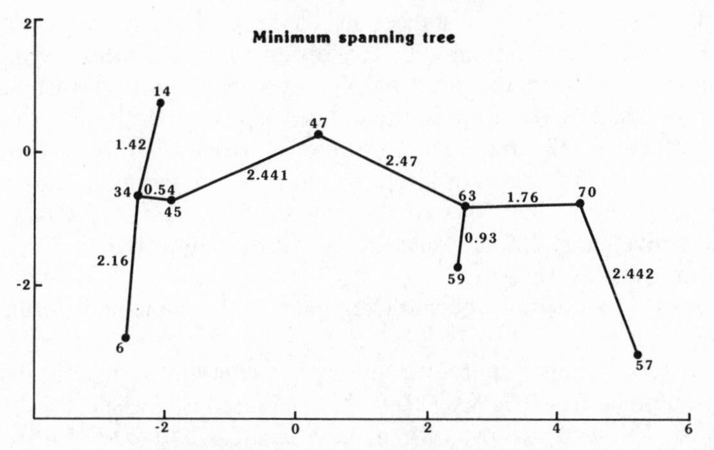

FIG. 10.9 Minimum spanning tree fitted to scatter of nine points in plane (see Fig. 10.1). The lengths of each link are given

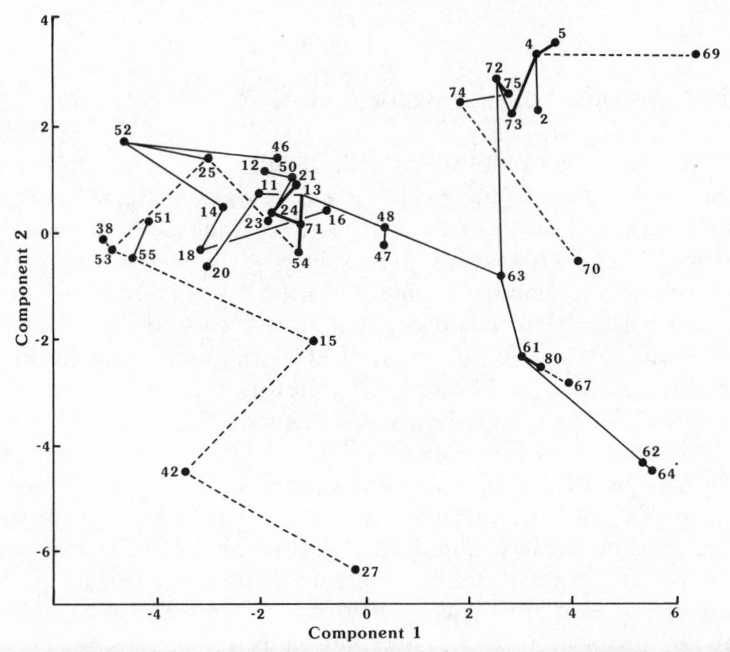

FIG. 10.10 Minimum spanning tree fitted to scatter of 40 west Oxfordshire profiles. The shortest links are drawn in bold, the longest are drawn with dashed lines

principal components (Fig. 10.6). In this graph the computed lengths of the links are indicated by different kinds of line. The shortest links are shown by bold lines, and the longest ones by dashed ones. The remainder are of intermediate length. The distances in a two-dimensional projection are more or less different from the true distances in the whole space. By drawing the MST on the principal component projection some of the distortion resulting from that projection is revealed. In this instance individuals 38 and 53, and 53 and 55 (far left), are shown to be considerably further apart than appears in the projection. So are individuals 67 and 80 (lower right). Individual 70 (right) is most similar to individual 74 and not 63; and individual 52 (far left) is more similar to 14 and 46 than would have been thought otherwise. Thus the MST is a useful way of exploring the distribution of individuals in character space and complements ordination analysis.

If a dendrogram is required to represent single-link grouping then it is readily obtained from the MST. Individuals joined by the shortest link are fused first, followed by those with the next shortest, and so on. Conversely, groups can be derived by dividing the tree through its longest link, followed by division through the link next in length, and so on.

Critical appraisal of hierarchical systems

In this chapter we have looked at the mechanism of hierarchical classification. Strict adherence to the geometric model was sometimes found to be unsatisfactory, and we saw how to obviate some of its shortcomings. We now consider the hierarchical strategies more generally.

First, what do the methods achieve? In the case of association analysis the answer is clear: all members of any class will be identical with respect to the attributes chosen for division. If the attributes are derived from continuous variables by dividing their scales then the individuals will not necessarily be identical, but they will be similar with respect to these. The classes will be separated by planes dividing the character space perpendicularly to the axes of the discriminating variables. When the attributes are chosen by the potential user of a classification to achieve some specific purpose the result is unexceptionable. However, when, as is usual, the aim is to derive more generally useful classes the result can be less happy. Some individuals are likely to be assigned to groups to which they bear little overall resemblance, simply because they happen to possess or lack, as the case may be, one or more of the discriminating attributes. Groups can be very heterogeneous with respect to attributes not used for division. Furthermore, if the population is clustered there is a risk that one or more clusters will be split (see Gower 1967 for an illustration). Clearly this is unsatisfactory, and it is the main reason why divisive numerical methods have not been used for classifying soil, and have been little used in most other fields. They suffer the same disadvantage as several modern systems of soil classification, though in the latter case misuse of Aristotelian logic is to blame (Webster 1968).

The agglomerative methods do not suffer from this defect since classes are formed by fusing individuals on their overall similarity. Each class contains individuals that are generally similar, and is 'natural' therefore in the sense of Gilmour (1937). However, the individuals comprising a group so derived, though generally similar, might not possess any attribute in common. Or, if the characters are continuous variables, then groups can overlap on all the original dimensions. Planes parallel to the original axes will not separate them, and divisions between them in character space need not even be straight.

Agglomeration will almost always produce polythetic groups, though some groups will be fully polythetic while others are only partially so; that is, groups in which every member possesses one or more attributes. Association analysis and similar divisive schemes are monothetic. As above, polythetic division is possible and theoretically attractive, and it is a pity that it takes prohibitively long for more than about twenty individuals.

The examples have illustrated some of the undesirable properties of hierarchical agglomeration, the effects of unequal group size and the tendency to chain, and how we can avoid them. We associated chaining with single-linkage grouping. However, we also saw (Chapter 7) how strict adherence to Euclidean geometry appears to exaggerate the difference between individuals that differ in one or a few properties. The effects extend into the agglomeration process and increase the tendency to chain in the centroid and group-average techniques (Moore and Russell 1967; Cuanalo and Webster 1970). The same effect is present in Fig. 10.7.

There is a further potential defect. When individuals are clustered in character space, i.e. when several portions of the space are more densely occupied than their surroundings, then small differences in similarities produced by calculating them in different ways will not appreciably alter the course of fusion. Suppose, for example, that three individuals P, Q, and R constitute a cluster. By one method P might fuse first with Q and then with R, whereas by another it might fuse with R first and then with Q. Such differences are of little consequence, and all the methods discussed are likely to isolate the same clusters. Similarly, if characters are measured by different techniques, or on different sub-samples by the same techniques, and the same similarity coefficient and fusion strategy applied, then the results will be much the same if the population is clustered. However, when individuals are more evenly spread small changes in the calculated similarity values can have more pronounced effects. The individual R, for example, might no longer be nearest to the group P + Q because of a small change in similarity, but to some other individual or group F and G. This can be especially serious if two similarities are exactly equal. Which fusion should take precedence? None of the strategies provides the answer, and only with single-linkage does it not matter. With this sort of population different procedures are likely to give different results, and a single analysis can lead to erroneous conclusions. Likewise, if the data describe a sample from a larger population that lacks clusters then sampling error is likely to have a major effect on the analysis.

The only safe course is to perform analysis by several methods and to compare results. If the study is based on a sample, then several samples should be analysed. If the results are much the same, then the population is almost certainly clustered, and those from any one method and sample will suffice. If the results are appreciably different, then this too gives useful information about the structure of the population. It probably means that the population either lacks structure, i.e. has a continuous distribution in property space, or is only weakly structured and any clusters are poorly defined. This can be verified by drawing one or more scatter diagrams after a principal component analysis or similar ordination procedure. The scatter of the sampling points in such a projection can often help to explain the results from classification. If the individuals do not occupy reasonably compact regions in the projection then it is likely that cluster-seeking techniques, such as hierarchical analyses, are unlikely to be useful. It can also be informative to plot the results of classification on such projections to see whether the classes occupy well-defined portions of the space.

On the other hand, agglomerative hierarchical analysis can help with the interpretation of the results from ordination by component (PCA) or coordinate analysis (PCO). As we have seen, the distances between individuals in a PCO projection are only approximations to the true distances between them. In particular, pairs that appear close can be far apart in the whole space, and therefore need to be treated with special caution. In an agglomerative analysis, on the other hand, individuals that fuse early are necessarily near one another: the technique is most reliable for identifying close neighbours. Thus, hierarchical agglomeration and ordination are complementary ways of exploring relationships among individuals spread in multidimensional space. It is best to regard the use of both sets of techniques as voyages of discovery rather than as scheduled sailings with assured destinations.

11

NUMERICAL CLASSIFICATION: NON-HIERARCHICAL METHODS

Le mieux est l'ennemi du bien.
Voltaire, *Dictionnaire philosophique*

Although hierarchical agglomerative methods dominate the literature of numerical classification, they are by no means the only ones. Nor are they always the most satisfactory and sensible, especially if the populations to be classified lack any inherent hierarchical structure. Non-hierarchical classification is likely to be a more appropriate alternative in many circumstances. Here we examine its underlying principles and its methods, and then illustrate it in operation with examples.

Non-hierarchical classification subdivides a set of individuals on which several properties have been measured into two or more disjoint groups. Each individual belongs to one and only one group. Any number of partitions of the population into different numbers of groups can be tried, but as the number of classes is reduced they need not combine to form nested classes as in a hierarchy. We could do this by dissection, but as we have seen (p. 168), this is unlikely to be satisfactory. Instead, we want a method that will identify any clusters that are present in the set and, given that these are expressed weakly in many populations, will optimize the subdivision in some sense. This usually means creating classes within which there is a minimum of variation and between which differences are maximized. These same ideas are embraced by the multivariate analysis of dispersion, described in Chapter 9. There we showed how the measures of within-group homogeneity or the separation between classes could be used to express the 'goodness of classification', and to compare different classifications. Here we embody them into the process of classification itself.

Edwards and Cavalli-Sforza (1965), mentioned in Chapter 10, attempted to do this. Their method was hierarchical, but it divided a set of individuals by minimizing the variation within the branches of the hierarchy at each division. In this sense it was optimal. Unfortunately, it was impractical because of the severe restrictions on the number of individuals that could be analysed at any one time: no more than about twenty. The same basic idea, namely minimizing the within-class variation, can be applied to classify many more individuals non-hierarchically. And if we will be satisfied with local optima the computing load can be modest. Several authors, e.g. Rubin (1967), Friedman and Rubin (1967), Demirmen (1969), and Marriott (1971), have shown how this can be done. The methods have been used in soil science: Crommelin and de Gruijter (1973) and de Gruijter (1977) were among the first to classify soil individuals in this way. More recently we used the method successfully to classify soil profiles and sampling sites (McBratney and Webster, 1981; Oliver and Webster, 1987a, 1989).

Method

The general procedure for non-hierarchical classification, sometimes known as *dynamic clustering*, is as follows. First we choose a mathematical criterion to judge the quality of the classification and which we can calculate. The criterion may be SS_W or Wilks's Criterion L to measure the dispersion within groups, or Trace $\mathbf{W}^{-1}\mathbf{B}$ to measure the separation between groups. Other less well-known test criteria include the number of agreements with binary class predictors—Gower's (1974) maximal predictive criterion—and the average entity stability (Rubin 1967). We consider the choice of measure later. The individuals are then divided in some convenient way into groups by cutting a dendrogram, by intuitive judgement of what seems reasonable, or even by random partition. The test criterion is calculated for this classification. Individuals are then moved from group to group and the criterion is calculated afresh after each reallocation. The changes are regarded as improvements if the criterion is itself improved, and they are retained. Otherwise they are not. The procedure continues iteratively until no further improvement seems possible.

The above procedure results in an optimum classification; but it is likely to be a local optimum rather than a global one, especially where the population is only weakly clustered. A local optimum occurs when none of the moves being considered improves the criterion further, yet there is some better division into the same number of groups if only we could find it. Another problem which arises for the same reason is deciding how many classes to choose. Unless we know how many groups there are beforehand, some rationale is needed for choosing the optimal number of groups.

The implementations of the general procedure may vary in detail, and we consider them now.

Test criteria

There are several test criteria that can be used for optimizing and judging the goodness of a classification. The one chosen depends on the nature of the problem and on the form of the data. Many derive from the general equation of the multivariate analysis of dispersion,

$$\mathbf{T} = \mathbf{W} + \mathbf{B},$$

where **T** is the total sums of squares and products (SSP) matrix and is fixed for any set of data, **W** is the SSP matrix within the groups, and **B** is the between-groups SSP matrix. Functions of **W** and **B** are sought as the test criteria for optimization. Wilks's Criterion L, proposed by Friedman and Rubin (1967) for the purpose, is an obvious choice since it measures the compactness of the groups on average. It is the ratio of the determinants: $|\mathbf{W}|/|\mathbf{T}|$ (Chapter 9). Minimizing L is equivalent to minimizing $|\mathbf{W}|$, since **T** is constant for a particular set of data. Wilks's Criterion takes account of the covariances as well as the variances in the groups and so preserves any within-group correlation present in the initial classification. This method makes no assumptions about the shape of any clusters in the population, and it is the appropriate criterion if there are elongated clusters present.

An alternative, but closely related, criterion to L is Trace $\mathbf{W}^{-1}\mathbf{B}$. It is a measure of the separation between group centroids, and its maximization was proposed by Friedman and Rubin (1967). Trace $\mathbf{W}^{-1}\mathbf{B}$ represents the sum of the Mahalanobis D^2, weighted by the group sizes, between the centroids of the groups and that of the whole set. Rao (1952) termed it the generalization of D^2 to more than two groups. If an unweighted criterion is preferred then unweighted forms of **W** and **B** can be used, and Tr $\mathbf{W}^{-1}\mathbf{B}$ represents the unweighted squared Mahalanobis distances between the groups and the centroid of the whole set of the data. Clearly, the best classification in this sense is that in which the groups are furthest apart, and so we shall want to find that one for which Tr $\mathbf{W}^{-1}\mathbf{B}$ is the largest.

The two criteria L and Tr $\mathbf{W}^{-1}\mathbf{B}$ are to a large extent complementary. The more compact the groups, and the smaller L is, the greater the Mahalanobis distances between them. The two are not equivalent, however. Their relation can be appreciated best if the two measures are expressed in terms of the latent roots, λ, of the matrix $\mathbf{W}^{-1}\mathbf{B}$. If there are g roots then

$$\text{Tr}\,\mathbf{W}^{-1}\mathbf{B} = \sum_{i=1}^{g} \lambda_i, \tag{11.1}$$

while

$$L=\frac{|\mathbf{W}|}{|\mathbf{T}|}=\frac{1}{|\mathbf{W}^{-1}\mathbf{B}+\mathbf{I}|}$$

$$=\frac{1}{\prod_{i=1}^{g}(1+\lambda_i)}. \tag{11.2}$$

Friedman and Rubin (1967) and Demirmen (1969) examined both measures empirically and found that L performed the better: Tr $\mathbf{W}^{-1}\mathbf{B}$ was somewhat unreliable. This seemed to be because it can be affected too readily by the variation in the largest eigenvalue, whereas L, based on products, reflects changes in the smaller eigenvalues better.

Both L and Tr $\mathbf{W}^{-1}\mathbf{B}$ remain the same for all linear transformations of the data. Further, using these criteria we can link the classification directly with the projection of the individuals in the plane of the first two canonical variates (see Chapter 9). If there are fewer individuals than variates then neither L nor Tr $\mathbf{W}^{-1}\mathbf{B}$ can be calculated since both $\mathbf{W}$ and $\mathbf{T}$ will be singular. In fact n, the number of individuals, must be larger than $p+g$, the number of variates plus groups, for the outcome to be sensible. If n is less than $p+g$ then either we must use SS_W or the number of variates must be reduced, by taking the leading principal components for example. If a few of the variates are strongly clustered then the optimal subdivision will depend very strongly on them with little regard to the values of the other variates.

Another criterion is the trace of $\mathbf{W}$ or equivalently SS_W, the sums of squares of deviations from the group means. It assumes ordinary Euclidean distance, and so the original variables must usually have been standardized to make the results sensible. It is a measure of compactness, though unlike L it takes into account only the variances and not the covariances. This criterion is appropriate where we know clustering is weak, because our main concern then will be that the variances within the final groups shall be small on average. In this case we should not usually want to create correlation within groups or to preserve any that happened to arise from an initial random partition. The criterion has a further advantage: it involves the least computing in a method that makes fairly heavy demands on computer time. Using SS_W tends to partition the individuals into hyperspherical groups and to split elongated clusters. So if we wish to isolate the latter then we should choose L. This criterion does not remain the same for all linear transformations of the data as do L and Tr $\mathbf{W}^{-1}\mathbf{B}$.

Finally, we note that using any of these criteria in the absence of well-defined clusters will tend to produce groups of roughly equal size.

Initiating clusters

As above, the user can provide an initial classification for the machine to improve. Alternatively, the individuals can be assigned to initial groups automatically. In the simplest schemes the n individuals are divided into k groups of approximately equal size in the order in which they are read from the file, as in Genstat (Genstat 5 Committee 1987). In MacQueen's (1967) method the first k individuals in the set are taken as seeds. Each of the other individuals is allocated to the seed to which it is nearest, provided the distance does not exceed some threshold value: otherwise it forms the seed of a new group, and the two existing groups that are closest to one another are merged. Thorndike (1953) chooses the k points that are mutually furthest apart as seeds for forming the groups. Beale's (1969) method starts with more groups than wanted. The centres of the groups are spaced at regular intervals of one standard deviation on each variable. The number of groups is then diminished until a criterion based on the residual sum of squares is satisfied. In all but the first of these methods individuals are allocated one at a time to the initial cluster centres to which they are nearest in the Euclidean sense. As each individual is added to the group the centroid is determined afresh.

Iterative reallocation

After all the points have been assigned to groups an attempt is made to improve the resulting classification by iteration. In the simpler procedures the centroids of the initial groups are calculated. Each individual is examined in turn and its distance to all of the group centroids is determined. If it is nearer to the centroid of some group other than its own then it is transferred to that group. Otherwise it remains where it is. Every individual may be treated in this way, after which the centroids are recalculated and the whole process repeated until there are no further changes. Alternatively, the group centroids may be recalculated after each transfer. Again, the process continues until there are no more transfers.

The more elaborate procedures are based on one of the test criteria mentioned above. Once all the individuals have been assigned to groups the test criterion is calculated. Individuals are then transferred one at a time to other groups, and after each transfer the test criterion is recalculated. If the criterion is diminished (if it is L or SS_W) or increased (if it is Tr $\mathbf{W}^{-1}\mathbf{B}$) then this transfer is regarded as an improvement and retained. If not, the individual is returned to its original group and the next individual is considered. As before, the process continues until the criterion shows no change.

The result of either of the above reallocation procedures is an optimum classification; but it is likely to be a local optimum rather than a global one. It is unlikely to be the best possible subdivision of the particular set of individuals. Even if it is the best possible, there is no ready means of knowing that it is so. More drastic means are needed for exploring other arrangements of the individuals by forcing more transfers to take place after a local optimum has been reached. One of the best of these is Banfield and Bassill's (1977) transfer algorithm embodied in Genstat. In this procedure pairs of individuals are exchanged between classes once simple transfers cease to produce an improvement. The criterion is recalculated after each exchange, which is retained if the criterion is improved, but not otherwise. Exchanges continue until no further improvement seems possible, and then the algorithm switches back to simple transfers. These two modes of improvement alternate until the criterion seems to have been optimized. We say 'seems to have been' because there is still no guarantee that the global optimum will have actually been achieved.

Another way of overcoming a local optimum is to make several starts from different randomly generated classifications and to retain the classification with the best optimum. This can be very time-consuming if repeated often, but according to Rubin (1967) it works well. Rubin describes a swifter alternative whereby a local optimum is overcome by transferring a substantial number of individuals from one group to another, followed by single transfers to an optimum as before. Rubin's experience suggests that with refinements of this kind a true optimum will often be found, and that it will usually be found if the data are at all clustered. Other ways include optimizing an already reasonable subdivision, and starting with more groups than is thought likely.

How many groups?

In principle, the above procedures can be used to classify a set of individuals into any reasonable number of groups. It can be repeated for as many groups as we wish. In some instances we may know *a priori* how many groups we want, but often we shall not. If a population is at all clustered we should want to identify the clusters as classes, of which there is an optimal number. Marriott (1971) proposed the following procedure for determining the number of classes and which of the several optimum classifications is best in this sense. He used Wilks's Criterion, L, as the measure of goodness, and found optimal classifications for different numbers of groups, g for $g = 2, 3, \ldots$. He then plotted g^2L against g. For a homogeneous population g^2L is expected to decline fairly steadily from a value of 1 when $g=1$. If, however, there is a sharp decline below the general trend for some value of g, say g_o, then this indicates that g_o clusters are present and have been identified.

Examples

To illustrate the application of these methods we explore data from two surveys in which a classification of the soil was desired, but where classification proved difficult or controversial. The first of these is Kelmscot, which we described in Chapter 9. The other is of the Wyre Forest in the Midlands of England (Oliver and Webster 1987a).

Kelmscot

At Kelmscot 21 properties of the soil at 84 sites had been recorded. The individuals, the sites, were not obviously clustered. This is apparent in Fig. 11.1, which is a scatter plot in the plane of the first two principal components. The analysis is summarized in Table 11.1, and the first and second principal components are seen to account for 21 and 10 per cent of the variation respectively. We cannot expect any of the hierarchical methods to perform well, and if we apply them we are likely to obtain widely different results. The dendrogram from a single-link analysis,

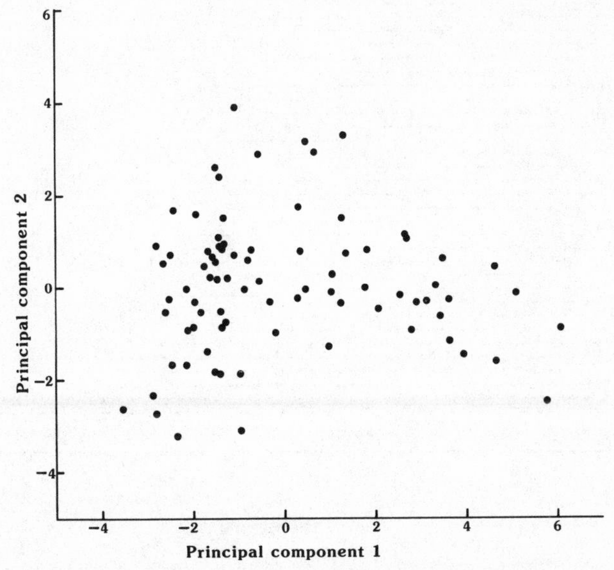

FIG. 11.1 Scatter of sampling sites at Kelmscot in the plane of the first two principal components

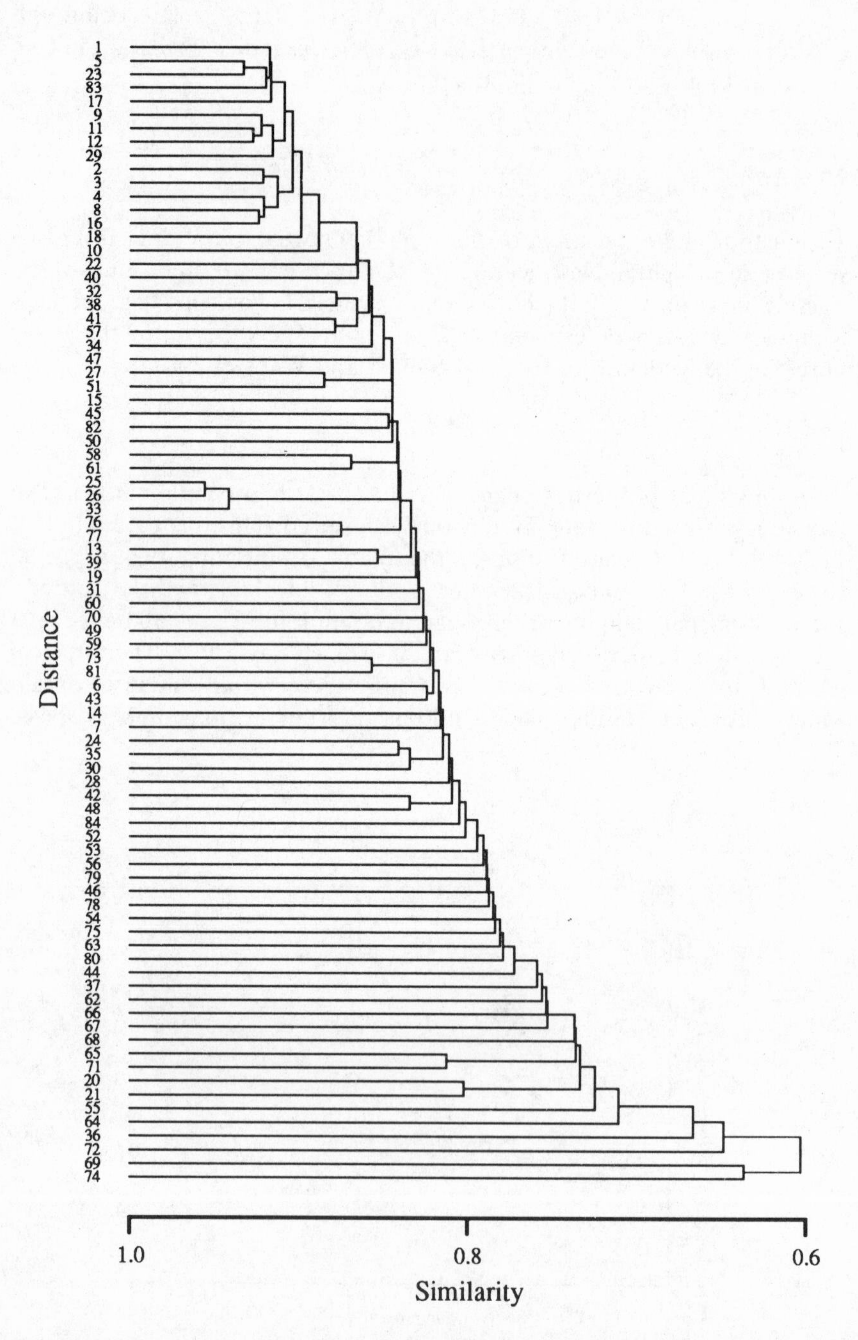

FIG. 11.2 Dendrogram formed by single-link agglomeration of the sampling sites at Kelmscot

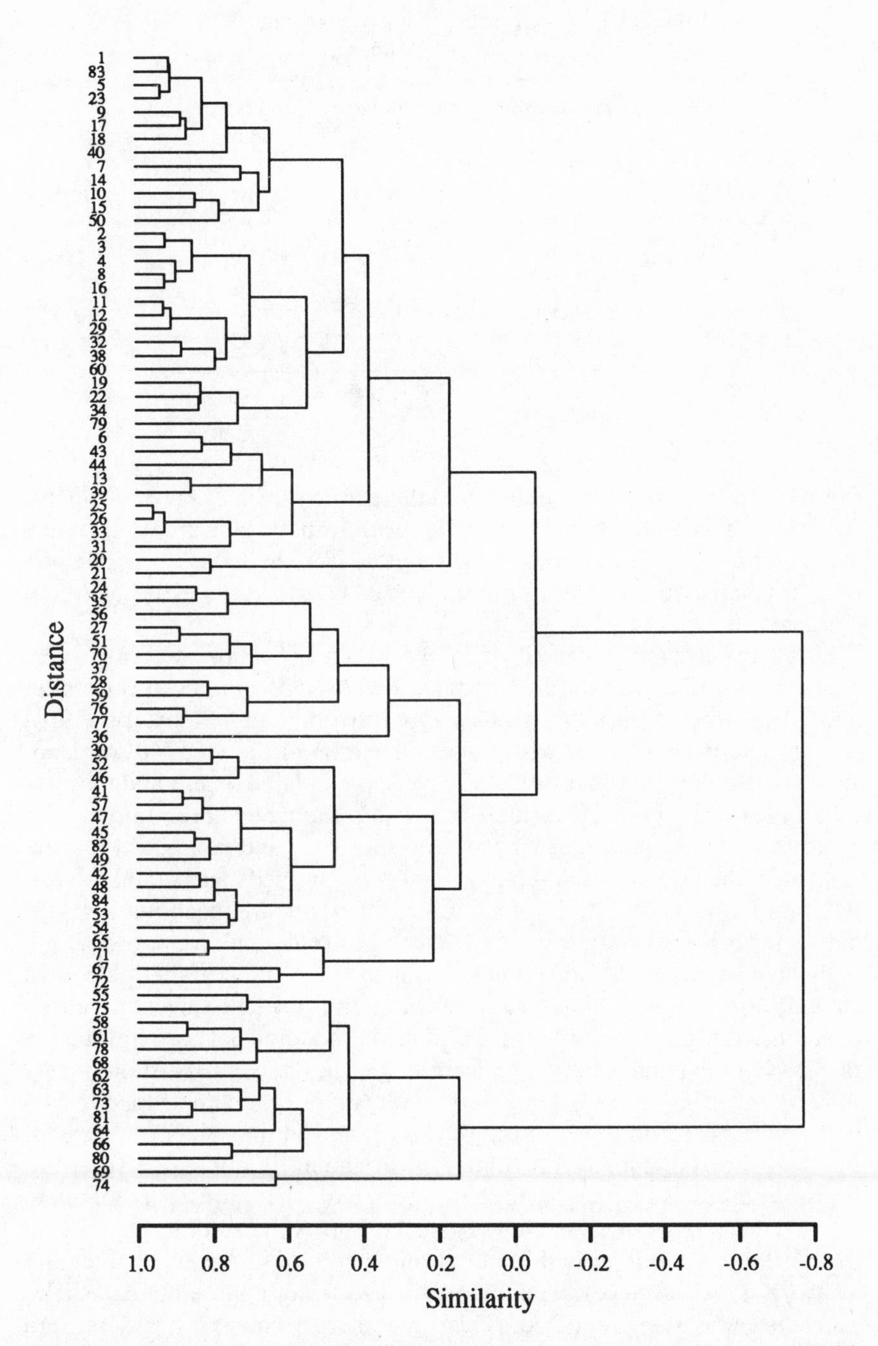

FIG. 11.3 Dendrogram of the Kelmscot sites formed by the flexible strategy with $\beta = -0.25$

TABLE 11.1 Latent roots of correlation matrix for 21 variates at Kelmscot

Order	Eigenvalue	Percentage variance	Cumulative %
1	5.422	25.82	25.82
2	2.098	9.99	35.81
3	1.922	9.15	45.96
4	1.457	6.94	52.90
5	1.346	6.41	59.31
6	1.179	5.62	64.93

Fig. 11.2, shows marked chaining, and there is no sensible way of using this result to subdivide the population. The result from flexible grouping with β set at -0.25, Fig. 11.3, appears much better. We shall assess its effectiveness in comparison with a non-hierarchical subdivision later using L and Trace $\mathbf{W}^{-1}\mathbf{B}$.

Using the principles described in the previous section, we can try to create a good classification by first taking an existing one, however crude, and attempting to improve it. This was Burrough and Webster's (1976) strategy. Suppose we start with the classification of the 1:63 360 soil map. We have already seen that its classes overlap to a large extent and that it is by no means the best. Nevertheless, it is a reasonable starting point. We perform a canonical variate analysis on the data, and compute the class centroids, the pooled within-class variance–covariance matrix, and the test statistics L and Tr $\mathbf{W}^{-1}\mathbf{B}$. Then we can calculate the Mahalanobis distance between each individual and each class centroid, and we transfer any individual to the class whose centroid is nearer to it than to that of its own class. In this instance the relevant information for the first five individuals at Kelmscot is given in Table 11.2. All of these individuals were originally in class 1. Individuals 1, 2, and 3 were nearer to the centroid of their class than to any other centroid and so they remain in that class. Individuals 4 and 5 were nearer to the centroid of class 3, and so they were transferred. Fig. 11.4 shows that many individuals are in similar situations. All these are transferred in this first stage of reallocation, and when transfer is complete L and Tr $\mathbf{W}^{-1}\mathbf{B}$ are recalculated. Table 11.3 gives the results. There were 27 transfers, L is diminished to about one-eighth of its original value, and Tr $\mathbf{W}^{-1}\mathbf{B}$ is increased correspondingly by a factor of approximately 8. The procedure was then repeated. This time there were five transfers, and modest changes in L and Tr $\mathbf{W}^{-1}\mathbf{B}$. The procedure was repeated again, this time resulting in only one transfer and only small changes in L and

TABLE 11.2 First five sites at Kelmscot, their Mahalanobis distances to the three class centroids, and their allocations in the first round of improvement

Individual	Initial group	Mahalanobis distance to: 1	2	3	New group
1	1	3.25	4.24	3.29	1
2	1	3.34	4.85	3.65	1
3	1	3.31	4.63	3.58	1
4	1	3.19	4.75	3.18	3
5	1	2.74	3.95	2.38	3

Tr $\mathbf{W}^{-1}\mathbf{B}$. A further attempt at improvement in this way achieved nothing: the classification had stabilized after the third round.

We can see what the reallocation achieved in terms of the separation between groups, or conversely the overlap of groups, by transforming the data to canonical variates. The latent roots and vectors of the matrix $\mathbf{W}^{-1}\mathbf{B}$ were extracted. The canonical variates both of the sampling sites and of the class centroids were then obtained by multiplying the data by the canonical vectors in a manner similar to finding principal component scores. Fig. 11.4 shows the scatter of the sampling sites projected on to the plane of the first two canonical variates for each of the three stages of reallocation. The circles of radius $\sqrt{\chi^2}$ with two degrees of freedom enclose the 90 per cent confidence regions for the groups, and they illustrate the increasing separation between the classes after each improvement in the order from part a to part d. Since there are only three classes, the group centroids are represented exactly in the two dimensions. The distances between them are the Mahalanobis distances.

Although the above procedure seems to have been successful, it is unlikely to have produced the best classification. We can try to improve the classification further by exchanging individuals between groups and alternating between exchange and simple transfer. We did this using Banfield and Bassill's (1977) algorithm, and the last entry in Table 11.3 shows the result. The best classification into three groups from this position involved only two further reallocations. It diminished L from 0.0291 to 0.0256 and increased Tr $\mathbf{W}^{-1}\mathbf{B}$ from 9.81 to 10.67.

As an alternative, we could apply the transfer and exchange algorithm to the initial 1:63 360 map classification. If we do so then we obtain a slightly different and slightly better classification with $L = 0.0250$ and Tr $\mathbf{W}^{-1}\mathbf{B} = 10.70$. However, we know that the 1:63 360 map classification was not the surveyor's best: the classification of the 1:25 000 was better, and we can use this as our starting point. This resulted in a classification that is better than the best we could achieve from the 1:63 360 classification. The Genstat algorithm produced two optima that it could not distinguish, with $L=0.0205$ and $L=0.0206$. The results are given in Table 11.4. We can obtain another starting classification by computing a dendrogram first and then

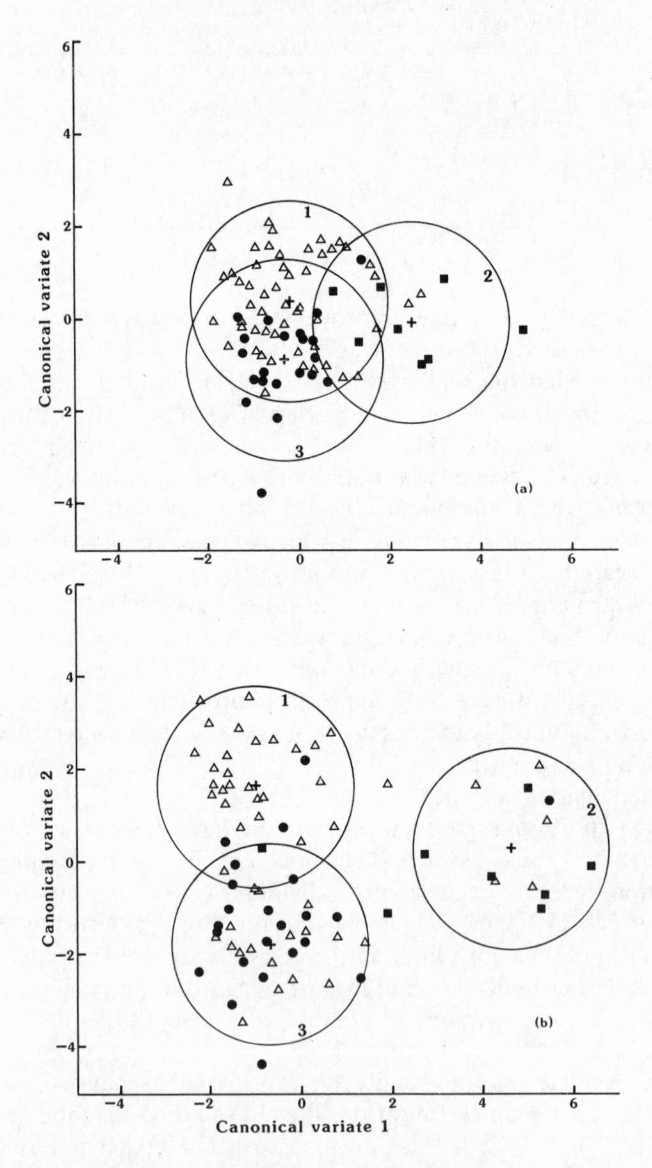

FIG. 11.4a & b Scatter of sampling sites at Kelmscot in the plane of the first two canonical variates for the original 1:63 360 map classification and the successive improvements

cutting it into three branches. Starting from the dendrogram in Fig. 11.3, we arrive at two optima, which again the algorithm does not distinguish (Table 11.5). These are slightly worse than the classification obtained by starting from the 1:25 000 map classification.

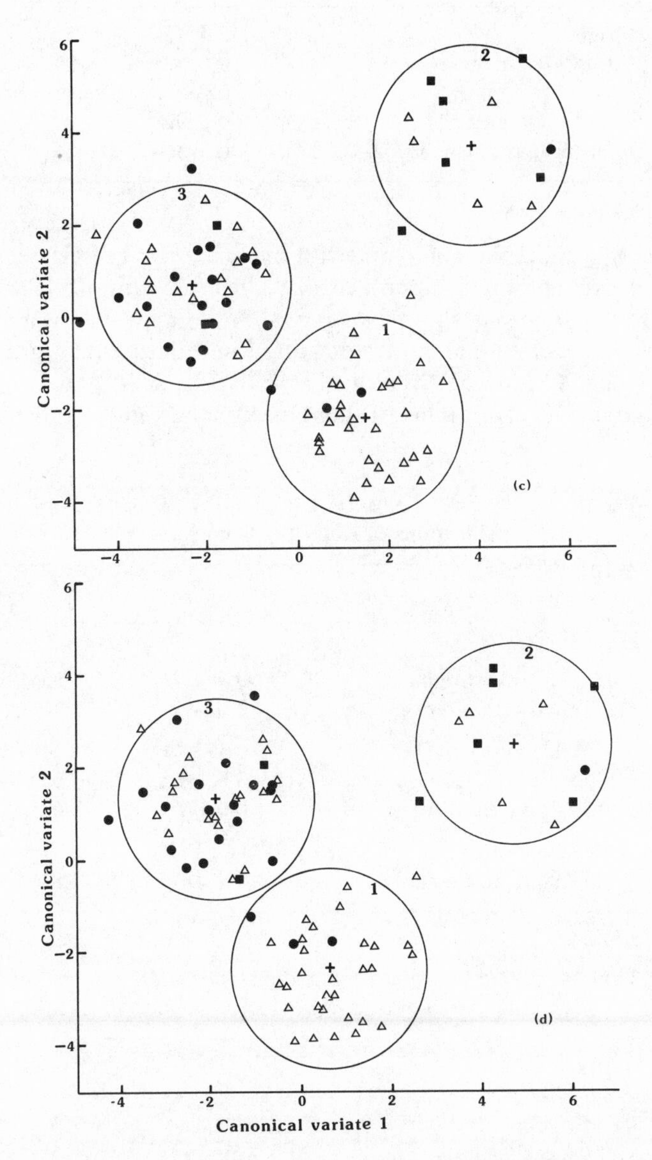

FIG. 11.4c & d Scatter of sampling sites at Kelmscot in the plane of the first two canonical variates for the original 1:63 360 map classification and the successive improvements

TABLE 11.3 Optimization from the 1:63 360 map classification of Kelmscot by discriminant analysis

	Number of transfers	L	Tr $\mathbf{W}^{-1}\mathbf{B}$
Original		0.4643	0.954
1st improvement	27	0.05594	6.590
2nd improvement	5	0.03276	9.096
3rd improvement	1	0.02908	9.809
Genstat improvement	2	0.02564	10.666

Instead of guiding the computer with our preconceived classes, we can take the more automatic approach. To illustrate this we allowed the machine to divide the sites into five classes. As it happened, that was done geographically because the data were stored in order by their grid coordinates. The classification was improved as above using L as criterion, and it too resulted in a local optimum: Table 11.6 lists the values of the criterion,

TABLE 11.4 Wilks's Criterion L and Tr $\mathbf{W}^{-1}\mathbf{B}$ for the optimized classifications of Kelmscot from various starts

	L	Tr $\mathbf{W}^{-1}\mathbf{B}$
Start		
1:63 360 map original	0.4643	0.954
1:63 360 map after optimization	0.02499	10.70
1:25 000 map original	0.1756	3.049
after 1st optimization	0.02052	13.64
after 2nd optimization	0.02062	13.67
Genstat optimization		
after start with 5 classes	0.02181	13.82

TABLE 11.5 Optimization from the dendrogram using the flexible strategy and $\beta = -0.25$ at Kelmscot

	L	Tr $\mathbf{W}^{-1}\mathbf{B}$
Dendrogram	0.05779	7.33
Mulvan reallocation	0.02875	11.62
Genstat optimization 1	0.02152	14.87
Genstat optimization 2	0.02175	15.21

TABLE 11.6 Genstat optimization starting at five classes

Number of classes	L	g^2L	Tr $\mathbf{W}^{-1}\mathbf{B}$
5	0.001537	0.0284	20.49
4	0.003932	0.0629	17.96
3	0.02181	0.1963	13.82
2	0.1586	0.6344	5.30

L, and also Tr $\mathbf{W}^{-1}\mathbf{B}$. The two most similar classes were joined, and the procedure repeated for four classes. The whole was repeated twice more for three and two groups. Table 11.6 summarizes the results. It shows among other things how L increases as the groups become fewer and therefore more heterogeneous, and how Tr $\mathbf{W}^{-1}\mathbf{B}$ decreases as the distances between the groups become less on average. At the three-class level the final classification is better than any of the results that started from the 1:63 360 map classification. In terms of the chosen criterion, however, it was not quite as good as those derived from the dendrogram or the 1:25 000 map classification. Interestingly, this optimum produced slightly more distinct groups with a larger value of Tr $\mathbf{W}^{-1}\mathbf{B}$ than that from the 1:25 000 map.

In this instance we have concentrated on classifications into three groups because that is the number that the soil surveyor originally recognized. We can see whether three groups conform to any weak clusters by plotting g^2L against g, as above. The result for $g = 1, 2, \ldots, 5$ is shown in Fig. 11.5: g^2L

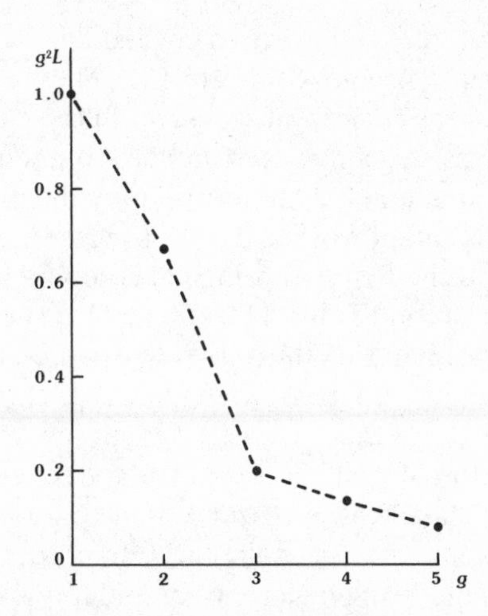

FIG. 11.5 Graph of g^2L against g for the optimal classifications of the Kelmscot sites

at $g=3$ falls below the general trend, and this confirms the original choice of three groups.

The exercise shows that we cannot identify a global optimum classification in this instance. We have found several local optima, all very similar to one another in their statistical character: given the evenly dispersed nature of the soil at Kelmscot, we should be satisfied with any one of them.

Wyre Forest

The second example of non-hierarchical classification derives from our study of the Wyre Forest in the Midlands of England (Oliver and Webster 1987a). The Wyre Forest is a region of semi-natural deciduous woodland to the west of the River Severn covering part of a dissected plateau. Its extent coincides closely with an outcrop of Middle Coal Measures of Upper Carboniferous age from which the soil has largely developed. Within this stratigraphic unit, the Old Hill Marl dominates in the east. Its lithology is complex with fairly thin beds of very variable grain size, typical of sediments deposited rapidly in deltas. The two main components of the Old Hill Marl are the Etruria Marl, which consists of red, purple, and mottled shales and siltstones with varying proportions of sand, and the Espley Sandstones, which vary considerably from fine micaceous sandstones to grits and conglomerates.

The soil in the eastern 6 km^2 of the Forest was sampled by dividing each kilometre square into 36 smaller squares and choosing a sampling point in each small square using random coordinates. This gave 201 fairly evenly distributed sites.

The properties of the soil were recorded at three constant depths, 0–5 cm (1), 15–20 cm (2), and 40–45 cm (3). They were percentage mottling, percentage stones, percentage sand, clay and silt, presence or absence of clay skins, structure, consistence, frequency of roots, Munsell hue, value and chroma, and for the topsoil only the presence or absence of bleached sand grains and of worms. There is moderate correlation between these properties, and the first two principal components account for 43 per cent of the total variation (Table 11.7). Fig. 11.6 shows the scatter of the sampling sites in the plane of these two axes; they are distributed fairly evenly in the projection. There is no evidence of clustering.

Little was known about the soil when this survey was undertaken. There was no clear indication of what classes of soil to recognize, and our aim was to create a sensible classification. An attempt to group the data hierarchically by single-linkage led to chaining. Other methods produced very different results, as did using different combinations of the variables. The dendrograms from centroid and weighted centroid methods, and a flexible grouping strategy with β equal to -0.25, showed quite a difference in the

TABLE 11.7 Latent roots of correlation matrix for 47 variates from the Wyre Forest

Order	Eigenvalue	Percentage variance	Cumulative %
1	15.82	33.66	33.66
2	4.43	9.43	43.09
3	3.46	7.36	50.45
4	2.93	6.23	56.68
5	1.66	3.53	60.21
6	1.36	2.90	63.11

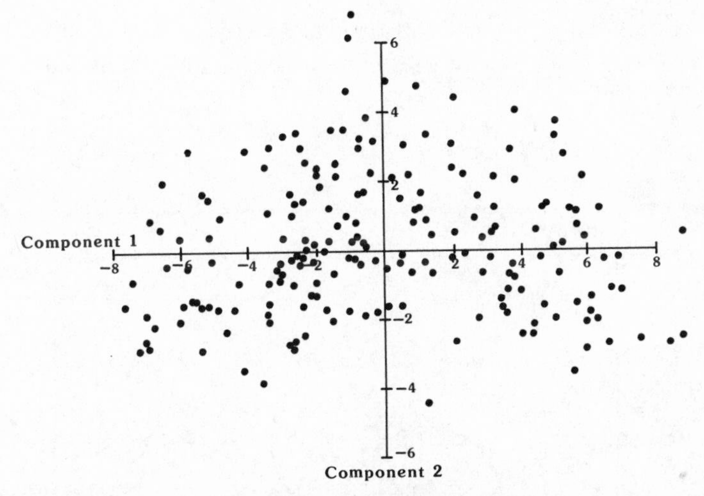

FIG. 11.6 Scatter of sampling sites from the Wyre Forest in the plane of the first two principal components

size of the different branches at any one fusion. We should expect such effects with unclustered data.

To classify the sites non-hierarchically we standardized the variates to zero mean and unit variance and selected SS_W as the criterion of optimality, which we wished to minimize. There was nothing to guide us on the number of classes to choose and so we started with more than we thought were likely in the hope of overcoming local optima. The sample was divided into ten equal classes by assigning the first 20 individuals in the data file to the first group, the next 20 to the second group, and so on. SS_W was computed, and then individuals were reallocated using Banfield and Bassill's transfer algorithm as before.

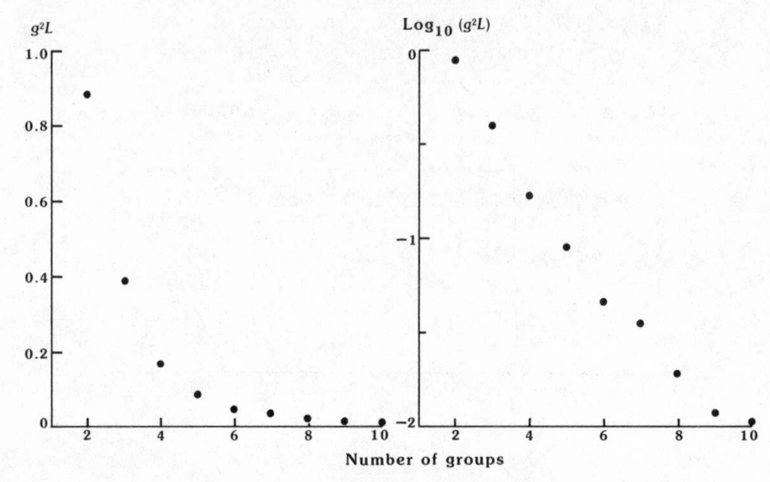

FIG. 11.7 Graph of g^2L against g for optimal classifications of the sites from the Wyre Forest

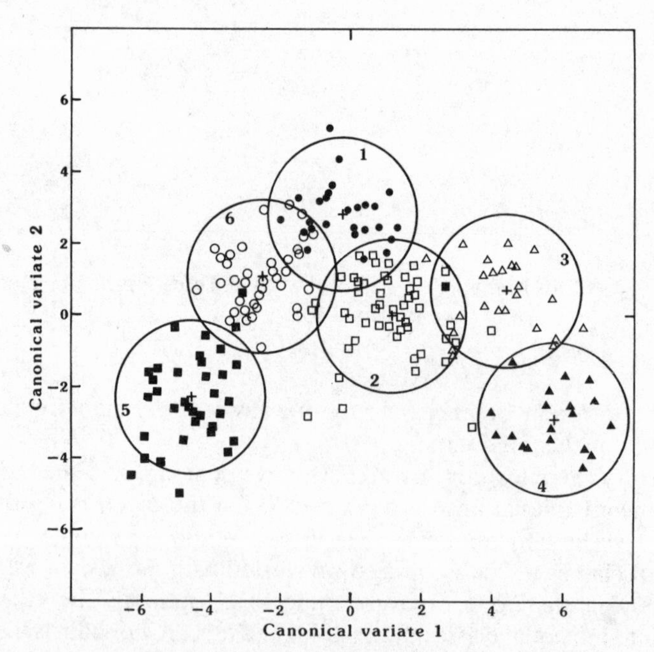

FIG. 11.8 Scatter of Wyre Forest sampling sites in the plane of the first two canonical axes for the non-hierarchical classification. The symbols show the classes, and the circles are the 90% confidence regions around each class centroid

TABLE 11.8 Wilks's Criterion L and Tr $\mathbf{W}^{-1}\mathbf{B}$ for a non-hierarchical classification and for the classification derived from the flexible sorting strategy with β equal to -0.25 of the soil of the Wyre Forest

Classification	L	Tr $\mathbf{W}^{-1}\mathbf{B}$
Non-hierarchical	0.001241	19.11
Flexible sorting strategy, $\beta = -0.25$	0.004939	12.91

Once an optimal classification for ten classes had been computed, the two most similar classes were combined, and the optimizing procedure was repeated taking the now nine classes as the start. The procedure was then repeated for 8, 7, . . ., 2 groups in turn, each time fusing the most similar pair of classes from the previous classification to give the new starting point.

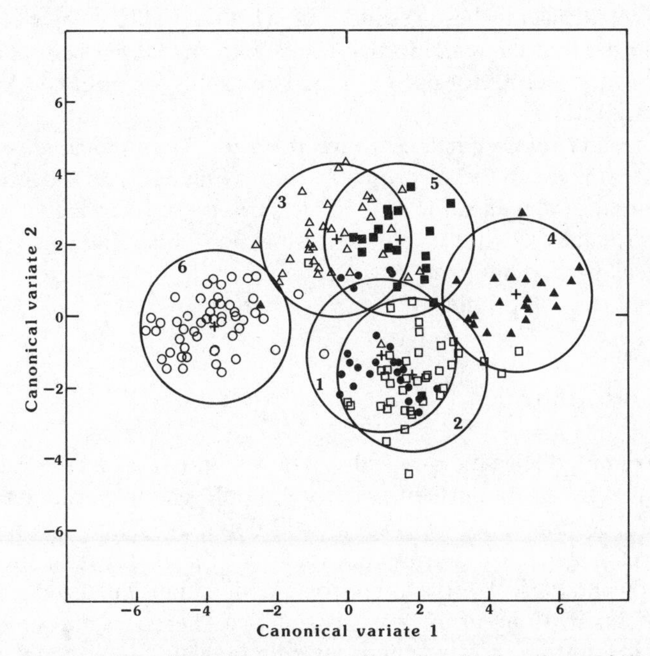

FIG. 11.9 Scatter of the Wyre Forest sampling sites in the plane of the first two canonical axes from cutting a dendrogram into six classes. Symbols indicate the different classes, and the circles are the 90% confidence regions around the group centroids

Following Marriott's (1971) procedure, we used Wilks's Criterion to determine the optimal number of classes. We computed L and plotted the values of g^2L against g. Fig. 11.7 shows the result. There is no trough, which confirms the lack of clusters inferred from the projection in the plane of the first two principal components. The graph changes slope most at $g=6$, however. This can be seen best when g^2L is plotted on a logarithmic scale, and it suggests that if there are even weak clusters then there are six of them.

The sampling sites were plotted in the plane of the first two principal components, and the symbols represent the classes to which they belong. We transformed the data to canonical variates for the optimal classification into six groups and computed L and Tr $\mathbf{W}^{-1}\mathbf{B}$. Fig. 11.8 shows the scatter of the sampling sites projected on to the plane of the first two canonical axes. The circles of radius $\sqrt{\chi^2}$ with two degrees of freedom enclose the 90 per cent confidence regions for the six groups. These classes occupy fairly distinct areas of the property space, and there is reasonable separation between them, suggesting that this is a good subdivision of the sample of soil profiles.

Having decided that six groups were optimal, we cut the dendrogram from the flexible sorting strategy at the six-group level and performed a dispersion analysis on these results. Fig. 11.9 shows the projection of the sites on to the first two canonical variates with the 90 per cent confidence circles drawn as before for each group. The values for L and Tr $\mathbf{W}^{-1}\mathbf{B}$ are given in Table 11.8.

These results show that the groups from the hierarchical classification are not as well separated in property space (Fig. 11.9) as those from the non-hierarchical classification (Fig. 11.8). Comparing the values of L and Tr $\mathbf{W}^{-1}\mathbf{B}$ (Table 11.8) for both classifications also indicates that the hierarchical method has been less effective than the non-hierarchical one in subdividing the soil profiles of the Wyre Forest.

Which technique?

Having examined both hierarchical and non-hierarchical methods of classification, now we can consider which we should use in particular circumstances. If the population consists of well-defined clusters, i.e. if parts of the character space are densely occupied with gaps between them, then most of the numerical procedures for classification will isolate them as classes. If, as in Chapter 10, we suspect that there is some hierarchical structure in addition then we might reveal it by a hierarchical technique. If on the other hand we wish to compare groups, we should do so by canonical variate analysis, and we should do best by optimizing Wilks's Criterion or Tr $\mathbf{W}^{-1}\mathbf{B}$ using a non-hierarchical technique.

Where a population lacks clusters a dendrogram will give a misleading picture of its structure, and it is very unlikely that we should be able to cut it to give as good a classification as we could create by non-hierarchical optimization. We might cut a dendrogram, however, to provide a sensible starting point for optimization.

To choose wisely we need to know whether a population, or at least a particular set of individuals, is clustered. One way of finding this out before attempting classification is by performing a principal component analysis or similar ordination (Chapter 8) and examining scatter diagrams in the plane of the leading components. Alternatively, we can create and examine a number of different levels of grouping using Wilks's Criterion and plotting g^2L against g as described above. We can also try using several different classification techniques. If they all produce similar results then the population will almost certainly be clustered; if the results are very different from one another then almost certainly it will not be. Many populations of soil, at least, are only weakly clustered. They can be divided in many equally reasonable ways.

Non-hierarchical classification has a further advantage over the hierarchical methods. Individuals are not irrevocably assigned to groups: they can be reallocated as groups change their character, and new individuals can be added following the same rules. The hierarchical methods do not allow this.

Non-hierarchical classification is not without its shortcomings. If clustering is weak then it might still be difficult to decide how many groups to choose, and which of several local optima to choose from. In the latter case it is unlikely to matter which of these is chosen. Groups tend to be of the same size and shape, so that weak clusters that are unequally represented may not be isolated as distinct classes. If we know that they occur and can identify them then it is worth using a technique that preserves them. In fact, in all cases we should use our prior knowledge to steer the computer to a sensible conclusion rather than treat the computer as a 'black box' that performs purely automatically.

With these ideas in mind, Webster and Burrough (1974) proposed a method for soil classification that exploited the human talent to recognize useful classes and create classifications intuitively. For each intuitively recognized class of soil several profiles were chosen as a representative sample. The sample contained only those profiles that seemed to belong to that class unequivocally, and were therefore easy to recognize. They may be thought of as the 'cores' of the classes. The centroids of the cores and the pooled variance–covariance matrix within the cores were calculated. Every other profile was then allocated to the core to whose centroid it was nearest in the Mahalanobis sense. If it was thought that some of the initial representatives were ill chosen then the Mahalanobis distances between them and the group centroids could be calculated and they could be reallocated if desired to improve the classification. Provided that the individuals chosen initially to represent the classes we have in mind do represent them truly, then the final classification will approximate to the most compact arrangement from that starting point. The method promised a particularly happy combination of man and machine, allowing them to do what they do best: the man to use his experience and intuition to choose the broad classes, the machine to handle large quantities of data and make routine decisions. All the hard decisions, the borderline issues that take a disproportionate amount of a person's time and attention, could be made by computer.

Whether we use a hierarchical method for classification or an optimizing non-hierarchical technique as in this chapter, the population actually classified is finite. That is, we classify a particular set of individuals on which we have data. When a population is infinite, as it is usually in the earth sciences, any analysis must be based on a sample. Either kind of method can then be used to create classes to which the remainder of the population can be allocated. If so the sample must be properly representative for a numerical classification to be sensible. Sound sampling, however, will not guarantee satisfaction for a population that lacks clusters because sampling variation can seriously affect the analysis. Different samples are likely to produce appreciably different results. Research is needed on the instability of classification resulting from sampling error and on how to achieve stability by suitable statistical and computing strategies. Despite immense effort, there is no certainty that the best classification will be found. A sound philosophy is to be satisfied with what seems to be a good classification and to use it until it proves to be inadequate.

12

SPATIAL DEPENDENCE

Slight not what's near, though aiming for what is far.
Euripides, *Rhesus*

Throughout most of the earlier chapters we have treated observations of the soil at different places as though they were independent. However, in Chapter 3, where we saw the effects of different sampling configurations, we hinted that they might be otherwise, and we deferred the explanation to this chapter. Most properties of the earth's surface, including those of the soil, vary continuously in space. As a consequence, the values at sites that are close together on the ground are more similar than those further apart; they depend upon one another in a statistical sense. Thus the observations cannot be regarded as independent, and a more advanced statistical treatment is required. This chapter and the next two consider this property known as spatial dependence, how to measure and model it, and its implications for sampling and estimation.

Viewed mathematically, the value of a soil property at any place on the earth's surface is a function of its position. One might expect there to be some mathematical expression to describe its variation from place to place. In practice, however, the variation is very irregular, and any adequate equation relating values of a property to position must be very complex. Fig. 12.1 shows the kind of variation to expect. Even this picture is incomplete; had the sampling interval, here 10 m, been shorter, then even more variation would have been revealed. Thus a complete mathematical description is not feasible. However, the trace is not wholly erratic; there is some structure in the variation in the sense that the values at positions near one another tend to be more similar than others. There may in addition be some trend of longer range, which may be deterministic, but it is by no means obvious. The only practicable approach is to regard such a property as a random variable and to treat its variation in space statistically. Such properties are known as *regionalized variables*.

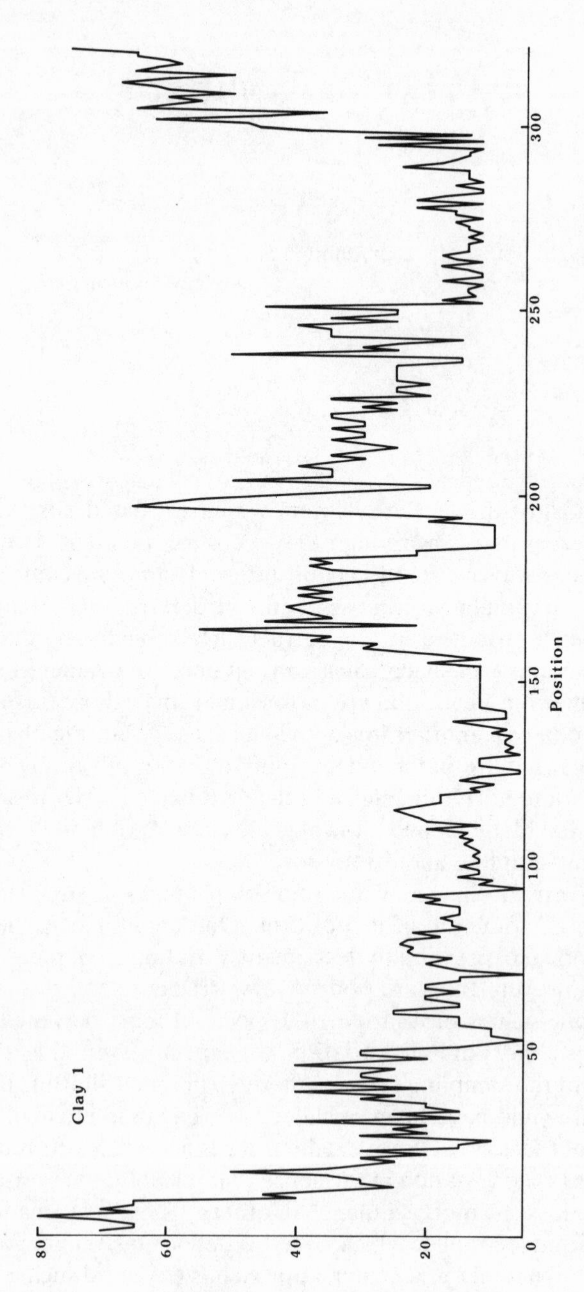

FIG. 12.1 Clay content in the topsoil along 3.2 km at Sandford St. Martin, England. (From Webster and Cuanalo 1975)

Regionalized variable theory

The development of spatial statistics, or *geostatistics* as it is known in the earth sciences, stems largely from mining and from two men in particular. G. Matheron (1965, 1969, 1971) and his colleagues in the French mining schools are largely responsible for the theory, i.e. the *Theory of Regionalized Variables*, while D. G. Krige (1966) developed and applied it empirically. Geostatistics seems to have evolved independently of the mainstream of statistics, and its terminology retains a strong mining flavour. It arose from the need to improve the estimates of ore concentrations in rock and of recoverable reserves from fragmentary information. Miners realized that the concentrations in the neighbourhood of a block were more like those within the block than those further away, and they found ways of taking this fact into account to improve their estimates.

Although regionalized variable theory had a late start compared with agricultural and multivariate statistics, its economic importance in mining, mineral prospecting, and reserve assessment has assured its rapid advance. The theory faces the real problems encountered in mining and, just as importantly, in the earth sciences generally. It provides the statistical tools for describing variation over the earth's surface, for estimating its attributes precisely, and for designing efficient sampling schemes. These form the subject of this and the following two chapters. Fuller accounts of the theory and its application are described in Matheron's source works. These are highly mathematical, and other texts, though aimed at the mining industry, are recommended for the earth and environmental scientist. Clark (1979) has provided a very practically orientated introduction. David's (1977) text is more advanced and theoretical. Journel and Huijbregts (1978) have written what is perhaps the best single account of the theory and its practical applications to date.

The semi-variance

We start with the simplest possible situation. Consider two places some distance apart at which a property Z has the values z_1 and z_2. The relation between the two values can be defined either by the algebraic difference in $z_1 - z_2$ or its absolute value. Alternatively, we can use the variance, equation (2.2), which is simply

$$\begin{aligned} s^2 &= (z_1 - \bar{z})^2 + (z_2 - \bar{z})^2 \\ &= \tfrac{1}{2}(z_1 - z_2)^2, \end{aligned} \tag{12.1}$$

where $\bar{z}$ is the mean of z_1 and z_2. The variance is to be preferred for the same reasons as given in Chapter 2.

This equation can be generalized for any two places. In doing so it will be helpful to change the notation from that of earlier chapters to avoid too many subscripts. Let the two places be $\mathbf{x}$ and $\mathbf{x}+\mathbf{h}$, where $\mathbf{x}$ denotes the coordinates of a place in one, two, or three dimensions, and $\mathbf{h}$ is a vector embracing both distance and direction, called the *lag*. Equation (12.1) then becomes

$$s^2 = \tfrac{1}{2}\{z(\mathbf{x}) - z(\mathbf{x}+\mathbf{h})\}^2, \qquad (12.2)$$

Suppose that Z has been measured at numerous places in a region and that there are m pairs separated by the vector $\mathbf{h}$. The average semi-variance at this lag can be calculated from

$$\bar{s}^2 = \frac{1}{2m}\sum_{i=1}^{m}\{z(\mathbf{x}_i) - z(\mathbf{x}_i+\mathbf{h})\}^2. \qquad (12.3)$$

The intrinsic hypothesis

A description of variation that is limited to a particular set of data is not very useful, and as in the early chapters we want to be able to extend it to the population as a whole. This can be achieved with spatial data by invoking the concept of stationarity. Formally, the actual value of a property at any place is regarded as one of the many values that might have been generated by some random process; i.e., it is a random variable $Z(\mathbf{x})$ with an actual value $z(\mathbf{x})$ at $\mathbf{x}$. This formality enables us to assign an expectation to Z at $\mathbf{x}$.

With this established, the following assumptions may be made. First, the expected value of the variable,

$$\mathrm{E}[z(\mathbf{x})] = \mu, \qquad (12.4)$$

is constant and does not depend on the position $\mathbf{x}$. When this holds the regionalized variable is said to be stationary in the mean. Second, the expected squared difference between values at places separated by the lag $\mathbf{h}$,

$$\mathrm{E}[\{z(\mathbf{x}) - z(\mathbf{x}+\mathbf{h})\}^2] = 2\gamma(\mathbf{h}), \qquad (12.5)$$

is finite and depends only on $\mathbf{h}$ and not on $\mathbf{x}$. In other words, the variance of the differences is stationary for any given lag. The quantity $\gamma(\mathbf{h})$ in this equation is the expectation of s^2 at lag $\mathbf{h}$. Notice that $\gamma(\mathbf{h})$ is half the expected square of difference between the two values. Regionalized variable theory focused attention on such differences, and $\gamma(\mathbf{h})$ became known as the semi-variance. Nevertheless it is the variance per site when sites are considered in pairs (Yates 1948). As above, γ depends on $\mathbf{h}$, and the function relating the two is known as the *semi-variogram*, or increasingly as just the *variogram*.

In Journel and Huijbregts (1978) these two assumptions are said to constitute the *intrinsic hypothesis*. However, Matheron (1965) originally defined the hypothesis somewhat more widely, with the first assumption (12.4) replaced by

$$\mathrm{E}[z(\mathbf{x}) - z(\mathbf{x}+\mathbf{h})] = u(\mathbf{h}),$$

the linear drift, which is not necessarily equal to zero. Whichever definition we take, we can expect the same degree of difference in the property between any two places $\mathbf{h}$ apart whatever the actual values of the property are.

We know that the soil and many other properties of the earth's surface change from one part of a large region to another. Nevertheless, the properties can usually be regarded as stationary within smaller neighbourhoods, say of size V, and this condition is known as *quasi-stationarity* in $z(\mathbf{x}) - z(\mathbf{x}+\mathbf{h})$. This leads to the following model of the variation:

$$z(\mathbf{x}) = \mu_V + \varepsilon(\mathbf{x}), \qquad (12.6)$$

where $z(\mathbf{x})$ is the value of the property Z at $\mathbf{x}$ within the region, μ_V is the mean value within the region, and $\varepsilon(\mathbf{x})$ is a spatially random component with a mean of zero and a variance defined by

$$\mathrm{var}[\varepsilon(\mathbf{x}) - \varepsilon(\mathbf{x}+\mathbf{h})] = \mathrm{E}[\{\varepsilon(\mathbf{x}) - \varepsilon(\mathbf{x}+\mathbf{h})\}^2] = 2\gamma(\mathbf{h}). \qquad (12.7)$$

In practice we have to limit the size of area over which we make our assumptions. It is for this reason that we used the subscript V in equation (12.6). This limitation is rarely serious because, as we shall see later in Chapter 14, we usually want estimates for areas spanning only small lag distances.

Equation (12.5) effectively defines the population semi-variance at lag $\mathbf{h}$. The semi-variances s^2 of (12.2) and (12.3) estimate γ without bias provided that the intrinsic hypothesis holds and the sampling itself is unbiased. To conform with geostatistical convention we shall denote the expectation or population parameter by $\gamma(\mathbf{h})$ and its estimate by $\hat{\gamma}(\mathbf{h})$. Likewise, we can distinguish between the sample variogram comprising the observed values and the variogram of the population that it estimates.

The covariance and autocorrelation

The covariance is an alternative means of describing the way in which values of a property vary in space. It has its origins in the analogous analysis of time series. For a lag **h** the covariance of Z is defined as

$$\begin{aligned} C(\mathbf{h}) &= \mathrm{E}[\{z(\mathbf{x}) - \mu\}\{z(\mathbf{x}+\mathbf{h}) - \mu\}] \\ &= \mathrm{E}[z(\mathbf{x})z(\mathbf{x}+\mathbf{h})] - \mu^2 \quad \text{for all } \mathbf{x}. \end{aligned} \tag{12.8}$$

When **h** is zero this expression defines the variance, σ^2:

$$\begin{aligned} C(0) &= \mathrm{E}[z^2(\mathbf{x})] - \mu^2 \\ &= \sigma^2. \end{aligned} \tag{12.9}$$

The ratio $C(\mathbf{h})/C(0)$ is the *autocorrelation coefficient* at lag **h**, denoted by $\rho(\mathbf{h})$. Like the product moment correlation coefficient between two variables (Chapter 5), $\rho(\mathbf{h})$ must lie between 1 and -1. A value of 1 indicates perfect correlation; i.e., for a particular lag the variable has identical values. Variation in space is rarely as regular as this, and in most instances the autocorrelation coefficient declines from near 1 at very short lags, where values of a property are similar, to near zero at long lags where there is little relation. Just as the variogram describes the relation between semi-variance and lag, so the covariance function, $C(\mathbf{h})$, and the *correlogram*, $\rho(\mathbf{h})$, relate the covariance and autocorrelation to **h**.

The definitions of the covariance and autocorrelation assume not only that the mean is constant but also that the variance is finite and the same throughout the region. These are the assumptions of second-order stationarity. Where they hold the covariance and the semi-variance are equivalent though complementary. The relation between the semi-variance, the covariance, and the autocorrelation can be shown by combining (12.5), (12.8), and (12.9) thus:

$$\begin{aligned} \gamma(\mathbf{h}) &= C(0) - C(\mathbf{h}) \\ &= \sigma^2\{1 - \rho(\mathbf{h})\}. \end{aligned} \tag{12.10}$$

It often happens in the earth sciences, however, that we cannot assume a constant or finite variance. The semi-variance often appears to increase without limit as the lag distance increases, at least within the range over which we can calculate it. The autocorrelation has no meaning then, whereas the semi-variance exists. Therefore, the semi-variance and the variogram may be regarded as generally more useful because they demand weaker assumptions. It is for these reasons that they, rather than the autocorrelation, are used in spatial analysis and form the core of regionalized variable theory.

Estimating the variogram

As we saw above, if we select several pairs of sampling sites **h** apart where we have measured a variable then, assuming that the intrinsic hypothesis holds, we can estimate $\gamma(\mathbf{h})$ using (12.3). To estimate the variogram we can very conveniently compute an ordered set of semi-variances from measurements made along a transect at regular intervals. This will give values $z(1)$, $z(2), \ldots, z(n)$ at n positions. We can then compute the semi-variance for the sampling interval by comparing all neighbouring pairs of observations as in Fig. 12.2. For the transect **h** becomes a scalar, h, equal to one sampling interval. The semi-variance for lag 1 is

$$\hat{\gamma}(1) = \frac{1}{2(n-1)} \sum_{i=1}^{n-1} \{z(i) - z(i+1)\}^2. \qquad (12.11)$$

All the observations are used twice except for those at the ends of the transect.

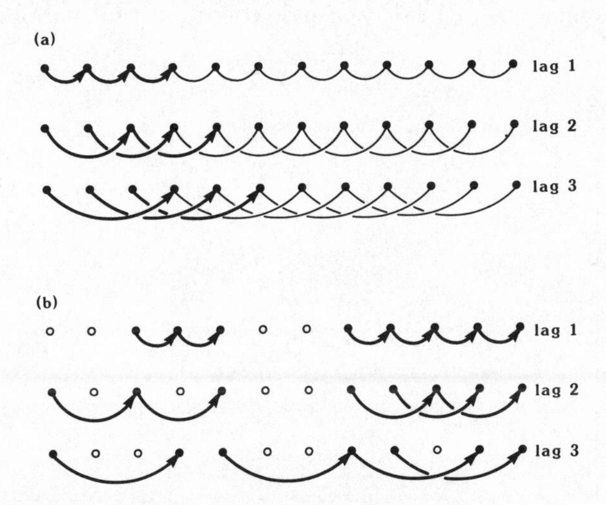

FIG. 12.2 Lagged comparisons for estimating semi-variances on linear transects, (a) for complete data, and (b) where some observations are missing and indicated by open circles

The calculation can then be repeated for any integral multiple of the sampling interval along the transect to obtain the semi-variances for increasingly long lags, $h=2, 3, 4, \ldots$ (Fig. 12.2a). Equation (12.11) generalizes to

$$\hat{\gamma}(h)=\frac{1}{2(n-h)}\sum_{i=1}^{n-h}\{z(i)-z(i+h)\}^2, \qquad (12.12)$$

and we obtain the ordered set of values $\gamma(1), \gamma(2), \gamma(3), \ldots$, which is the sample variogram. Fig. 12.3 shows some examples.

More often than not the computation is less tidy than this. Where transects cross roads, rivers, bare rock, or buildings there will be gaps in the record, and the summation for lag h is made over the actual number of pairs $m<n-h$ (Fig. 12.2b). Note that there is no need to estimate missing data to fill the gaps, as is often the case in multivariate analysis. Often several transects are surveyed because it is more convenient or because they cover the region of interest more evenly. The semi-variances can then be computed by pooling the individual sums for each lag from all the transects and dividing by the total m comparisons.

A more serious difficulty arises where transects have been sampled irregularly. This is overcome most easily by choosing a set of lags h_i, $i=1, 2, \ldots$ at arbitrary but constant increments d, and then associating with each h_i a class of width d and limits $(i-1)d$ and id and mid-point $h_i=(i-\frac{1}{2})d$. Each pair of observations separated by the lag $h_i \pm d/2$ is then used to estimate $\gamma(h_i)$, and each pair contributes to one and only one estimate.

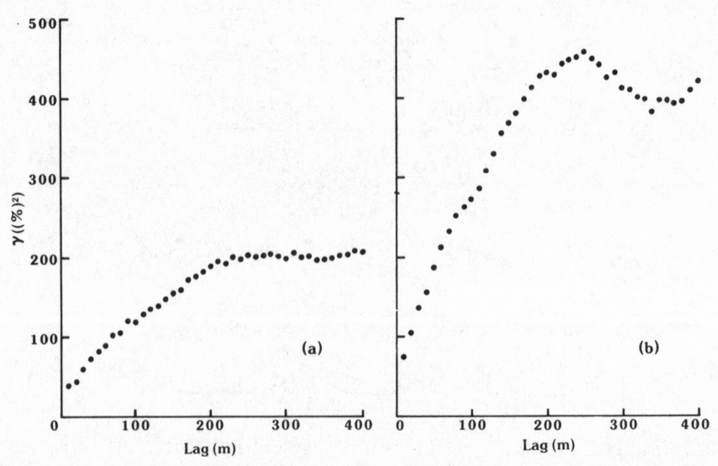

FIG. 12.3 Sample variograms of clay content, (a) for the topsoil at Sandford, and (b) for the subsoil. (From Webster and Cuanalo 1975)

One effect of this procedure is to smooth the variogram: the larger the increment is, the smoother the result will be. Therefore, the increment must be chosen with care. If it is small there might be few comparisons from which to compute the semi-variance and so $\gamma(h)$ will be estimated crudely. If on the other hand d is large then detail may be lost by unnecessary smoothing. The best compromise will depend on the number of data, the evenness of the sampling, and the form of the underlying variogram. It is better still to sample at regular intervals initially. The soil and many other properties of the environment are accessible almost everywhere, and there is rarely good reason to sample irregularly.

Two dimensions

Properties of the earth's surface vary in two lateral dimensions, and in regional survey we are interested generally in variation in a plane, which may be different in different directions. In other words, we must be prepared for anisotropy. In two dimensions the variogram is itself a two-dimensional function.

Where measurements have been recorded at regular intervals on a rectangular grid the sample variogram can be calculated straightforwardly by an extension of equation (12.6). Suppose the grid has m rows and n columns; we estimate the semi-variances as follows:

$$\hat{\gamma}(p,q)=\frac{1}{2(m-p)(n-q)}\sum_{i=1}^{m-p}\sum_{j=1}^{n-q}\{z(i,j)-z(i+p,j+q)\}^2$$

$$\hat{\gamma}(p,-q)=\frac{1}{2(m-p)(n-q)}\sum_{i=1}^{m-p}\sum_{j=q+1}^{n}\{z(i,j)-z(i+p,j-q)\}^2, \quad (12.13)$$

where p and q are the lags in the two dimensions. These equations enable half of the variogram to be computed for lags from $-q$ to q and from 0 to p. The variogram is symmetrical about its centre, so the full set of semi-variances is readily obtained by computing the remainder as

$$\hat{\gamma}(-p,q)=\hat{\gamma}(p,-q)$$

and

$$\hat{\gamma}(-p,-q)=\hat{\gamma}(p,q). \quad (12.14)$$

As in the one-dimensional case, there will often be missing data, and perhaps more importantly, the region of interest may have an irregular shape. Where this is so the quantity $(m-p)$ $(n-q)$ in the denominators of

(12.13) must be replaced by the actual numbers of comparisons in each sum.

It is very common for data in two dimensions to be unevenly scattered. Each pair of observations is then separated by a potentially unique lag in both distance and direction. In these circumstances the variogram is computed best by grouping the separations by both distance and direction. A series of lag distances and directions is usually chosen to form regular progressions as in the one-dimensional case. A range in each is chosen, usually equal to the class interval between successive lags, and applied so that the nominal lag lies in the centre of the range. Each squared difference then contributes to the average semi-variance for the class of lag into which it falls by virtue of the separation between its two sampling points. Fig. 12.4 shows the geometry of the grouping. The nominal lag is represented by the line OL of length l and direction θ. The range in distance is w and that in direction is α.

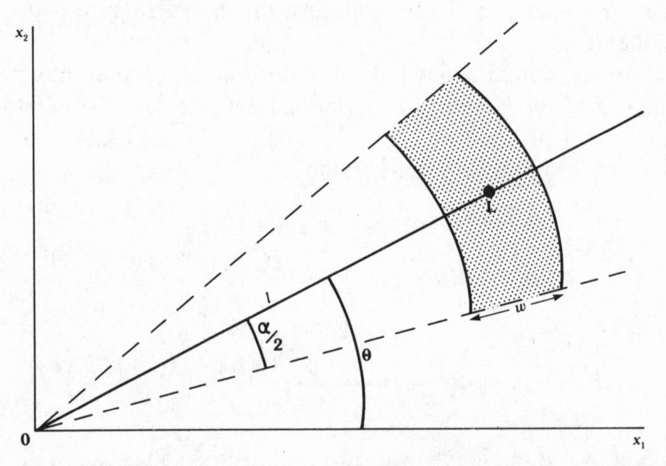

FIG. 12.4 Grouping of lag by distance and direction. The shaded area shows the extent of one group (see text)

Confidence limits of variograms

Estimated semi-variances are subject to error, the size of which depends on the number of comparisons at each nominal lag and the statistical distribution of the population. Unfortunately there is no easy means of obtaining confidence limits analytically for the variogram. In general, the larger the sample, the more precisely are the semi-variances estimated. For transect surveys good working practice is to have at least 100 sampling points and to compute the variogram to no more than about one-fifth of the length, as is recommended in the analysis of time-series (Jenkins and Watts 1968). One hundred points should also be regarded as a minimum for two dimensions, but where variation is anisotropic several hundred sites are likely to be needed to estimate the anisotropy satisfactorily.

Confidence limits are also wider for sample data that are not normally distributed. The simple formulae for estimating the variogram, equations (12.12) and (12.13), are sensitive to highly skewed distributions, but they can be used safely provided that the data can be transformed simply to normality. The result will be variograms of the transformed data, of course. Otherwise a robust estimator of the kind devised by Cressie and Hawkins (1980) can be used.

General characteristics of variograms

Features

The two simple examples of variograms in Fig. 12.3 illustrate the way in which the semi-variance changes with increasing lag distance. In both, the semi-variance is small at short lags and increases steadily with increasing lag distance. Fig. 12.3a, computed from the data in Fig. 12.1, has an initial slope indicating that differences in clay content increase with increasing distance. The steeper initial slope of Fig. 12.3b for clay content in the subsoil shows that there is more change with increasing separation. In each, the dissimilarity increases until eventually it reaches a maximum at which the graph flattens and which is known as the *sill variance*. The sill variance estimates a quantity known as the *a priori* variance of the random variable. The lag at which the variogram reaches its sill is the *range*. This is the limit of spatial dependence; beyond it the variance bears no relation to the separating distance. Variograms with sills represent second-order stationarity.

Fig. 12.5 shows two other common forms of variogram in which the semi-variance appears to increase indefinitely. This might seem surprising, but in many instances it accords intuitively with experience; as the area of interest increases, so more sources of variation are encountered.

These variograms have another characteristic feature. Although by definition the semi-variance at lag zero is itself zero, these measured semi-variances tend towards some positive value as the lag distance approaches zero. This value is known as the *nugget variance*, and the phenomenon, which is widely recognized, is the *nugget effect*. The terms derive from gold mining. Gold nuggets are small; their diameters are very much smaller than the spacing between sampling cores. They are also sparse; most cores contain none, a very small proportion contain one each, and only very rarely does a core contain more than one nugget. Thus, finding a gold nugget in a drill core is regarded as a purely random event.

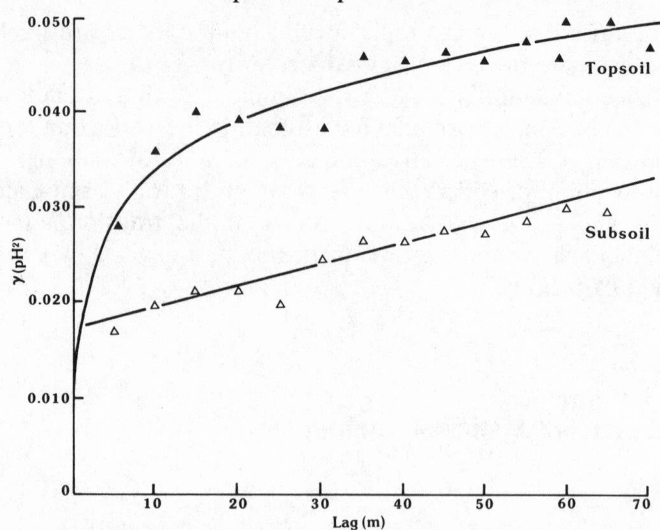

FIG. 12.5 Variograms of the pH of the soil in the Wyre Forest, England. (From Oliver and Webster 1987b)

Few soil properties, or indeed other regionalized variables, are like this; most should be regarded as continuous variables. Yet their sample variograms usually appear to have appreciable nugget variances, and in extreme cases all the variation appears to be nugget, i.e. the variogram appears flat. There are now several examples of this behaviour in the literature (e.g. Campbell 1978; McBratney and Webster 1981; Voltz 1986). See also the example for the Wyre Forest (Fig. 13.4) given in the next chapter. With continuous variables the nugget variance may arise partly from measurement error, though this is usually small in relation to the spatial variation. The principal cause is usually spatially dependent variation that occurs over distances much smaller than the shortest sampling interval. The true shape of the variogram in this range can be identified only by denser sampling.

The variation in two dimensions can be different in different directions, and the variogram should be computed in at least three directions to detect any anisotropy. If the directional variograms for a property in a region differ substantially from one another then this might signal anisotropy. In Fig. 12.6 the variograms for stone content at Plas Gogerddan have quite different gradients, indicating that the rate of change in stone content of the soil varies with direction. The gradient of the variogram is greatest in the direction $3\pi/4$ and least perpendicular to this in the direction $\pi/4$. Differences in the rate of variation according to direction occur on river levées and beaches, where the particle size distribution varies more gradually parallel to the water line than at right angles to it. Similarly,

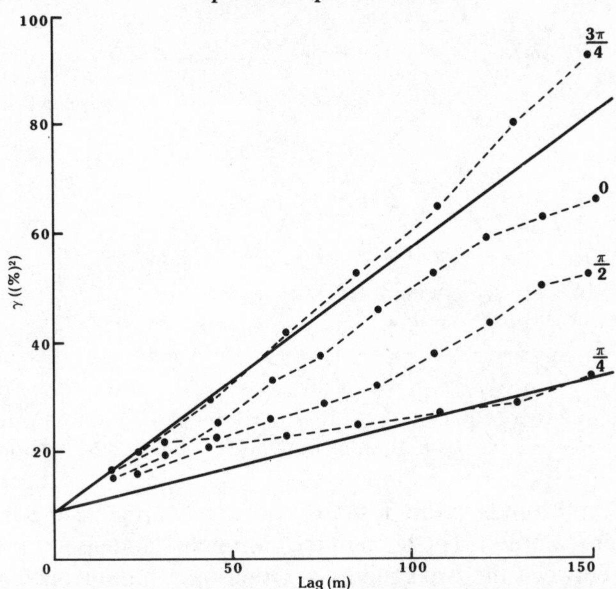

FIG. 12.6 Variogram of stone content at Plas Gogerddan, Wales, computed in four different directions. The solid lines are the envelope of the anisotropic model (see text). (From Burgess and Webster 1980)

where the land surface bevels dipping sedimentary strata the soil and rock along the strike are more likely to be similar than those at places the same distance away in the direction of the dip. In other instances the total amount of variation is different from one direction to another, and there we encounter real differences in the sill height of the variograms for different directions. This signifies *zonal anisotropy*. Interpreting differences in sill height as evidence of anisotropy needs care because they could arise as a proportional effect (Guarascio 1976). This is a sampling effect which can be eliminated by dividing the semi-variances for each direction by the variance of the data for the particular direction. So in an analysis of two-dimensional variation anisotropy must be allowed for, and in a later section we describe how to take account of it when fitting models to the variogram. If the variation in two dimensions is isotropic then the semi-variances can be computed using equation (12.3) without regard to direction. The result appears as a one-dimensional variogram; Fig. 12.7 is an example.

The kinds of variogram that we have described above have approached the origin in either a linear or a convex-upward form, and some have intercepted the ordinate at a positive value. A less common form is one that approaches the origin and is concave upwards. Such a variogram may derive from intrinsic variation obeying (12.4) and (12.5). Alternatively, it may signal a departure from the intrinsic hypothesis. The latter can arise

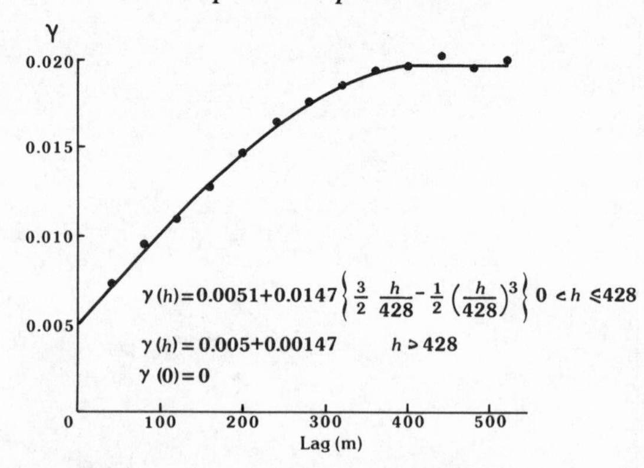

FIG. 12.7 Variogram of potassium at Broom's Barn representing isotropic variation in two dimensions. (Based on a study by Webster and McBratney 1987)

because of local trends in the property and steady progressions in the data when they are sampled. The form of the variogram alone is not sufficient to distinguish between the two situations. Other information, such as general knowledge of the situation in the field, is needed. Such trends must be taken into account when estimating the variogram, and we take the matter a little further under the heading 'Drift' later in this chapter.

Scale. The observed variogram of any property is to some extent a function of the scale of the investigation. It will change as the extent of the region covered becomes either smaller or larger. It also depends on the *support* of the sample, i.e. the size, shape, and orientation of the volumes of material on areas of ground on which individual measurements are made, and which we defined in Chapter 3. The larger is the support, the more variation each measurement encompasses, and the less there is in the intervening space. This has a smoothing effect on the variogram. Investigators should realize that the results refer specifically to the particular support on which their measurements were made. The support should remain constant throughout any one investigation therefore and should be reported.

A variogram on one support can be related, at least theoretically, to that on another. Suppose that $\gamma(\mathbf{h})$ is the variogram for the support on which the measurements were made, and let us regard it as the punctual variogram. The variogram for larger blocks of, say, size B can be determined from it by a procedure known as *regularization*. Let the regularized variogram be denoted by $\gamma_B(\mathbf{h})$. It can be shown that for any lag $\mathbf{h}$

$$\gamma_B(\mathbf{h}) = \bar{\gamma}(B, B_{\mathbf{h}}) - \bar{\gamma}(B, B), \tag{12.15}$$

where $B_{\mathbf{h}}$ denotes another block the same size and shape as B separated from it by the vector $\mathbf{h}$, $\bar{\gamma}(B, B_{\mathbf{h}})$ is the average semi-variance between two blocks, and $\bar{\gamma}(B,B)$ is the average semi-variance within the blocks, i.e. the within-block variance. If $|\mathbf{h}|$ is large relative to the distances across the block then $\bar{\gamma}(B,B_{\mathbf{h}})$ is approximately the punctual semi-variance at lag $\mathbf{h}$, and

$$\gamma_B(\mathbf{h}) \simeq \gamma(\mathbf{h}) - \bar{\gamma}(B, B). \tag{12.16}$$

So for $|\mathbf{h}| \gg \surd$ (area of B), the regularized variogram is derived from the punctual variogram by subtracting the within-block variance.

This relation is important when considering bulking. If the variogram of a property measured at points is known then that for samples bulked over a larger area, the regularized variogram, can be determined from it.

Models for variograms

Many variograms have essentially simple forms, and much of the apparent complexity is sampling fluctuation. It is natural, therefore, to seek simple functions to describe their underlying form and to pass closely through the sample values. In Chapter 14 we shall need such functions for optimal estimation.

Recapitulating, the function we choose must describe in many instances at least three features: an intercept on the ordinate, a portion of increasing semi-variance, and a sill. In two dimensions it must also be able to describe anisotropy. There is another condition: the function must be conditional negative semi-definite, or CNSD for short. We explain this below.

Suppose that $Z(\mathbf{x})$ is a second-order stationary random function giving rise to the values of the regionalized variable $z(\mathbf{x}_i)$, $i = 1, 2, \ldots, n$, at sites $\mathbf{x}_1, \mathbf{x}_2, \ldots, \mathbf{x}_n$, and that its variogram is $\gamma(\mathbf{h})$. Consider the linear sum

$$Y = \sum_{i=1}^{n} \lambda_i z(\mathbf{x}_i), \tag{12.17}$$

where λ_i, $i = 1, 2, \ldots, n$, are any arbitrary weights. The quantity is itself a random variable with variance

$$\text{var}[Y] = \sum_{i=1}^{n} \sum_{j=1}^{n} \lambda_i \lambda_j C(\mathbf{x}_i, \mathbf{x}_j), \tag{12.18}$$

where $C(\mathbf{x}_i, \mathbf{x}_j)$ is the covariance of Z between $\mathbf{x}_i$ and $\mathbf{x}_j$. The variance of Y may be positive or zero, but it may not be negative. The right-hand side of (12.18) must ensure that this is so. Thus, the covariance function, $C(\mathbf{h})$, must be positive semi-definite. This means that the matrix of covariances that (12.18) represents must be positive semi-definite—its determinant and all its principal minors must be positive or zero.

We must consider, however, those situations where the variance appears to increase without limit and the properties do not have definable covariances. Provided that the intrinsic hypothesis holds, we can make use of the following relation. We rewrite (12.18) as

$$\operatorname{var}[Y] = C(0) \sum_{i=1}^{n} \lambda_i \sum_{j=1}^{n} \lambda_j - \sum_{i=1}^{n} \sum_{j=1}^{n} \lambda_i \lambda_j \gamma(\mathbf{x}_i, \mathbf{x}_j), \qquad (12.19)$$

where $\gamma(\mathbf{x}_i, \mathbf{x}_j)$ is the semi-variance of the property between $\mathbf{x}_i$ and $\mathbf{x}_j$. The first term on the right-hand side of this equation can be eliminated if we choose the weights to sum to zero, i.e. if $\Sigma\lambda_i = 0$. The variance of Y is then

$$\operatorname{var}[Y] = -\sum_{i=1}^{n} \sum_{j=1}^{n} \lambda_i \lambda_j \gamma(\mathbf{x}_i, \mathbf{x}_j). \qquad (12.20)$$

This may not be negative either. On removing the sign the variogram function must be negative semi-definite with the added condition that the weights in (12.20) sum to 0. Such functions are often referred to as *authorized* functions or models in the French literature.

As it happens, there are just two main families of functions that describe the simple forms of variograms: they are essentially those that are bounded by a sill and those that are not. We describe them below in their isotropic form, and in which, therefore, the lag $\mathbf{h}$ becomes the scalar $h = |\mathbf{h}|$.

Bounded variograms

Bounded linear model. The simplest form of bounded variogram is one in which the semi-variance increases linearly from zero with increasing h until it reaches its sill and thereafter remains constant. Its formula is

$$\begin{cases} \gamma(h) = c\left(\dfrac{h}{a}\right) & \text{for } h \leq a \\ \gamma(h) = c & \text{for } h > a, \end{cases} \qquad (12.21)$$

where c and a are parameters of the model. The quantity c is the *sill variance*, and a is the *range* of the model, which marks the limit of spatial dependence. This function is CNSD in one dimension only. It is not authorized for more, however well it might appear to fit the sample values. Figure 12.8a shows an example of the model, actually with an added intercept, fitted to a sample variogram.

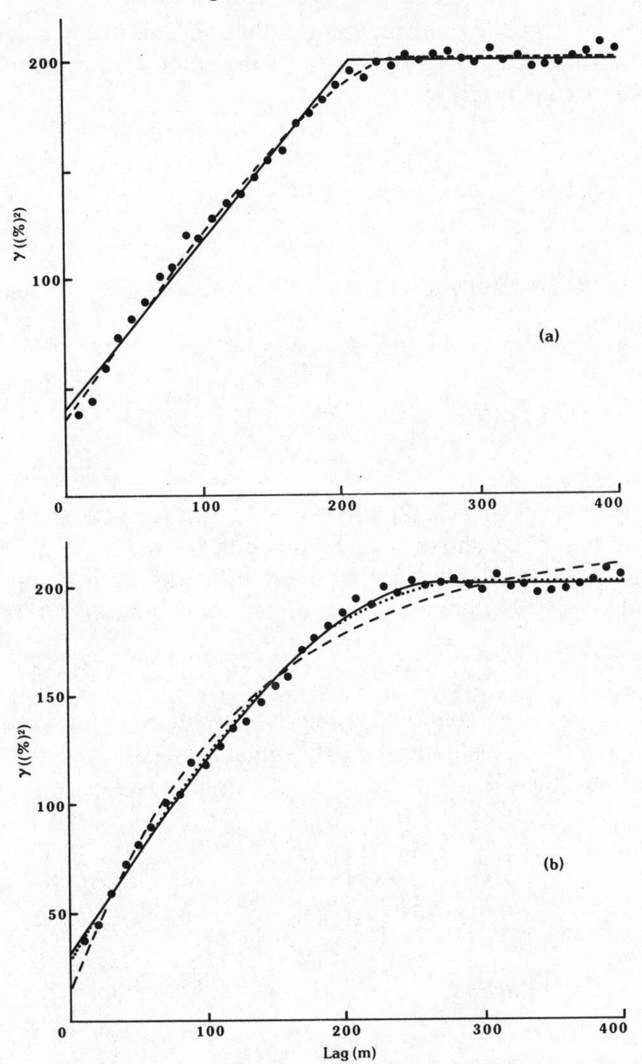

FIG. 12.8 Variogram of clay content in the topsoil at Sandford with (a) a bounded linear model (solid line) and circular model (dashed line) fitted, (b) spherical model (solid line), a penta-spherical model (dotted line), and an exponential model (dashed line) fitted

Circular model. The formula of the circular model is

$$\begin{cases} \gamma(h) = c\left\{1 - \dfrac{2}{\pi}\cos^{-1}\dfrac{h}{a} + \dfrac{2h}{\pi a}\sqrt{(1 - h^2/a^2)}\right\} & \text{for } h \leqslant a \\ \gamma(h) = c & \text{for } h > a, \end{cases} \tag{12.22}$$

where c and a have the same meaning as before. This formula derives from the area of intersection of two discs of diameter a whose centres are h apart. This area is given by

$$A=\frac{a^2}{2}\cos^{-1}\frac{h}{a}-\frac{h}{2\pi}\sqrt{(a^2-h^2)} \qquad \text{for } h \leqslant a. \tag{12.23}$$

Expressing this as a function of the area of the disc gives the autocorrelation function

$$\rho(h)=\frac{2}{\pi}\left\{\cos^{-1}\frac{h}{a}-\frac{h}{a}\sqrt{(1-h^2/a^2)}\right\}, \tag{12.24}$$

and the variogram of (12.22) follows from the relation in (12.10). The example in Fig. 12.8a shows that the function follows fairly closely that of the bounded linear model, but that it curves markedly as h approaches a. This model is CNSD in one and two dimensions, but not in three.

Spherical model. The spherical model is the natural extension into three dimensions of the ideas underlying the circular model. It derives from the volume of intersection of two spheres. Its formula is

$$\begin{cases} \gamma(h)=c\left\{\dfrac{3h}{2a}-\dfrac{1}{2}\left(\dfrac{h}{a}\right)^3\right\} & \text{for } h \leqslant a \\ \gamma(h)=c & \text{for } h > a. \end{cases} \tag{12.25}$$

This is one of the most commonly fitted models. It is much used in mining, and although defined for three dimensions it often fits two-dimensional variograms better than its two-dimensional analogue, the circular model. Figure 12.8b shows an example of its fitting to the same data as in Fig. 12.8a. Note that it curves more gradually than the circular model.

Penta-spherical model. Extending the above to more than three dimensions can give a series of functions with ever more gradual curvature. We mention just one more that we are finding increasingly apt (Webster and

Oliver 1989). This is the five-dimensional analogue of the circular function with formula

$$\begin{cases} \gamma(h) = c\left\{\dfrac{15h}{8a} - \dfrac{5}{4}\left(\dfrac{h}{a}\right)^3 + \dfrac{3}{8}\left(\dfrac{h}{a}\right)^5\right\} & \text{for } h \leqslant a \\ \gamma(h) = c & \text{for } h > a. \end{cases} \qquad (12.26)$$

Fig. 12.8b shows its form and fitting to the experimental values.

Exponential model. Some variograms approach their sills asymptotically, and these are often well represented by a simple exponential function:

$$\gamma(h) = c\{1 - \exp(-h/r)\}, \qquad (12.27)$$

where c is the asymptote, the sill, and r is a distance parameter controlling the spatial extent of the function. Fig. 12.8b shows the best fitting exponential curve for the variogram of clay content at Sandford. There are many other examples in the literature now. Strictly, there is no limit to the spatial dependence of a property described by an exponential variogram. For practical purposes, however, a limit of $a' = 3r$ at which $\gamma(a')$ is approximately $0.95c$ is often used as the effective range.

The exponential function has an important place in statistical theory. It represents the essence of randomness in space. It is the variogram of both first-order auto-regressive and Markov processes. It is the form to expect where changes in soil type are the main contributors to soil variation and where such changes, i.e. soil boundaries, occur as a Poisson process. Burgess and Webster (1984), Webster and Burgess (1984a), and Müller and Böttcher (1987) found many instances where the spacings between soil boundaries on transects had Poisson distributions. If the intensity of the process is α then the mean distance between the boundaries is $d = 1/\alpha$, and the variogram is

$$\begin{aligned} \gamma(h) &= c\{1 - \exp(-h/d)\} \\ &= c\{1 - \exp(-\alpha h)\}. \end{aligned} \qquad (12.28)$$

Oliver (1984) found that the variograms for several soil properties in the Wyre Forest in the English Midlands were exponential with approximately equal values of r. The above process is a plausible explanation for this.

Unbounded variograms

Linear model. Again, the linear model is the simplest in this group with equation

$$\gamma(h) = wh. \tag{12.29}$$

It has been much used to describe experimental variograms. This is partly because it is easy to fit by simple regression. Also, many variograms are approximately linear over short lag distances, and the semi-variances become progressively less well estimated as the lag increases. Further, as we shall see in Chapter 14, a variogram used for kriging is usually needed only for short lags.

Power function. The linear variogram may be regarded as a special case of of a more general family of functions:

$$\gamma(h) = wh^{\alpha} \tag{12.30}$$

where $0 < \alpha < 2$. Clearly, where $\alpha = 1$ we have the linear variogram, equation (12.29). If α is greater than 1, however, then the form of the function is a curve that is concave upwards, while a curve that is convex upwards has a value of $\alpha < 1$. Fig. 12.9 shows some examples.

The limiting values of 0 and 2 are excluded. If $\alpha = 0$, then clearly $\gamma(h)$ is constant with a semi-variance of w at $h = 0$, which we know cannot be. At the other extreme, when $\alpha = 2$ we have a parabola, which can arise only from variation that is smooth and differentiable. In other words, we should no longer have random variation.

Pure nugget variation

Although we may preclude $\gamma(h) = wh^0$ as a variogram, we do need some function that is positive and flat for all positive values of h, but is zero at $h = 0$. We can do this formally using the Dirac function, $\delta(h)$, which takes the value 1 when $h = 0$ and zero otherwise. The variogram is

$$\gamma(h) = c_0\{1 - \delta(h)\}. \tag{12.31}$$

Such a variogram is possible only for variables that have discrete values or occur at only discrete positions. It is impossible for continuous variables such as most soil properties. Nevertheless, it appears in our data, either because the shortest sampling interval is larger than the range of spatial dependence, or because there is measurement error, or both. In any event, we must make formal provision for it, and (12.31) does this.

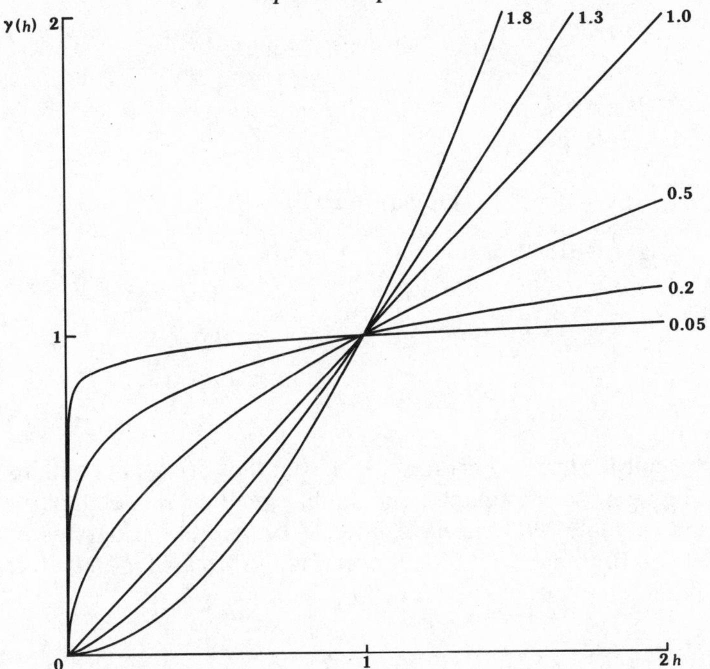

Fig. 12.9 Graphs of power functions for $\gamma(h) = \mathrm{w}\,h^{\alpha}$ for several values of α between 0 and 2 including the linear form when $\alpha = 1$

Combining models

All of the above models describe straight lines or simple curves. In many instances, however, we find that the variogram appears more complex, and a more elaborate model is needed to describe it sensibly. This is best done by combining two or more of the simple functions listed above. This is allowed because any linear combination of authorized models is itself authorized. So we may have

$$\gamma(\mathbf{h}) = \gamma_1(\mathbf{h}) + \gamma_2(\mathbf{h}) + \gamma_3(\mathbf{h}) + \ldots \qquad (12.32)$$

The most common combination by far is that of one of the increasing functions with the Dirac function to describe variograms that show both spatial dependence and a nugget variance. Thus, the full equation for the exponential variogram in Fig. 12.8b is

$$\gamma(h) = c_0\{1 - \delta(h)\} + c\{1 - \exp(-h/r)\}, \qquad (12.33)$$

where c_0 is the nugget variance and c is the sill variance of the spatially dependent component. The equation for the variograms in Fig. 12.5 is

$$\gamma(h) = c_0\{1 - \delta(h)\} + wh^\alpha,$$

where wh^α describes the spatially dependent part. These are often expressed in the form

$$\begin{cases} \gamma(h) = c_0 + c\{1 - \exp(-h/r)\} & \text{for } h > 0 \\ \gamma(0) = 0 \end{cases}$$

and

$$\begin{cases} \gamma(h) = c_0 + wh^\alpha & \text{for } h > 0 \\ \gamma(0) = 0, \end{cases}$$

respectively.

In geochemical surveys it is often found that the variogram contains two recognizable spatial components. The double spherical model in particular has proved valuable, and we have fitted it to several variograms of soil properties (McBratney *et al.* 1982; Webster and Nortcliff 1984). Its formula formula is

$$\begin{cases} \gamma(h) = c_1\left\{\dfrac{3h}{2a_1} - \dfrac{1}{2}\left(\dfrac{h}{a_1}\right)^3\right\} + c_2\left\{\dfrac{3h}{2a_2} - \dfrac{1}{2}\left(\dfrac{h}{a_2}\right)^3\right\} & \text{for } h \leqslant a_1 \\ \gamma(h) = c_1 + c_2\left\{\dfrac{3h}{2a_2} - \dfrac{1}{2}\left(\dfrac{h}{a_2}\right)^3\right\} & \text{for } a_1 < h \leqslant a_2 \\ \gamma(h) = c_1 + c_2 & \text{for } h > a_2. \end{cases} \qquad (12.34)$$

It may also have a nugget variance, $c_0\{1 - \delta(h)\}$. Fig. 12.10 shows an example of this model with a nugget component fitted to the experimental variogram of available copper in the topsoil of south-east Scotland. The dashed lines show the individual simple components and the solid line their combination. The coefficients are given in the figure caption.

Anisotropic models

The models above describe isotropic variation, and the variogram function depends only on the modulus $|\mathbf{h}|$ of the lag. There are numerous situations where the variation is not the same in every direction: it is anisotropic. In these circumstances the variogram function depends not only on the distance but also on the direction, θ, of the lag.

If the anisotropy in a region can be taken into account by a simple linear transformation of the rectangular coordinates then it is known as *geometric* or *affine* anisotropy (David 1977). The formula for such a transform-

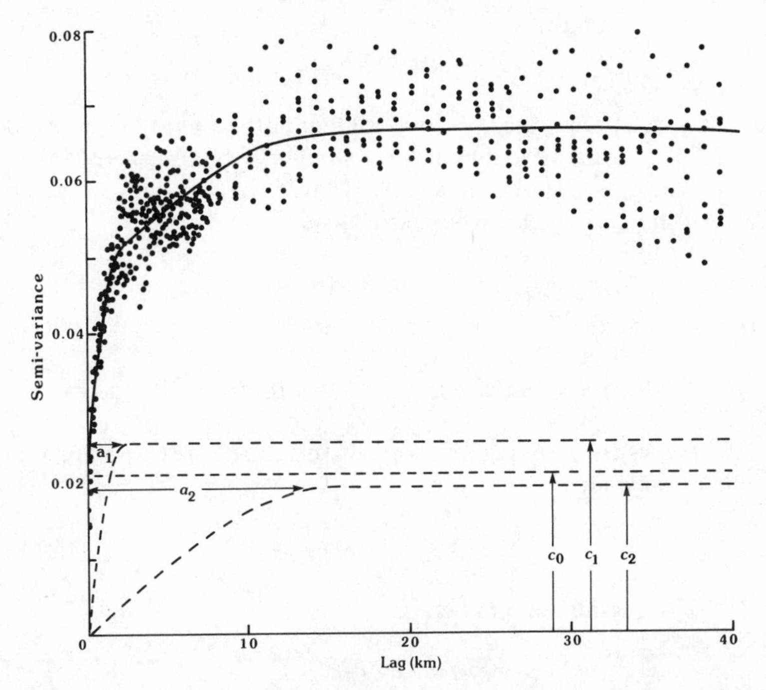

FIG. 12.10 Variogram of available copper in the topsoil in south-east Scotland with a nugget variance and two spherical components. The coefficients are $c_0 = 0.0213$, $c_1 = 0.0257$ and $c_2 = 0.0196$ with units of $(\log_{10}\,\text{mg Cu/kg soil})^2$ and $a_1 = 2.26$ km and $a_2 = 15.5$ km. The full formula is given in the text. (From McBratney *et al.* 1982)

ation is

$$\Omega(\theta) = \sqrt{\{A^2\cos^2(\theta - \phi) + B^2\sin^2(\theta - \phi)\}}. \tag{12.35}$$

In the unbounded models A is the gradient of the variogram in the direction of maximum variation, B is the gradient in the direction of minimum variation, and ϕ is the angle of the maximum gradient. In the bounded models A is the distance parameter in the direction of maximum range and B is the distance parameter in the direction of minimum range. In both instances the proportion $A:B$ can be regarded as the anisotropy ratio.

The transformation of geometric anisotropy can be appreciated by imagining the land on a rubber sheet. By stretching the sheet in the direction ϕ in the proportion $A:B$ the semi-variogram in that direction will have the same gradient as that in the perpendicular direction. In this way the variation is made isotropic.

The function $\Omega(\theta)$ is a factor that can be applied to either the gradient of an unbounded model or to the distance parameter of a bounded model.

For a power function the model is

$$\gamma(h, \theta) = \Omega(\theta)|\mathbf{h}|^{\alpha}. \tag{12.36}$$

The unbounded variograms of stone content in the soil at Plas Gogerddan (Fig. 12.6) for four directions have a similar nugget variance but quite different gradients. Burgess *et al.* (1981) fitted the following linear model modified to take account of geometric anisotropy:

$$\gamma(h, \theta) = c_0 + \Omega(\theta)|\mathbf{h}|,$$

and they obtained

$$\gamma(h, \theta) = 9.06 + \sqrt{\{0.472^2 \cos^2(\theta - 1.00) + 0.158^2 \sin^2(\theta - 1.00)\}}|\mathbf{h}|.$$

Fig. 12.6 shows the defining envelope of this model as two oblique lines. The line of maximum gradient is

$$\gamma_1(h) = 9.06 + 0.472h,$$

and that of the minimum gradient is

$$\gamma_2(h) = 9.06 + 0.158h.$$

This form of display contains all the information in the variogram except the orientation, which can be stated.

Buraymah and Webster (1989) give an example of bounded anisotropic variation for the pH of their irrigated plot. The best-fitting model was exponential with the formula

$$\gamma(|\mathbf{h}|, \theta) = c_0 + c\{1 - \exp[-|\mathbf{h}|/\Omega(\theta)]\}, \tag{12.37}$$

with $\Omega(\theta)$ defined as in (12.35). The coefficients were

$$A = 185.7\text{m}$$

$$B = 67.7\text{m}$$

$$\phi = 2.84 \text{ rad}.$$

The anisotropy ratio, $A{:}B$, was about 3, and the direction of maximum variation is ENE–WSW. Fig. 12.11 shows the variogram for eight directions and the envelope of the fitted model.

Any observed anisotropy that cannot be accounted for by a simple linear transformation of the coordinates requires a model of zonal anisotropy. Zonal anisotropy may be expected when the directional variograms have quite different sills (Fig. 12.12). In other words, the actual amount of

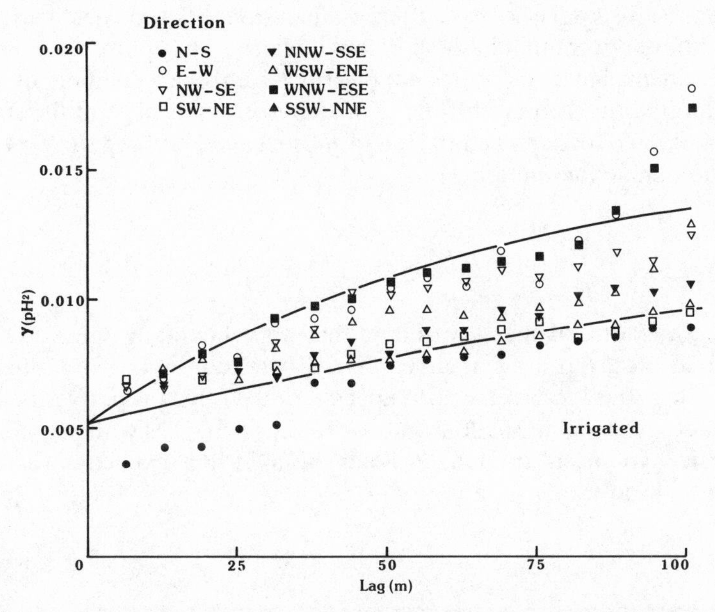

FIG. 12.11 Variogram of pH in the topsoil of an irrigated plot of land in the Gezira of Sudan. The plotted points are the estimated semi-variances in eight directions and the dashed lines show the envelope of the exponential model. (From Buraymah and Webster 1989)

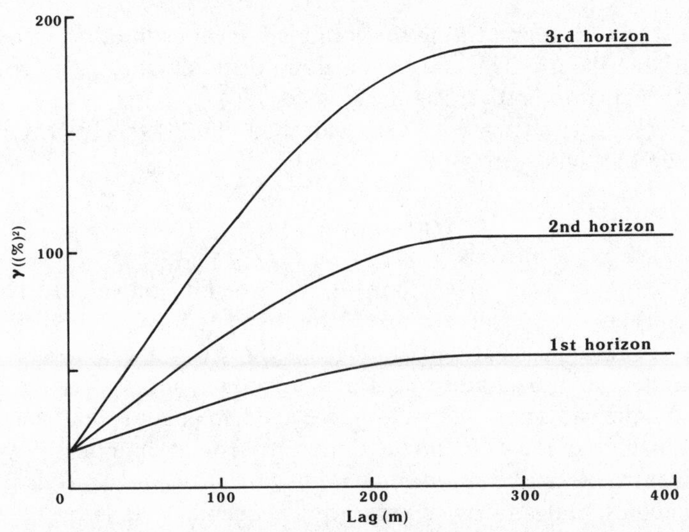

FIG. 12.12 Variograms of a property with different sills in different directions indicating zonal anisotropy

variation differs according to direction. This situation would almost certainly arise in studies of soil in three dimensions. The vertical variation is very different from that in the horizontal planes. The appropriate model in these circumstances is a nested one in which each component of the structure has its own anisotropy. Zonal anisotropy is beyond the scope of this book, and readers are referred to Journel and Huijbregts (1978) if they wish to pursue the matter.

Drift

Drift is a feature of spatial variation that we deliberately deferred earlier. The intrinsic hypothesis assumes that all variation is random. However, where there are local trends the expected value of the property is no longer constant, even within small neighbourhoods: it changes with position. The intrinsic hypothesis no longer holds because the expected value is a function of position:

$$\mathrm{E}[z(\mathbf{x})] = u(\mathbf{x}). \tag{12.38}$$

The model that we used earlier to describe the variation, equation (12.6), must be modified by replacing μ_V, the mean, by the more general term $u(\mathbf{x})$, giving

$$z(\mathbf{x}) = u(\mathbf{x}) + \varepsilon(\mathbf{x}). \tag{12.39}$$

The quantity $u(\mathbf{x})$ representing the trend is known as the *drift* in regionalized variable theory. This use of the term 'drift' should not be confused with drift in agglomerative classification (p. 176).

Where the drift changes we can rearrange (12.39) to give $\varepsilon(\mathbf{x})$ as the deviation from the drift:

$$\varepsilon(\mathbf{x}) = z(\mathbf{x}) - u(\mathbf{x}). \tag{12.40}$$

In these circumstances (12.5) and (12.7) are not equivalent. The raw semi-variances no longer estimate the expected squared differences between the residuals at two places. We need to know the values of $u(\mathbf{x})$ to compute them. Unfortunately the drift is unknown, and to determine it we must know the variogram. Therefore, we need an analysis that will enable us to estimate both the drift and the deviations from it simultaneously. This apparent impasse can be resolved by a full *structural analysis* (Olea 1975). This combines trial and error with good judgement. It is by no means straightforward and is beyond the scope of this chapter. Nevertheless, it is important because it forms the basis of *universal kriging*. Readers who wish to pursue the matter should read Olea's account, which contains a full description of the analysis and for which he provides a computer program (Olea 1977). Webster and Burgess (1980) summarize the procedure and give an example of its application to soil.

Fitting models

The above sections show how to estimate the variogram at particular lags and some of the simple functions that we can use to describe the variation. There remains the task of choosing a suitable function and fitting it to the experimental or sample values. This is important because, as we shall see in Chapter 14, such a function and estimates of its parameters are needed to estimate local values of the variable itself. Some geostatisticians fit models by eye, but the practice is unreliable: a statistically based procedure is much to be preferred.

Of the statistical methods, least squares fitting is the most common. As in regression analysis (Chapter 6), a model is fitted such that the sum of the squares of the differences between the observed semi-variances and the model's predictions is least. For the linear model this is a simple regression with semi-variance as the dependent variable and lag as the predictor. Other power functions can also be fitted by regression simply by taking logarithms. Thus, $\gamma(h) = wh^{\alpha}$ becomes $\log \gamma(h) = \log w + \alpha \log h$, which is linear in the logarithms. Other models, in particular the bounded models, that have nonlinear parameters cannot be made linear by transformation. Their parameters must be estimated, therefore, by iteration.

Simple least-squares approximation still leaves room for improvement. The observed semi-variances are usually based on different numbers of comparisons, and it makes sense to give them weight in proportion to the number of comparisons when fitting models. Thus, the weight for the estimate of γ at the jth lag is $m(\mathbf{h}_j)$, where m is the number of comparisons. We also know that the confidence intervals on the variogram widen as the semi-variance increases. The weight may be proportional to the observed semi-variance, or to that expected, or to functions of them. Thus, Cressie (1985) has suggested the weight $m(\mathbf{h}_j)/\gamma^{*2}(\mathbf{h}_j)$, where $\gamma^{*}(\mathbf{h}_j)$ is the value predicted by the model at $\mathbf{h}_j$. G. M. Laslett, quoted by McBratney and Webster (1986), proposed an improved weighting: $m(\mathbf{h}_j)\hat{\gamma}(\mathbf{h}_j)/\gamma^{*3}(\mathbf{h}_j)$. Both of these usually give substantially more weight at the shortest lags than weighting on the number of comparisons alone. Overall, however, the improvement seems to be small.

Least-squares fitting assumes that the residuals from the fitted model are normally distributed and independent of one another and that the estimated semi-variances all have the same variance. It is often criticized, mainly on the grounds that the assumptions are unrealistic. We have

TABLE 12.1 Coefficients of the bounded models fitted to the sample variograms of clay content at Sandford by weighted least squares approximation. The weights were proportional to the number of comparisons in each estimate.

Model	Coefficients			Residual mean square
	Variances		Distance parameter, m	
	c_0	c		
Bounded linear	38.91	160.82	209	6572
Circular	33.55	167.82	242	3766
Spherical	30.69	170.47	273	3648
Penta-spherical	28.13	173.86	327	4418
Exponential	11.94	205.38	121	13 805

already remarked that the variances are not all the same. It is not clear whether the residuals are normally distributed, and they are unlikely to be completely independent, though the degree of dependence is unknown. Nevertheless, there is no general agreement on a single method that is better. In our experience the method with weights proportional to the numbers of comparisons in the estimated semi-variances works reliably. We can recommend it, though we do not discourage investigators from exploring the merits of other weighting schemes or other sound methods of fitting models to variograms.

Allied to choosing values of the parameters that give a good fit is the choice of the model itself. This can also be done by minimizing the sums of the squares of the residuals from the model. Table 12.1 lists the residual mean squares (RMS) for several plausible models for the variograms of clay content in Fig. 12.8. The spherical model has the smallest RMS, while those for the exponential and bounded linear models are much larger. The last two can also be seen to be a poorer fit. The residual sum of squares can always be diminished by increasing the complexity of the models, but unless an investigator has good physical grounds for choosing a complex model he should choose the best fitting of the simpler ones.

The whole question of choosing and fitting models is still controversial and a matter of statistical research. McBratney and Webster (1986) and Webster and McBratney (1989) give further guidance, especially on the use of the Akaike Information Criterion. As above, many of the models are nonlinear in their parameters, and so a good computer program is needed for the fitting. We have used MLP, the Maximum Likelihood Program written by Ross (1987), but Genstat V (Genstat 5 Committee 1987) contains the same algorithms and is more widely available and may be preferred.

13

NESTED SAMPLING AND ANALYSIS

As I was going to St. Ives
I met a man with seven wives.
Each wife had seven sacks,
Each sack had seven cats,
Each cat had seven kits . . .

Nursery rhyme

In Chapter 12 we introduced the theory of regionalized variables and its central tool, the variogram. We return now to a method embodied in classical statistics that long predates regionalized variable theory, and yet can produce a first approximation to the variogram. The method involves analysing the variance from a multi-stage or nested sampling design. In Chapter 4 we described the simplest form of analysis of variance, i.e. the partition of variance into that present within classes and that attributable to differences between classes. With more complex sampling schemes, and nested ones in particular, it is possible to identify other sources of variation within a population.

Much of the credit for this approach to spatial analysis is due to Youden and Mehlich (1937). They realized the importance of knowing the spatial scale of the variation in soil when designing an efficient sampling scheme for estimating mean values. They adapted classical multi-stage hierarchical sampling for this purpose, with each stage in the hierarchy representing a distance between sites. Such a design identifies the distances within which most of the variation occurs. The particular merit of this method is that a wide range of spatial scales can be covered in a single analysis. This is especially valuable where variation occurs on spatial scales that differ by several orders of magnitude simultaneously, i.e. where the variation is nested. Youden and Mehlich's adaptation links the classical nested analysis of variance and regionalized variable theory, though it was Miesch (1975) who formalized the connection and demonstrated the equivalence of the two approaches for estimating the variogram.

In this chapter we describe the theory of nested sampling and analysis, its adaptation to describe spatial variation, and its role in a complete spatial investigation.

Theory

The model for nested variation is based on the idea that a population can be divided into classes at a number of distinct stages to form a hierarchy. An initial division of a population at stage 1 into classes can be divided at stage 2 into sub-classes to form a two-stage nested or hierarchical classification. These classes can be subdivided further at stage 3 to give finer classes, and so on. Each stage constitutes a category, and any individual observation belongs to one and only one class in each category. The underlying theory is that a single observation embodies variation contributed from every stage in the hierarchy, including an unresolved variance in the smallest subdivision.

Since we regard the values of any one property of the earth's surface as just one realization of the corresponding random process, as described in Chapter 12, and are not interested in contrasts between any particular pairs of sites, the appropriate mathematical model is the random effects model of the analysis of variance, Model II of Marcuse (1949). For a design with *m* stages the model of variation is

$$z_{ijk\ldots m} = \mu + A_i + B_{ij} + C_{ijk} + \ldots + \varepsilon_{ijk\ldots m}, \qquad (13.1)$$

where $z_{ijk\ldots m}$ is the value of the *m*th unit in . . . the *k*th class at stage 3, in the *j*th class at stage 2, and in the *i*th class at stage 1. The general mean is μ; A_i is the difference between μ and the mean of class *i* in the first category; B_{ij} is the difference between the mean of the *j*th subclass in class *i* and the mean of class *i*; and so on. The final quantity $\varepsilon_{ijk\ldots m}$ represents the deviation of the observed value from its class mean at the last stage of subdivision. The quantities A_i, B_{ij}, C_{ijk}, . . . , $\varepsilon_{ijk\ldots m}$ are assumed to be independent random variables associated with stages 1, 2, 3, . . . , *m*, respectively, having means of zero and variances $\sigma_1^2, \sigma_2^2, \sigma_3^2, \ldots, \sigma_m^2$. The latter are components of variance (Chapter 4). The individual component for a given stage measures the variation attributable to that stage, and together they sum to the total variance:

$$\sigma^2 = \sigma_1^2 + \sigma_2^2 + \sigma_3^2 + \ldots + \sigma_m^2, \qquad (13.2)$$

Nested design

A multi-stage or nested sampling design with replication at each level enables the components of variance to be estimated by a hierarchical analysis of variance. Sampling schemes that subdivide a population into classes at two or more distinct stages are often used in medicine, agriculture, and manufacturing. Whatever the basis of the subdivision, an essential characteristic of the design is that there are small classes within larger classes at each stage apart from the lowest. The hierarchy may comprise the stages in a systematic classification scheme such as order, main group, subgroup, and family of soil; or distinct physical units such as farm, field, plot; or tree, branch, leaf; or simple arbitrary subdivisions such as large, medium, and small.

Youden and Mehlich (1937) were the first to apply this kind of sampling to spatial variation. The originality of their contribution was that they saw, for the values of a variable distributed in space, that the stages could be represented by different sampling intervals, and that provided these are suitably nested the hierarchical model is valid. In this instance the components of variance represent the variation associated with different separating distances and form the foundation for designing more efficient sampling in the future. Despite this, the method lay largely dormant for almost forty years. Hammond *et al.* (1958) used the technique to design efficient methods for sampling soil within individual fields, and in geology Olson and Potter (1954) and Krumbein and Slack (1956) used it to describe variation in current bedding and radioactivity in shales respectively. More recently Webster and Butler (1976), described below, and Nortcliff (1978) have used it to measure the contribution to the variance of soil properties from different distances.

Link with regionalized variable theory

Miesch (1975) further established the merit of nested survey and analysis in which the stages represent distances by drawing attention to the link between the components of variance and the spatial autocorrelation of regionalized variable theory. He showed that, if the components of variance were accumulated, starting with the smallest spacing, they were equivalent to the semi-variances obtained by equation (12.6) over the same range of distances. In practice, they have been only rough estimates of the true semi-variances because each is based on a few degrees of freedom.

Suppose we have m stages of subdivision and distances $d_1, d_2, \ldots, d_m$ where d_1 is the shortest distance at the mth stage and d_m the largest distance at the first stage. The equivalence is given by

$$\sigma_m^2 = \gamma(d_1)$$

$$\sigma_{m-1}^2 + \sigma_m^2 = \gamma(d_2)$$

$$\sigma_{m-2}^2 + \sigma_{m-1}^2 + \sigma_m^2 = \gamma(d_3), \text{ and so on.} \qquad (13.3)$$

The values $\gamma(d_i)$ are the equivalent semi-variances. When plotted against distance they provide an approximate description of the way in which the property varies in space in a given region: graphs such as Fig. 13.2 are first approximations to the variograms of these properties.

This link is important because a nested survey and analysis can form the first stage in a complete analysis of the spatial variation of particular attributes in a region where little or nothing is known about their scales of variation. As mentioned in Chapter 12, the variogram depends on scale, and when it is estimated conventionally it rarely encompasses much more than one order of magnitude. A nested survey allows a rough variogram to be obtained over several orders of magnitude of distance in a single analysis. Sampling to estimate the variogram precisely can then be concentrated in the range of distance that encompasses most of the variation. We show how to use this two-step approach for determining spatial scale later with an example from the Wyre Forest.

Youden and Mehlich's survey

Youden and Mehlich's (1937) original example illustrates the principle of nested survey and analysis. In their survey of the soil in part of Broome County in New York State they devised a sampling scheme with four stages. They applied it to two soil series: the Culvers and the Sassafras series. On each of these two types of soil they selected nine primary stations approximately 1.6 km apart. At each station two sub-stations were chosen 305 m apart, and from these two sampling areas were selected 30.5 m apart. In each area two sampling points were located 3.05 m apart. This survey was balanced in the sense that all the classes at each particular stage were subdivided equally to give a hierarchy of sampling points as shown in Fig. 13.1. The progression of the spacings was geometric, and the components of variance might reasonably be regarded as independent, thereby allowing confidence limits to be determined. In terms of the

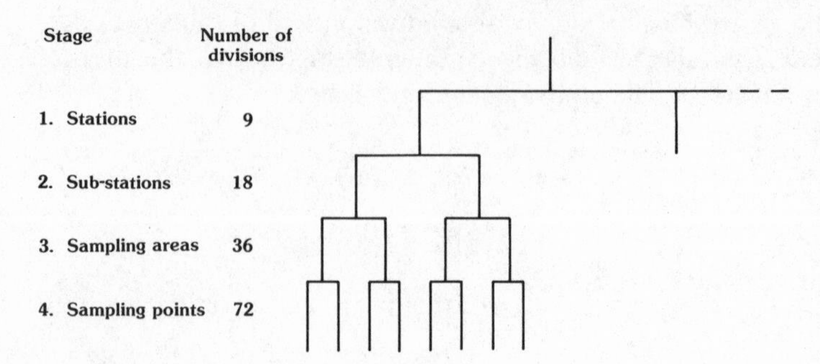

FIG. 13.1 Balanced hierarchical sampling scheme for a survey of the soil of part of Broome County, New York State

analysis in Table 13.1, $m=4$, $n_1=9$, and $n_2=n_3=n_4=2$. Thus there were 72 sampling points in all at which soil was taken from a depth of 0–15 cm, and its pH was determined.

For each series the variation associated with each sampling interval was determined by a nested analysis as described by Table 13.1. The analysis proceeds by first computing the sums of squares of deviations of the means of the classes in stage 1 from the general mean and multiplying each by the number of observations that make up the class mean. For each class in stage 2, the difference between its mean and the mean of the class to which it belongs in stage 1 is squared and multiplied by the number of observations in that class. The sum of these values is the appropriate sum of squares. This is repeated for each stage, and the sums of squares of the individual stages sum to the total sum of squares. The mean squares are obtained by dividing the sums of squares of each stage by the appropriate degrees of freedom. The latter are apportioned for each set of measurements as follows (Table 13.2). The number of degrees of freedom (d.f.) between stations is 8, i.e. n_1-1. There are 18 $(=n_1n_2)$ sub-stations and therefore 17 d.f. Since the differences between stations already accounts for 8, only 9 d.f. remain for differences between sub-stations within stations (stage 2). Thus, each pair of sub-stations contributes one degree of freedom from which to estimate the variance owing to differences between sub-stations within stations. Similarly, there are 18 d.f. for differences between areas within sub-stations (stage 3) and 36 d.f. for differences between sampling points within areas (stage 4). The mean square at each stage apart from the lowest contains a unique contribution to the variance from that stage, plus contributions from the components in all stages below (Table 13.1). For instance, the unique contribution to the variance at stage 2 (Table 13.1) is $n_m n_{m-1} \ldots n_3 \sigma_2^2$. This enables each component of variance to be determined separately from its mean square. For a balanced

TABLE 13.1 Derivation of the components of variance for a balanced design

Source	Degrees of freedom	Parameters estimated by mean squares
Stage 1	n_1-1	$\sigma_m^2+n_m\sigma_{m-1}^2+\cdots+n_m n_{m-1}\ldots n_3\sigma_2^2$ $+n_m n_{m-1}\ldots n_3 n_2\sigma_1^2$
Stage 2	$n_1(n_2-1)$	$\sigma_m^2+n_m\sigma_{m-1}^2+\cdots+n_m n_{m-1}\ldots n_3\sigma_2^2$
Stage 3	$n_1 n_2(n_3-1)$	$\sigma_m^2+n_m\sigma_{m-1}^2+\cdots+n_m n_{m-1}\ldots n_4\sigma_3^2$
⋮	⋮	⋮
Stage $m-1$	$n_1 n_2 n_3 \ldots (n_{m-1}-1)$	$\sigma_m^2+n_m\sigma_{m-1}^2$
Stage m (residual)	$n_1 n_2 n_3 \ldots n_{m-1}(n_m-1)$	σ_m^2
Total	$n_1 n_2 n_3 \ldots n_{m-1} n_m-1$	

TABLE 13.2 Components of variance of pH in two soil series in Broome County, New York

Stage	Spacing (m)	Degrees of freedom	Culvers series 0–15 cm		Sassafras series 0–15 cm	
			Estimated component	Percentage of variance	Estimated component	Percentage of variance
1	1600	8	0.02819	39.7	0	0
2	305	9	0.02340	32.9	0.04440	60.3
3	30.5	18	0.00552	7.8	0.00698	9.5
4	3.5	36	0.01391	19.6	0.02225	30.2

design the values of each component can be tested to judge whether it is larger than zero by computing the *F* ratio:

$$F = \frac{\text{mean square at stage } m}{\text{mean square at stage } m+1}.$$

The results of the analyses of topsoil pH for the Culvers and Sassafras soil series are given in Table 13.2. The largest component of variance for the topsoil of the Culvers series derives from the largest spacing, 1.6 km, and it accounts for almost 40 per cent of the total variance. The between-sub-stations (stage 2, 30.5 m) component accounts for a third of the variance. The variance in stage 4 from points only 3.05 m apart accounts for 20 per cent of the variance, which is the third largest. For the Sassafras series stage 4 contributes even more variance, 30.2 per cent, whereas the 1.6 km spacing contributes nothing. In this case most of the variance derives from stage 3 over distances between 30 and 300 m. This can be appreciated most clearly by plotting the accumulated components of variance against distance, as in Fig. 13.2, in which the distance is on a logarithmic scale. The graph for the Culvers series, again a rough variogram, shows that the variance increases substantially as the distance between the sampling points is increased, and it seems to continue to increase without limit. It is another example of the unbounded variation described in Chapter 12 (Fig. 12.9). However, the variance for the Sassafras series reaches a maximum, its sill, at $h = 305$ m. The full extent of the variation at this range of scales seems to have been encompassed, and we may judge the Sassafras series to be second-order stationary. Another feature of these graphs is that as the separating distance approaches zero the variance seems to approach a finite value. So again we have examples of the 'nugget effect'. It is difficult to imagine that adjacent samples are not

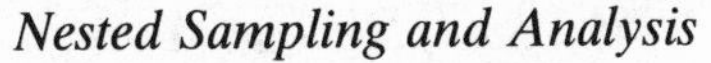

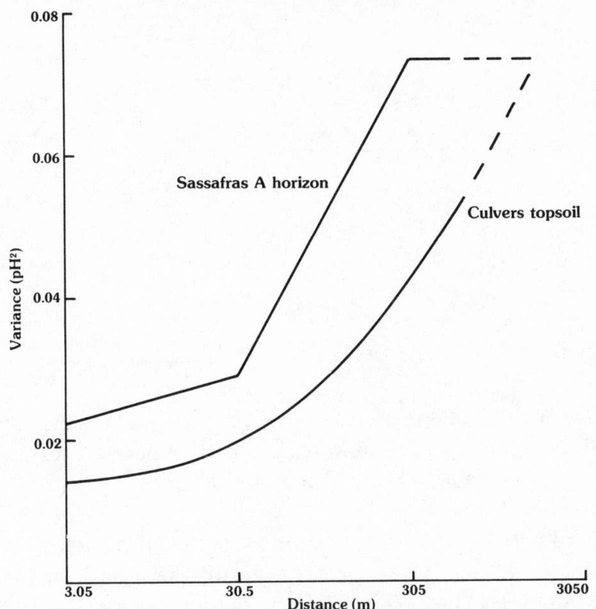

FIG. 13.2 Graphs of accumulated variance of pH of the topsoil against separating distances on a logarithmic scale for the Sassafras and Culvers soil series, Broome County. (From Youden and Mehlich 1937)

very similar, yet the sampling evidence often suggests that they are not. The true situation can be revealed only by sampling more closely, which means incorporating more stages in the design.

Ginninderra Experiment Station

In the example above the scale of variation of just one property was determined to improve future sampling for estimating its regional mean. Webster and Butler (1976) used a nested survey for an *a posteriori* examination of a general purpose soil classification to see whether they could explain the difficulties encountered in using it to map the 400 ha station in the Australian Capital Territory. Their nested design of four stages comprised eight primary stations at 180 m intervals, with two-fold subdivisions at each of 80, 18, and 5 m, to give 64 sampling points. At each sampling point several soil properties were measured on 10 cm diameter cores of topsoil. Data for larger spacings were also available, and these have been added to the analysis. The estimated components of variance for four properties of the topsoil were accumulated and plotted against sampling distance (Fig. 13.3). It is clear that most of the variation in potassium content is contributed from spacings between 50 and 180 m,

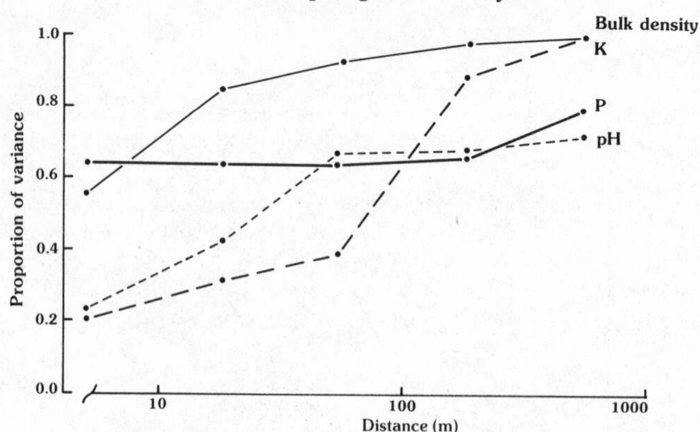

FIG. 13.3 Graphs of accumulated variance of properties of the topsoil against separating distance on a logarithmic scale for Ginninderra, Australian Capital Territory. (From Webster and Butler 1976)

whereas for pH it derives from distances between 5 and 50 m. For bulk density most of the variation is present within 18 m, and for phosphorus it is largely within 5 m. These results show distinct differences in the scale of variation of the different properties, and they explain why it had been difficult to make a generally useful soil map of the region. This example also shows the value of the method for comparing the patterns of variation of several soil properties over a region.

Unequal sampling

Youden and Mehlich's (1937) sampling design was fully balanced, and the analysis was straightforward. Surveyors cannot always ensure that there are the same number of individuals in every class; and even if they can they may wish to vary the numbers. For instance, some classes may cover more ground than others, and it is reasonable that the more extensive classes are better represented in the sample. Furthermore, to achieve good spatial resolution over a wide range of distances demands many stages. Since the sample size doubles for each additional stage in a balanced hierarchy, nested sampling could readily become prohibitively expensive if many stages were required. Youden and Mehlich's design with just four stages had $9 \times 2 \times 2 \times 2 = 72$ sampling points. A fifth and sixth stage would have required 144 and 288 sampling points, respectively. As it happens, however, full replication is unnecessary because with this number of data the mean squares for the lower stages are estimated much more precisely than those for the higher stages. Economy can be achieved by replicating only a proportion of the sampling centres at one or more stages once there are sufficient degrees of freedom.

Thus sampling might be unequal either by chance or by design. The penalty for this lack of balance is somewhat more complex estimation and interpretation. The analysis of variance of a hierarchy of classes is much the same whether the classes in any one level are sampled equally or not. With n observations there are $n-1$ d.f. The number of degrees of freedom at any stage in the hierarchy equals the number of classes at that stage less the number of classes at the stage above. The sums of squares are exactly as described earlier in the chapter. However, the numbers of observations m in the classes at any one level are not all the same. So, taking stage 2 again as an example, the factors $n_m, \ldots, n_m n_{m-1} \ldots n_3$ in Table 13.1 must be replaced by independently determined values $u_{2,m-1}, \ldots, u_{2,2}$. The mean squares are calculated as usual. The general scheme is given in Table 13.3. The table also shows that the coefficients required to estimate the components of variance contributing to the mean squares at each level are no longer simply the sample sizes in each mean. Nor are they the same for every mean square, and this complicates inferential testing. Thus the F ratio of the mean squares cannot test whether $\sigma_1^2=0$ or $\sigma_2^2=0$.

Gower (1962) and Gates and Shiue (1962) have provided computational procedures for calculating the coefficients for an unbalanced design. It is also presented in standard texts such as Snedecor and Cochran (1980), so we merely summarize it here. Suppose that there are C_i groups at the ith level and that within the kth group at the ith level there are c_{jk}^i subgroups at level j, each containing n_{pj}, $p=1, 2, \ldots, c_{jk}$, where $i \leq j$. Then the coefficient u_{ij} is given by

$$u_{ij} = \left\{ \sum_{k=1}^{C_i} \sum_{p=1}^{c_{jk}^i} \frac{n_{pj}^2}{n_{ik}} - \sum_{k=1}^{C_{i-l}} \sum_{p=l}^{c_{jk}^{i-1}} \frac{n_{pj}^2}{n_{i-1,k}} \right\} \Big/ d_i, \tag{13.4}$$

where d_i is the number of degrees of freedom at level i. The formula holds for $i=1$ by denoting the whole sample in one group as level 0. At the other

Table 13.3 Derivation of components of variance for an unbalanced design

Source	Degrees of freedom	Parameters estimated by mean squares
Stage 1	f_1	$\sigma_m^2 + u_{1,m-1}\sigma_{m-1}^2 + \cdots + u_{1,2}\sigma_3^2 + u_{1,2}\sigma_2^2 + u_{1,1}\sigma_1^2$
Stage 2	$f_2 - f_1$	$\sigma_m^2 + u_{2,m-1}\sigma_{m-1}^2 + \cdots + u_{2,3}\sigma_3^2 + u_{2,2}\sigma_2^2$
Stage 3	$f_3 - f_2$	$\sigma_m^2 + u_{3,m-1}\sigma_{m-1}^2 + \cdots + u_{3,3}\sigma_3^2$
⋮	⋮	⋮
Stage $m-1$	$f_{m-1} - f_{m-2}$	$\sigma_m^2 + u_{m-1,m-1}\sigma_{m-1}^2$
Stage m	$N - f_{m-1}$	σ_m^2
Total	$N-1$	

extreme, in the design that we describe with m levels, the coefficients of σ_m^2 are all unity because the numerators on the right-hand side of (13.4) equal the degrees of freedom.

Wyre Forest Survey

A survey of the soil in the Wyre Forest, in the English Midlands, illustrates the application of an unbalanced design (Oliver 1984). It also shows how the information it affords can be used to determine the variogram more precisely. We knew from an earlier survey that all the variation in the range of properties examined occurs within 167 m, the average distance between sampling sites in that survey. Fig. 13.4 shows the variograms for the content of sand in the topsoil and subsoil from this survey; clearly, they are both entirely nugget.

The nested survey was designed to discover how the variation is distributed over distances less than 167 m. It had five stages covering the range 6–600 m, which was expected to encompass most of the spatial variation. The sampling intervals were increased in a geometrical progression of approximately three-fold increments (Table 13.4) so that the components could be regarded as independent. The design incorporated a sampling interval of 600 m in case there were long-range spatial structures present.

Nine primary centres were located at the nodes of a 600 m square grid oriented randomly over the region. All other points were then located on

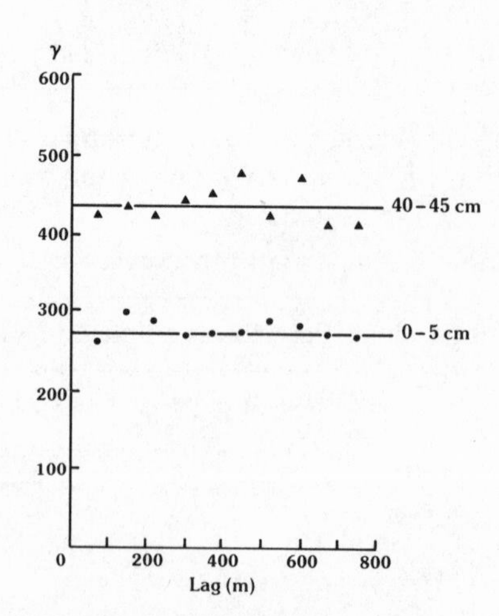

Fig. 13.4 Sample and model variograms of the content of sand in the topsoil and subsoil for the first survey of the Wyre Forest, England. (From Oliver and Webster 1987b)

TABLE 13.4 Nested sampling design for determining the scale of spatial variation in the soil of the Wyre Forest

Stage	Sampling interval (m)	Number of sampling points
1	600	9
2	190	18
3	60	36
4	19	72
5	6	108

random orientations from these as follows. From each grid node a second site was chosen 190 m away to provide the second stage. From each of the now 18 sites another point was chosen 60 m away (stage 3). The procedure was repeated at stage 4 to locate points 19 m away from those of stage 3, giving 72 points. At the fifth stage just half of the fourth stage points were replicated by sampling 6 m away. This gave a sample of 108 sites rather than 144 for a fully balanced survey. Table 13.4 summarizes the design, Fig. 13.5 illustrates the hierarchical structure used for one centre, and Fig. 13.6 shows the configuration of sampling points for one first-stage centre. This design achieved a 25 per cent economy in sampling effort, and Fig. 13.7 shows the economy possible with even more stages. At each sampling point several properties of the soil were recorded at four fixed depths in the soil profile: 0–5 cm (1), 10–15 cm (2), 25–30 cm (3), and 50–55 cm (4).

Each variable was analysed according to the scheme outlined in Table 13.3. The estimated components of variance for sand content at the four depths are listed in Table 13.5. Fig. 13.8 shows the accumulated components of variance for each depth in the profile plotted against separating

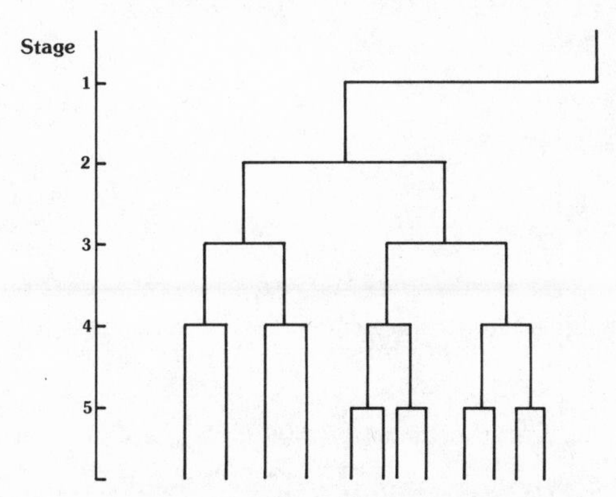

FIG. 13.5 The unbalanced hierarchical sampling scheme for one of the nine centres in the nested survey of the Wyre Forest, England

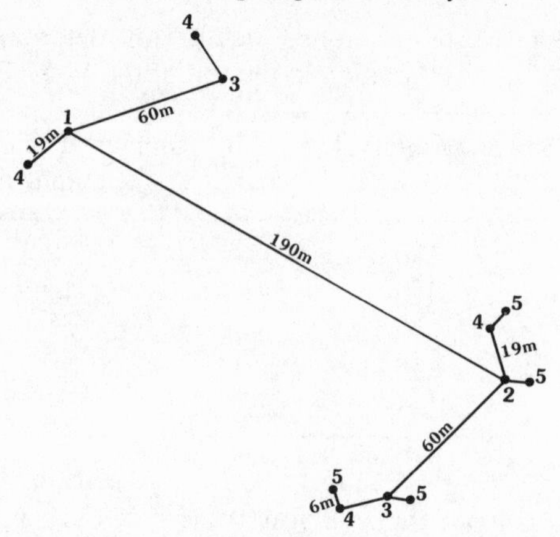

FIG. 13.6 The spatial configuration of the sampling points for one of the nine centres in the nested survey of the Wyre Forest, England. (From Oliver and Webster 1986)

distance on a logarithmic scale to give first approximations to the variograms.

The most important feature of the results is that at least 80 per cent of the variation for all properties occurs within 60 m. Stages 1 and 2, i.e. distances of 190–600 m and 60–190 m, respectively, account for less than 20

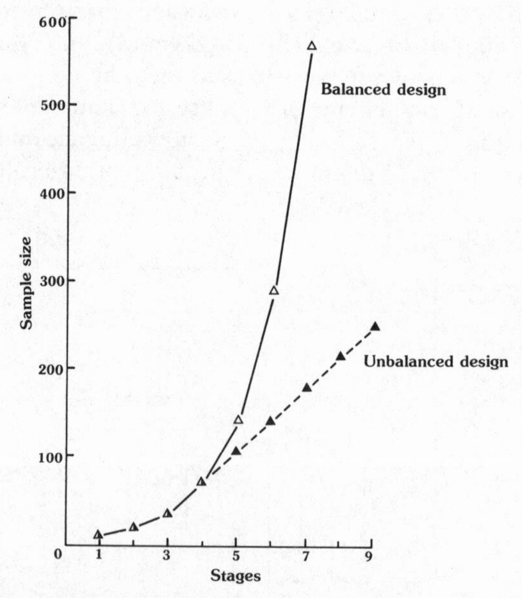

FIG. 13.7 A comparison of the number of samples required for balanced and unbalanced nested designs. (From Oliver and Webster 1986)

TABLE 13.5 Components of variance and percentage variance of sand content of the soil at four depths contributed by each stage in the survey of the Wyre Forest.

	Component of variance at depth (cm)				Percentage variance explained at depth (cm)			
Stage	0–5	10–15	25–30	50–55	0–5	10–15	25–30	50–55
1	37.45	21.26	34.17	40.40	12.0	6.5	7.6	7.5
2	−47.25	−60.49	−80.94	−104.70	0.0	0.0	0.0	0.0
3	85.79	139.90	171.20	317.10	27.5	43.0	37.9	51.5
4	147.40	113.80	155.40	0.70	47.2	35.0	3.4	0.1
5	41.69	50.21	90.46	251.70	13.3	15.4	20.0	40.9

per cent of the variation. The estimated components of variance for stage 2 were generally negative. This suggests that either there is some repetition in soil character at that distance, or the components estimate zero because there is truly no contribution to the variance at this stage. The confidence limits of the components are wide, and therefore we cannot be sure how to interpret these negative values. Even at stage 5 there is still a considerable contribution to the total variance. This represents the unresolved variation within 6 m plus errors of measurement. The former could have been investigated further by adding more stages.

In Chapter 12 we pointed out that the observed variogram depends on the spatial scale over which we measure it. If a large extent is covered with wide sampling intervals then all the variance might appear as nugget, as in the initial survey of the Wyre Forest (Oliver and Webster 1987b).

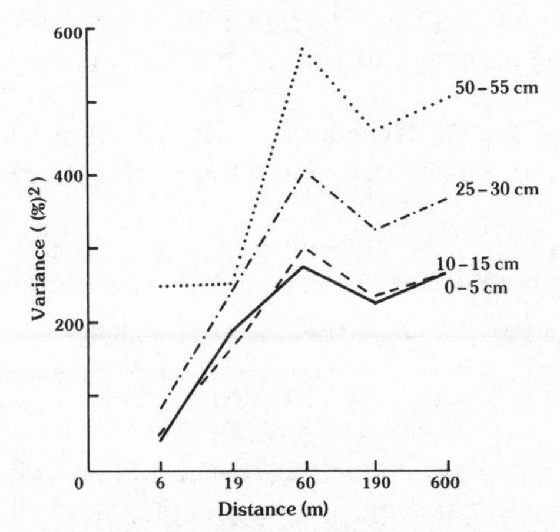

FIG. 13.8 Accumulated components of variance plotted against distance on a logarithmic scale for sand content in the soil of the Wyre Forest, England

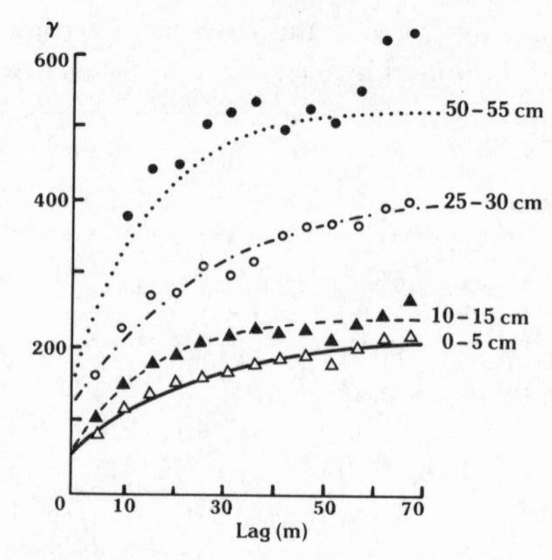

FIG. 13.9 Sample and model variograms for sand content from the linear survey of the soil of the Wyre Forest, England

Alternatively, if very close intervals are chosen to resolve the short-range variation then the sampling required to estimate the contributions at the larger distances would be too costly. A nested survey enables us to choose a suitable sampling scheme by identifying the scale at which most of the variation occurs. For instance, once rough variograms to 600 m for the soil properties of the Wyre Forest had shown that most of their spatial variation occurred over distances less than 60 m, we could estimate the variogram precisely over that range by linear sampling.

Ten transects 100 m long and one of 500 m were sampled at 5 m intervals to ensure that the residual variance would be no greater than in the nested survey. The same properties of the soil were measured as for the nested survey. The semi-variances were estimated by equation (12.12) for integer multiples of the sampling interval from 5 to 70 m.

The experimental semi-variances for the sand content of the subsoil are given in Fig. 13.9. An exponential model provided the best fit. The effective range of spatial dependence of 41.7 m, three times the distance parameter of the model, is well within the limit of 60 m indicated by the nested sampling.

This result illustrates the place of nested survey as the first stage in a spatial analysis where little or nothing is known about a region.

Improving the efficiency of sampling

In Chapters 3 and 4 we considered the sampling effort required, the efficiency of different sampling designs, and the effects of classification for estimating the mean value of a property within a given region. We can now show how the results from a nested survey and analysis can be used to plan an efficient survey to estimate a regional mean. We do so using the components of variance already estimated. In Broome County (Youden and Mehlich 1937) the mean for each area (stage 3 of the design) was estimated from two sampling points, so the variance of an area mean is estimated by

$$V_{\text{area}} = \frac{s_4^2}{2}. \tag{13.5}$$

The mean of a sub-station (stage 2) was determined from two area means, each composed of values from two sampling points. Its estimated variance is

$$V_{\text{sub-station}} = \frac{s_4^2}{2\times 2} + \frac{s_3^2}{2} = \frac{s_4^2 + 2s_3^2}{4}. \tag{13.6}$$

By similar reasoning, the estimated variance of a station (stage 1) mean is

$$V_{\text{station}} = \frac{s_4^2}{2\times 2\times 2} + \frac{s_3^2}{2\times 2} + \frac{s_2^2}{2} = \frac{s_4^2 + 2s_3^2 + 4s_2^2}{8}, \tag{13.7}$$

and the variance of the mean of the whole region is

$$V_R = \frac{s_4^2}{9\times 2\times 2\times 2} + \frac{s_3^2}{9\times 2\times 2} + \frac{s_2^2}{9\times 2} + \frac{s_1^2}{9} = \frac{s_4^2 + 2s_3^2 + 4s_2^2 + 8s_1^2}{72}. \tag{13.8}$$

Thus the variance of the mean at any level in the hierarchy is equal to the mean square within that stage divided by the number of observations that go to make that mean. In particular, the estimated variance of the mean for the whole region is the between-stations mean square divided by the sample size.

Once the components of variance have been determined using the formula above, it is possible to compare the efficiency of other sampling schemes. For a nested design with four stages, the estimated variance of the mean for any sampling scheme based on those four stages will be

$$V_{\bar{z}} = \frac{\sigma_4^2}{n_1', n_2', n_3', n_4'} + \frac{\sigma_3^2}{n_1', n_2', n_3'} + \frac{\sigma_2^2}{n_1', n_2'} + \frac{\sigma_1^2}{n_1'} \tag{13.9}$$

where n_1', n_2', n_3', n_4', are the numbers of sub-units to be chosen for each class in the four stages 1, 2, 3, and 4, respectively. It is clear that, other things being equal, we should increase the sampling at those stages where the contribution to the variance is largest. This should be at the station stage for the Culvers soil and at the sub-station stage for the Sassafras soil, for which the estimated components are largest.

The estimated variance for the pH of the Sassafras series from the 72 sampling points was 0.00297 with a standard error of 0.0545. If we doubled the number of substations so that $n_1 = 9$ and $n_2 = 36$ and did not replicate at the lower stages, then

$$V'_{\bar{z}} = \frac{s_1^2}{9} + \frac{s_2^2}{36} + \frac{s_3^2}{1} + \frac{s_4^2}{1}$$

$$= \frac{0 + 0.04440 + 0.00698 + 0.0225}{36}$$

$$= 0.002052.$$

The standard error is 0.0453. Thus, a smaller standard error can be obtained with half the original sample size by sampling at intensities that reflect the scales of variation.

Bulking

Some properties of soil, rocks, or sediments are expensive to measure. If there is no objection to determining these properties on material disturbed by mixing then the cost of measurement can be reduced by bulking. Bulking can also improve the precision of a regional estimate, for instance where the property is very variable locally. Inasmuch as the analysis of a bulked sample approximates the mean of analyses of its component subsamples, bulking maintains something of the precision of replicated observations. Here we show how the nested analysis of variance can be used to determine the mean for a region from a bulked sample, and in the next chapter we consider how to use measures of spatial dependence to optimize the spacing of the samples that will form the bulked sample.

The principle of bulking is simple. If n cores of material are taken from an area B and the property Z measured on each then the value of Z in B is estimated by the sample mean $\bar{z}$, with variance σ^2/n. If, however, the material from the n cores is mixed thoroughly and carefully sub-sampled for analysis in the laboratory then Z can be determined on the resultant mixture. For simple quantities such as the total amounts of elements, calcium carbonate, particles of different sizes, and organic matter, the value obtained, $\bar{z}_B$, will approximate the true mean of Z apart from laboratory error, and the variance of z_B will be σ^2/n. Often this is approximately true even when the quantities are not obviously additive, as for example with available nutrients or pH.

Bulking means that σ^2 cannot be estimated from a single value of the bulked sample, and this implies that it is only the mean value that is of interest in area B. If there are many areas like B, representing subdivisions of a region, and we sample a small proportion of them, then we can take the mean of the measurements on the bulked samples as the mean for the soil or other attribute of the region and obtain an estimate of its standard error. In multi-stage sampling the estimation variance of a sample mean is the mean square at the highest level divided by the total number of observations. If there are just two levels and the primary units are sampled equally, then the variance is

$$V_{\bar{z}} = \frac{1}{n_1 n_2} \frac{\sum_{i=1}^{n_1} n_2(\bar{z}_i - \bar{z})^2}{n_1 - 1} = \frac{\sum_{i=1}^{n_1} (\bar{z}_i - \bar{z})^2}{n_1(n_1 - 1)}. \tag{13.10}$$

To obtain $V_{\bar{z}}$ we need know only the means of the first-stage units ($\bar{z}_i$). This result can be applied usefully when samples are bulked within the primary units. We simply replace the means, $\bar{z}_i$, by the values for the bulked samples (z_i^*), so that

$$V_{\bar{z}} = \frac{\sum_{i=1}^{n_j} (z_i^* - \bar{z})^2}{n_1(n_1 - 1)}. \tag{13.11}$$

Also, we recall that

$$V_{\bar{z}} = \frac{s_2^2}{n_1 n_2} + \frac{s_1^2}{n_1} = \frac{1}{n_1}\left(\frac{s_2^2}{n_2} + s_1^2\right).$$

The contribution of the first term to the true variance can be diminished by increasing n_2. Alternatively, we can achieve the same end by bulking. In fact, if sufficient cores are bulked s_2^2/n_2 can be made so small that it may be ignored. If so, the variance of the bulked values about their mean then approximates closely the component of variance represented by differences between the primary units:

$$\frac{\sum_{i=1}^{n_i} (z_i^* - \bar{z})^2}{n - 1} \simeq s_1^2. \tag{13.12}$$

So by bulking enough cores within, for instance, each field, the between-field component of variance can be estimated economically and with little bias. In the example of the Culvers soil where $s_2^2=0.04283$ at 305 m spacing and $s_1^2=0.02819$ at stations 1.6 km apart, bulking 20 cores at each station would lead s_1^2 to be overestimated by about 7.5 per cent. If 50 cores were bulked the bias would be about 3 per cent.

Finite population correction

In the example above we calculated the sampling error as though we were choosing at every stage a small sample from a large population of classes, and we ignored the finite population correction (f.p.c.). If the sampling stages are represented by distinct areas, for instance districts, farms, and fields, then the numbers of divisions in each of the first two stages are finite, and the sampling fraction is likely to be appreciable. In these circumstances it is wise to apply finite population corrections. We shall consider the matter here for two stages only.

Suppose we can divide a survey area into N_1 primary regions (stage 1) of equal area, and each primary region is divided further into N_2 sub-regions; we might divide a 100 km^2 map sheet into 100 1 km^2 squares and divide each of these into 100 1 ha squares. We then choose n_1/N_1 primary regions randomly, and within each of those choose n_2/N_2 1 ha squares. The f.p.c.s at the two stages are then

$$\text{Stage 1: } \frac{N_1-n_1}{N_1}$$

$$\text{Stage 2: } \frac{N_2-n_2}{N_2}.$$

If the mean squares in the analysis of variance are M_1 and $M_2=s_2^2$ for stages 1 and 2 respectively then the stage 1 component of variance can be estimated as

$$s_1^2=\frac{M_1}{n_2}-\frac{N_2-n_2}{N_2}\frac{s_2^2}{n_2}. \tag{13.13}$$

The estimated variance of the mean for the whole area is

$$V_{\bar{z}}=\frac{N_1-n_1}{N_1}\frac{s_1^2}{n_1}+\frac{N_2-n_2}{N_2}\frac{s_2^2}{n_1n_2}. \tag{13.14}$$

Comparable expressions can be applied if there are more than two stages. Hammond *et al*. (1958) describe several examples of sampling using f.p.c. in a three-stage scheme in which the first two stages were area subdivisions. Incidentally, a very large proportion of the total variance of several soil properties within large fields was present within areas of a few square metres.

For most soil sampling the f.p.c. of the final lowest stage will be very small. It is likely to be appreciable only for properties that are measured over areas, for example the proportion of exposed rock, or the proportion of soil surface covered by stones.

Two other extreme cases should be taken into account. If in the two-stage sampling $n_1 = N_1$, i.e. if we sample every first-stage group, then the first term in (13.14) is zero, and the sampling error depends solely on s_2^2. The result is stratified sampling with first-stage groups as strata. If $n_2 = N_2$ the whole of each chosen primary group would be recorded and would result in simple random sampling.

14

LOCAL ESTIMATION: KRIGING

The best things are most difficult.

Plato, *Cratylus*

In the early chapters of this book we stressed estimation because the aim of many quantitative surveys, whether of soil or mineral resources or vegetation, is to estimate the amounts or concentrations of some attribute of the terrain. Chapter 3 was concerned with averages or totals for whole regions. Chapter 4 recognized that such values were often of little interest and not necessarily sensible. They could be made more meaningful, however, by stratifying a region into classes according to the type of soil, rock, or physiography. Estimates could then be made for each type separately, and they would also be more precise since variation within each class would be less than in the region as a whole. They would also be more local if each stratum were represented by just one sub-region or a few close sub-regions.

Spatial classification as the basis for estimation has a long history in the earth sciences. However, it is a somewhat inflexible approach because classification into soil series or stratigraphic units, for example, tends to be a once-for-all exercise. Once the classes are defined and their boundaries mapped, little can be conveyed about their variation in composition from place to place. All sites within a spatial class are assumed to be similar in terms of the properties of interest, and there is no means of indicating intergrades between classes. Classification also provides only two kinds of estimate, one the mean for each class and the other the prediction for any point within a class. For any one class the mean and predicted values are the same: only their variances differ. Recalling the result from Chapter 4, we have for class j its mean μ_j estimated from n_j observations by $\hat{\mu}_j$ with a variance of σ_j^2/n_j, and the prediction for any point, also estimated by $\hat{\mu}_j$, but with its variance $\sigma_j^2 + \sigma_j^2/n_j$. Thus local variation has little bearing on the estimate of the mean, prediction at points may be crude, and one cannot estimate values for areas smaller than the extent of the mapped class with known confidence. The procedure takes little cognizance of the spatial configuration of the data. Furthermore, a classification designed to predict values of one group of properties might be of little worth for estimating other properties. Classification involves arbitrary decisions, and these are increasingly subjective and difficult the more gradual the variation is and the more detailed the investigation.

The shortcomings of classification have long been recognized. Thiessen (1911), the climatologist, devised one of the early alternatives to classification for local estimation. Having obtained meteorological records from weather stations in a region, he divided the region into tessellating polygons such that there was just one station within each polygon and all points within the polygon were nearer to that station than to any other. Such polygons are often known as *Thiessen polygons*; other names are *Voroni polygons* and *Dirichlet tiles*. The merit of tessellation is that the predicted value of any point within a polygon is the recorded value for the station within it. Intuitively this is the most appropriate value because it is the nearest. The major weaknesses of the procedure are that each prediction is based on just one measurement, there is no estimate of error, and pertinent information from other nearby sites is ignored.

Many other techniques such as trend surface analysis and arbitrarily weighted averages of data have been proposed for estimating values of properties between sampling points. These have been primarily for mapping. Essentially they interpolate, and as computers have become more powerful, so these methods have tended to become more elaborate. Most are intuitively reasonable, but they are usually more or less biased, their errors are unknown, and some are frankly cosmetic. They are mathematical but not statistical.

One method, perhaps better described as a set of methods, works locally and has all the desirable qualities for statistical estimation. In the earth sciences it is known as *kriging* after D. G. Krige (1966) who devised it empirically for estimating the gold content of ore in the South African goldfields. It is firmly grounded in the theory of regionalized variables and its development is largely due to Matheron (1965, 1971, 1976), although its principles had already been recognized by A. N. Kolmogorov (1941) and Russian meteorologists since the 1940s (Gandin 1965). Not only does the method provide statistically sound estimates, but it can also be used to plan sampling in a rational way. The advantages of kriging are such that we devote the remainder of this chapter to it.

Kriging

At its simplest, kriging is no more than a method of weighted averaging of the observed values of a property Z within a neighbourhood V. Suppose that we wish to know the average value, $z(B)$, of the variable Z over a block of land B. Its estimate is a weighted average of the data $z(\mathbf{x}_1)$, $z(\mathbf{x}_2)$, $\ldots$, $z(\mathbf{x}_n)$, which may be either inside or outside the block:

$$\hat{z}(B)=\sum_{i=1}^{n}\lambda_i z(\mathbf{x}_i). \tag{14.1}$$

The λ_i, $i=1,2,\ldots,n$, are the weights assigned to the sampling points. A desirable property of an estimator is that it be free of bias, and to assure this the weights are chosen to sum to 1:

$$\sum_{i=1}^{n}\lambda_i=1. \tag{14.2}$$

The estimation variance of $\hat{z}(B)$, or kriging variance, is

$$\begin{aligned}\sigma^2(B)&=\mathrm{E}[\{\hat{z}(B)-z(B)\}^2]\\&=2\sum_{i=1}^{n}\lambda_i\bar{\gamma}(\mathbf{x}_i,B)\\&\quad-\sum_{i=1}^{n}\sum_{j=1}^{n}\lambda_i\lambda_j\gamma(\mathbf{x}_i,\mathbf{x}_j)-\bar{\gamma}(B,B),\end{aligned} \tag{14.3}$$

where $\gamma(\mathbf{x}_i,\mathbf{x}_j)$ is the semi-variance of Z between the ith and the jth sampling points, $\bar{\gamma}(\mathbf{x}_i, B)$ is the average semi-variance between the block and the ith sampling point, and $\bar{\gamma}(B, B)$ is the average variance within the block, i.e. the within-block variance. Subject to the non-bias condition of (14.2), the value of $\sigma^2(B)$ in (14.3) is least when

$$\sum_{i=1}^{n}\lambda_i\gamma(\mathbf{x}_i,\mathbf{x}_j)+\psi=\bar{\gamma}(\mathbf{x}_j,B)\quad\text{for all } j. \tag{14.4}$$

This introduces a Lagrange multiplier, ψ, to achieve minimization.

These equations are the kriging equations. Their solution provides the weights for estimating $z(B)$ using (14.1) and also enables us to estimate the kriging variance by

$$\hat{\sigma}^2(B)=\sum_{i=1}^{n}\lambda_i\bar{\gamma}(\mathbf{x}_i,B)+\psi-\bar{\gamma}(B,B). \tag{14.5}$$

In this way we obtain estimates that are unbiased and have minimum variance. For this reason simple kriging, or *ordinary* kriging as it is known technically, is often called BLUE, Best Linear Unbiased Estimation. Since the estimation variances are also determined, kriging has all the desirable qualities of a statistical procedure.

The equation (14.4) can be written succinctly in matrix form as

$$\mathbf{A}\begin{bmatrix}\boldsymbol{\lambda}\\ \psi\end{bmatrix}=\mathbf{b}, \tag{14.6}$$

where

$$\mathbf{A}=\begin{bmatrix}\gamma(\mathbf{x}_1,\mathbf{x}_1) & \gamma(\mathbf{x}_1,\mathbf{x}_2) & \dots & \gamma(\mathbf{x}_1,\mathbf{x}_n) & 1\\ \gamma(\mathbf{x}_2,\mathbf{x}_1) & \gamma(\mathbf{x}_2,\mathbf{x}_2) & \dots & \gamma(\mathbf{x}_2,\mathbf{x}_n) & 1\\ \cdot & \cdot & \cdot & \cdot & \cdot\\ \cdot & \cdot & \cdot & \cdot & \cdot\\ \cdot & \cdot & \cdot & \cdot & \cdot\\ \gamma(\mathbf{x}_n,\mathbf{x}_1) & \gamma(\mathbf{x}_n,\mathbf{x}_2) & \dots & \gamma(\mathbf{x}_n,\mathbf{x}_n) & 1\\ 1 & 1 & \dots & 1 & 0\end{bmatrix},$$

$$\begin{bmatrix}\boldsymbol{\lambda}\\ \psi\end{bmatrix}=\begin{bmatrix}\lambda_1\\ \lambda_2\\ \cdot\\ \cdot\\ \cdot\\ \lambda_n\\ \psi\end{bmatrix}, \text{ and } \mathbf{b}=\begin{bmatrix}\bar{\gamma}(\mathbf{x}_1,B)\\ \bar{\gamma}(\mathbf{x}_2,B)\\ \cdot\\ \cdot\\ \cdot\\ \bar{\gamma}(\mathbf{x}_n,B)\\ 1\end{bmatrix}.$$

Matrix **A** is inverted, and the weights are given by

$$\begin{bmatrix}\boldsymbol{\lambda}\\ \psi\end{bmatrix}=\mathbf{A}^{-1}\mathbf{b}. \tag{14.7}$$

As above, this solution leads to an estimate of $\sigma^2(B)$, which in matrix notation is simply

$$\hat{\sigma}^2(B)=\mathbf{b}^T\begin{bmatrix}\boldsymbol{\lambda}\\ \psi\end{bmatrix}-\bar{\gamma}(B,B). \tag{14.8}$$

The importance of the variogram in practical geostatistics will now be clear. Equation (14.4) contains semi-variances, and these are obtained

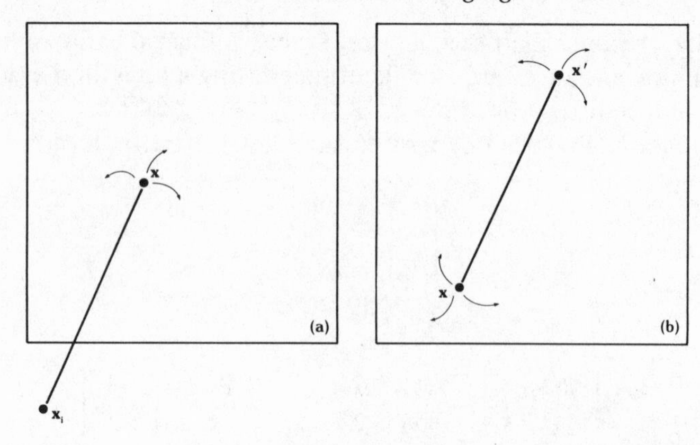

Fig. 14.1 Integration of the variogram (a) between a sampling point and a block, and (b) within a block

from a suitable model fitted to the experimental values. The variogram is essential for kriging.

The average semi-variances $\bar{\gamma}(\mathbf{x}_i, B)$ and $\bar{\gamma}(B, B)$ that appear in (14.3)–(14.8) might require a little explanation. Formally these are integrals of the variogram. The integral between the ith sampling point and the block B is defined by

$$\bar{\gamma}(\mathbf{x}_i, B) = \frac{1}{B} \int_B \gamma(\mathbf{x}_i, \mathbf{x}) d\mathbf{x}, \qquad (14.9)$$

where $\gamma(\mathbf{x}_i, \mathbf{x})$ denotes the semi-variance between the sampling point $\mathbf{x}_i$ and a point $\mathbf{x}$ describing the block. The integration is illustrated in Fig. 14.1a. The within-block variance is the double integral

$$\bar{\gamma}(B, B) = \frac{1}{B^2} \int_B \int_B \gamma(\mathbf{x}, \mathbf{x}') d\mathbf{x}\, d\mathbf{x}', \qquad (14.10)$$

where $\gamma(\mathbf{x}, \mathbf{x}')$ is the semi-variance between two points $\mathbf{x}$ and $\mathbf{x}'$ that sweep independently over B. This integration is shown in Fig. 14.1b.

Formulae for these integrals have been worked out for rectangular blocks and certain types of variogram and are available as auxiliary functions; see, for example, Clark (1976), Journel and Huijbregts (1978), and Webster and Burgess (1984b). Nowadays the integration is performed numerically by computer, and it is so swift that these auxiliary functions have lost their appeal.

A block may be of any reasonable size and shape. At its smallest it may be a 'point', $\mathbf{x}_0$, having the same dimensions as the support on which the original measurements were made. In these circumstances we have punctual kriging, and the equations simplify somewhat: the semi-variances in **b** are no longer averages but single semi-variances $\gamma(\mathbf{x}_i, \mathbf{x}_0)$. The quantity $\bar{\gamma}(B, B)$ becomes $\gamma(\mathbf{x}_0, \mathbf{x}_0)=0$, and hence disappears from (14.3), (14.5), and (14.8). This can have interesting consequences where there is a sizeable nugget variance, as we shall show later.

The weights, too, are interesting, as the following examples show. Figure 14.2a gives the punctual kriging weights for potassium at Broom's Barn based on the variogram of Fig. 12.7. The target point in the example has coordinates 16 m east and 12 m north of a sampling point near the centre of the sampling grid. The nearest sampling point carries a quarter of the weight, and the nearest four account for nearly three-quarters of it. The second 'shell' of sampling points has small weights, and sampling points beyond them have weights so close to zero that they may be disregarded in many instances. They are effectively screened by the nearer points. Figure 14.2b illustrates the effect of anisotropy. The stone content at Plas Gogerddan varies anisotropically, and this was expressed in the variogram (Fig. 12.6). This carries through to the kriging weights. Spatial dependence is strongest in the direction approximately $3\pi/4$, and so the weights are largest in directions close to this.

In the extreme, a target point, $\mathbf{x}_0$, can coincide with a sampling point. In these circumstances the estimation variance, equation (14.3), is reduced to nothing by taking the sample value as the estimate there. When the kriging equations are solved, the weight at $\mathbf{x}_0$ is found to be 1 and all the other weights are zero.

The above results are important. First, they demonstrate that a kriged estimate is local. This seems to be right intuitively and to accord with the sense of spatial dependence expressed in the variogram. They are also of practical significance. If only close data carry weight then matrix **A** need never be large and its inversion will be swift. This does not matter when making just one estimate. In mapping, however, where many such estimates may be needed, it can make a very big difference to the feasibility of kriging because the time required to invert a matrix is approximately proportional to the cube of its order. For data on square grids the 25 nearest observations will almost certainly be enough, and in many instances the nearest 16 may be. For irregular sampling, Olea (1975) suggests using the nearest two sampling points within each octant around $\mathbf{x}_0$, making 16 in total. A further advantage of selecting only the nearest 16 or 25 points is that the variogram need be estimated accurately only over short distances, i.e. to the lag of the furthest point to carry effective weight, and where there is the greatest confidence. The confidence limits to the variogram widen considerably at larger lag distances unless the sampling itself is massive. Such a variogram might not apply over the whole region, anyway.

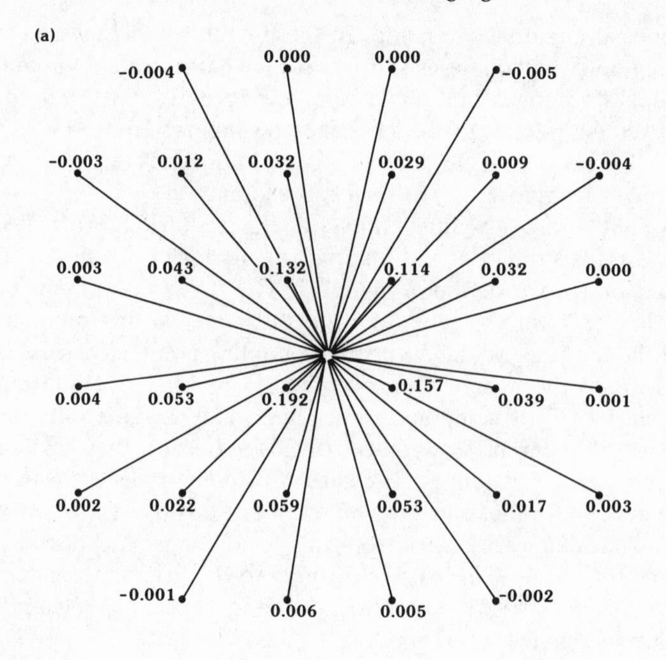

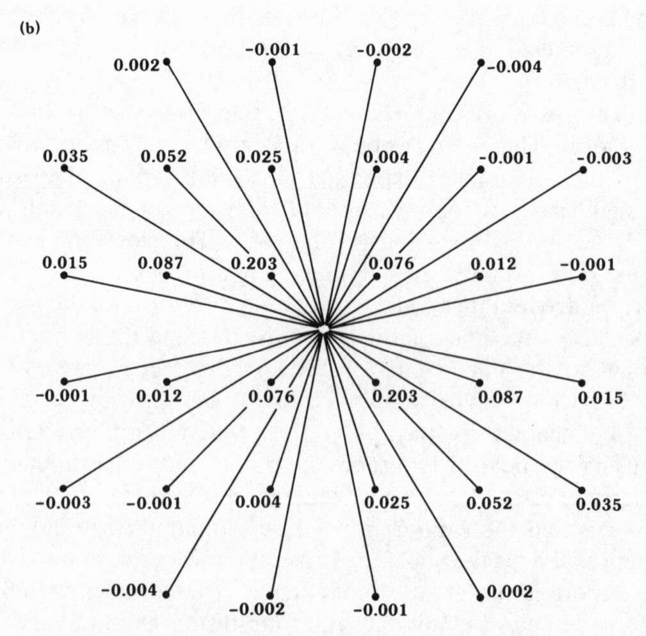

FIG. 14.2 Examples of weights for punctual kriging (a) for potassium at Broom's Barn Farm, England, and (b) for stone content at Plas Gogerddan, Wales

Mapping

Kriging was developed in mining for estimating the amounts of metal in blocks of rock. In these circumstances every block of rock is of interest: each will be treated individually. If it contains sufficient metal to return a profit then it will be mined and sent for processing. If not then it may either be left in place or removed and sent to waste. In studies of the soil, and in environmental science more generally, the interest has so far been mainly in spatial distributions and their displays on maps. Kriging has been used mainly for interpolation. We therefore devote a section to kriging in this context.

The most practicable way of using kriging to map a variable is to estimate its values at the nodes of a fine grid. The results can be displayed by threading isarithms through the resultant grid on the screen of a monitor or by drawing them using a pen-plotter. Alternatively, the statistical surface so constructed can be displayed by layered shading, colouring, or as a perspective diagram. There are many computer programs and packages available for this. We used Surface II (Sampson 1978) to produce the examples in this chapter. Most of these programs perform a further interpolation. They find the positions where the chosen isarithms intersect the sides of grid cells, usually by linear interpolation, and then join these positions either by straight lines or less often by curves. Although this second interpolation is not optimal in the kriging sense, its accuracy can be increased by making the grid mesh finer. In most instances a mesh with an interval of 2 mm on the finished map will be adequate. The map should be graphically acceptable even when the segments of the isarithms across the grid cells are straight. Fig. 14.3 shows such a map of exchangeable potassium at Broom's Barn made by block kriging from sample data and the variogram of Fig. 12.7. The result may also be shown as a perspective diagram (Fig. 14.4). These are based on the original study by Webster and McBratney (1987).

Mapping in two stages in this way has a further advantage. Corresponding to the grid of kriged values there is a grid of the kriging variances, and these variances or their square roots can also be mapped isarithmically to give a map of the errors. It can be regarded as a reliability map. This is perhaps the single most important advantage of kriging for mapping. The kriging variances can be displayed as perspective diagrams, and Fig. 14.5a shows this representation of the block kriging variances of potassium at Broom's Barn. Note how the variances increase markedly near the margins of the farm and where the farm buildings and access roads are.

Interpolating a grid of values by kriging can involve heavy computation. The core of the difficulty is the matrix inversion. In principle, all the estimates and their variances could be found from a single inversion of

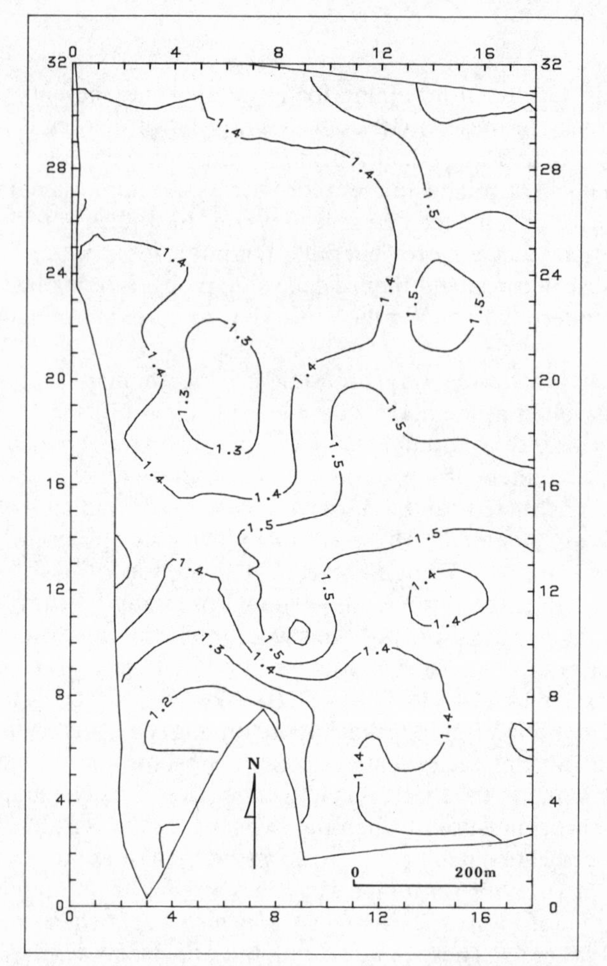

FIG. 14.3 Map of the exchangeable potassium at Broom's Barn Farm made by kriging 50 m × 50 m blocks on a fine grid. The units are $\log_{10}$ (mg K/kg soil)

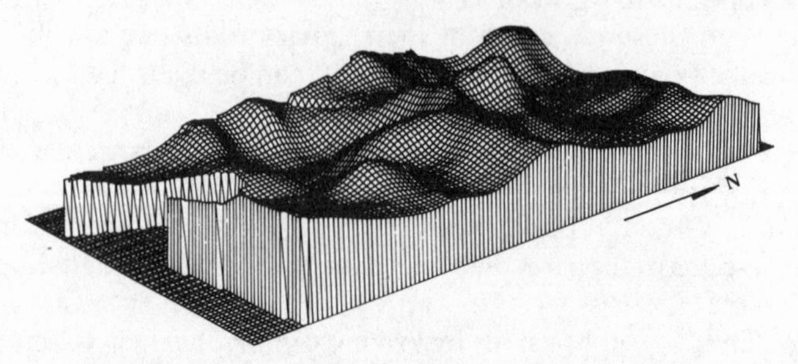

FIG. 14.4 Perspective diagram of exchangeable potassium at Broom's Barn Farm made by kriging 50 m × 50 m blocks on a fine grid

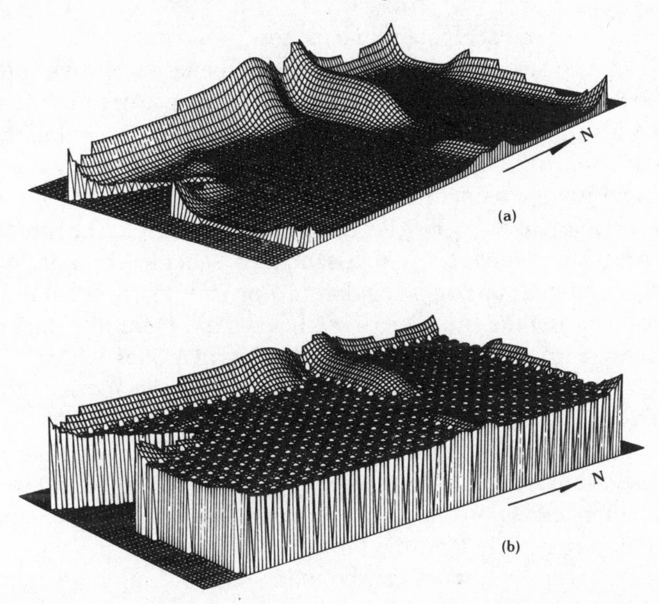

FIG. 14.5 Perspective diagrams of the kriging variances of exchangeable potassium at Broom's Barn Farm (a) for block kriging over 50 m × 50 m, and (b) for punctual kriging

matrix **A** in (14.6) containing all semi-variances between the sampling sites. If there were many data this would be formidable, but as we have already seen, this is neither necessary nor desirable. The alternative is to use matrices of local data around each grid node. Each matrix is small, but now there are many of them to invert. Fortunately, there are substantial economies to be made.

The matrix inversions can be speeded up by working with covariances instead of semi-variances. This works because in the usual method of

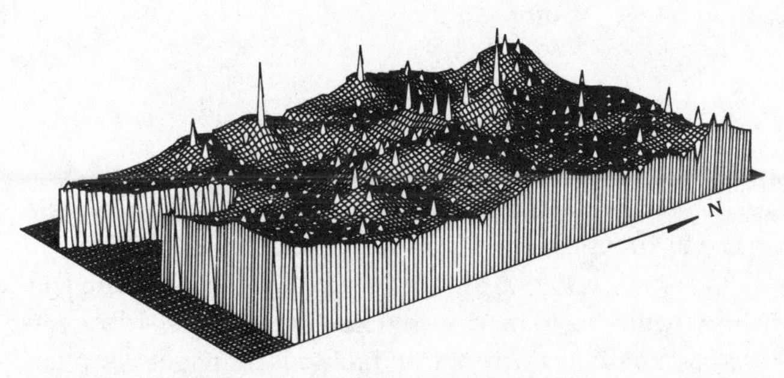

FIG. 14.6 Perspective diagram of exchangeable potassium at Broom's Barn Farm made by punctual kriging

inversion the largest element in each row of the matrix has to be sought as a pivot. In a covariance matrix the largest element is always in the diagonal, and so no search is needed. For variables that are second-order stationary all the formulae for finding the weights from the variogram also apply to the covariance function with only changes of sign. For variables that are intrinsic only the technique can still be used by taking the sample variance as the covariance at $|\mathbf{h}| = 0$. Other economies can be made, but their precise form depends on the sampling scheme, and in particular on whether the data are on a regular grid or irregularly scattered.

Consider first the situation of scattered data. Here the same few data will often serve to estimate $z(\mathbf{x})$ at several grid nodes within a small area. Furthermore, the finer the interpolation grid the more nodes will it be possible to interpolate from the same observations. For all of these matrix **A** remains the same, and so needs inverting just once. To take advantage of this the computer must be programmed to perform the interpolations for small compact regions in turn and to arrange the data in small blocks so that they are readily identified for each region.

Much larger economies can be made where sampling is regular. As we have seen, the kriging weights depend on the variogram and on the configuration of the sampling in relation to the interpolated point or block. They do not depend on the measured values. With regular sampling and interpolation on a regular grid the same configuration recurs many times. Not only does the variogram remain constant but so does the matrix **A** for any given configuration. Therefore, only one matrix inversion is needed for each configuration. If sampling is performed on a square grid and the interpolation grid fits on to it with interval $1/r$ times that of the sampling grid then there are only r^2 possible configurations except near the edge of the map. Where variation is isotropic the spatial relations have four-fold symmetry, so even fewer solutions are needed.

Thus by skilful programming, using devices such as those above, it is possible to krige a large figure field in a quite acceptable time. There is no justification for using crude or inappropriate methods of interpolation because of shortage of time on a computer.

Effects of nugget variance

In Chapter 12 we noted that the nugget component of the variance, defined using the Dirac function as $\gamma(\mathbf{h}) = c_0\{1 - \delta(\mathbf{h})\}$, represented discontinuity in the underlying variable. Although we believe most properties of the soil and of the environment to be continuous generally, we often have to accept that our best estimate of a variogram includes a nugget variance in addition to the spatially dependent component. In these circumstances it may be helpful to envisage the estimate as comprising two parts. The one deriving from the autocorrelated variation is continuous and varies according to the measured values in the neighbourhood. The other derives from the nugget variance. It is the classical estimate with the same mean value everywhere except at the sampling points. Combining the two produces an undulating surface with spikes where there are data. Fig. 14.6 shows a good example. The effect of the nugget component therefore is to create discontinuities at the sampling points and smooth the surface between them, and this effect increases as the proportion of the nugget variance increases. Fig. 14.5b shows the estimation variances, and in particular it shows how the kriging variance diminishes to zero at the sampling points. Incidentally, if all the variance is nugget then the kriged surface is flat between the sampling points. There must be some spatial dependence for interpolation to be sensible and profitable whatever method is used.

When an investigator wishes to estimate values at particular points this feature of kriging is immaterial; the estimates will be the best everywhere. If, however, he aims to map the variation of a property then the discontinuities may distract his attention from the spatially dependent variation that is of interest. Further, the positions of the discontinuities depend on the original sampling configuration, and if this were different then the spikes, whose heights above or below the local average are random, would also be very different. This feature of punctual kriging is a sampling effect. It can be avoided either by not kriging at sampling points, which will usually be the case with irregular sampling, by choosing a grid for interpolation in such a way that none of its nodes coincides with any sampling point, or by omitting each original value when kriging at its position.

This problem does not arise with block kriging, and investigators may choose to map from block estimates for this reason. Fig. 14.4 shows the block-kriged surface of exchangeable potassium. It lacks the spikes and shows the regional pattern of the variation more clearly. However, it would be better to choose block kriging because such estimates are intrinsically of interest.

A further feature of the nugget variance is that it sets a lower limit to the kriging variance, σ_0^2, at points: this variance cannot be less than the nugget variance. It is contained entirely in the within-block variance, however, and so it does not contribute to the block-kriging variance. As a result block estimates are more precise.

The reliability of kriging depends on how accurately the true variogram is represented by the chosen model. As we saw in Chapter 12, a model is found by fitting a curve or surface to a set of estimated semi-variances. The nugget variance is taken as the intercept of this curve or surface on the ordinate. We accept this interpretation as a conservative or safe one because we do not know anything of the shape of the variogram at lag distances less than the shortest sampling interval. This means that in practice the nugget variance is likely to be overestimated and as a result so is the punctual kriging variance. Therefore the estimates from a survey will themselves be conservative.

More care is needed with block kriging. The estimation variance, equation (14.3), comprises three terms, one of which is the within-block variance, which is estimated by integrating the variogram from $|\mathbf{h}| = 0$ to the limit of the block. If the semi-variance is overestimated at short lags then the within-block variance will also be overestimated, at least for small blocks. Subtracting an overestimated within-block variance in (14.3) will cause the kriging variance to be underestimated. Caution is needed, therefore, when interpreting the kriging variances for blocks whose sides are smaller than the shortest sampling interval of the variogram: the estimates may be less reliable than they seem. Estimates for larger blocks should be reliable since the contribution to the within-block variance from the short lags will be a small proportion of the whole.

The above has important implications for a survey. It means that the shortest sampling interval used to determine the variogram should be less than the lengths of the smallest blocks of land that will be of interest. This should be borne in mind when planning sampling to estimate the variogram, as discussed in Chapter 13.

Designing sampling schemes

We return now to the matter of efficient sampling explored in Chapter 3, but whose explanations we had to leave for want of adequate theory. Here we show how the theory of regionalized variables can be used to design sampling schemes.

Earlier in the chapter, equation (14.4), we saw that the kriging weights for any particular estimate depend on the configuration of the sampling points in relation to the point or block to be estimated and on the variogram. They do not depend on the observed values at those sites. The same applies to the kriging variances, equation (14.5). So, if the variogram is known then the errors of estimates can be determined for any particular configuration *before the sampling is actually performed*. It should be possible, therefore, to design a sampling scheme to meet a specified tolerance or precision.

The examples in Chapter 3 showed that the most precise estimates were obtained from surveys on regular grids. We can now say why. Where a variable is spatially dependent information from an observation pertains to the neighbourhood surrounding it. If its variogram is bounded then this neighbourhood is limited by the range of the variogram. If two observations are made within overlapping neighbourhoods then they duplicate information to some extent. Clustered points mean that the same information is replicated several times while some parts of a region may be left unrepresented. Redundancy can be minimized by placing the sampling points as far away from their neighbours as possible, which, at the same time, also minimizes the area that is unrepresented if the sampling is sparse. This can be achieved by sampling on a regular equilateral triangular grid. For a grid with one node per unit area neighbouring sampling points are 1.0746 units of distance apart, and no other point is more than $d_{max}=0.6204$ away from a sampling point. Any other sampling scheme will have some neighbours that are closer and some other points for which d_{max} is greater.

We should expect triangular sampling configurations to be most efficient, therefore. Indeed, Matérn (1960) and Dalenius *et al.* (1961) showed them to be so for estimating mean values of areas where the variograms are exponential. The same is true with unbounded variograms and in most circumstances with bounded variograms with finite ranges. With the latter, however, a hexagonal grid can be the most efficient in certain very restricted circumstances (Yfantis *et al.* 1987).

Triangular grids are rather inconvenient, and for practical reasons rectangular grids are preferred. The rectangular grid that maximizes the distances between neighbours is square. For unit sampling density the grid interval is 1, and the maximum distance between a sampling point and any other point is now $d_{max}=1/\sqrt{2}=0.7071$. This is more than that for a triangular grid, and will result in somewhat larger estimation errors. For a hexagonal grid $d_{max}=0.8773$ for unit sampling density.

Having established the general principle that regular sampling is the most efficient, we are left only to determine the sampling intensity. The precise procedure for this, and the results, differ somewhat between estimating for a single isolated region and estimating values for isarithmic mapping. Since we have emphasized the latter, we deal with mapping first.

Designs for mapping

We have already seen that when making a map the error of interpolation is not the same everywhere. With punctual kriging there is no error at the sampling points, and in general the further a target point is from data the larger the error is. So if we sample on a regular grid we minimize d_{max}, which is the distance between a target point at the centre of a grid cell and its nearest sampling point at a grid node. Except near the margins of the map, we also minimize the maximum kriging error.

Following Burgess *et al.* (1981) (see also McBratney *et al.* 1981), we can exploit this to plan our sampling. We decide that the error of our map should nowhere exceed some tolerable threshold. Knowing the variogram and assuming sampling to be on a regular grid, we calculate the kriging

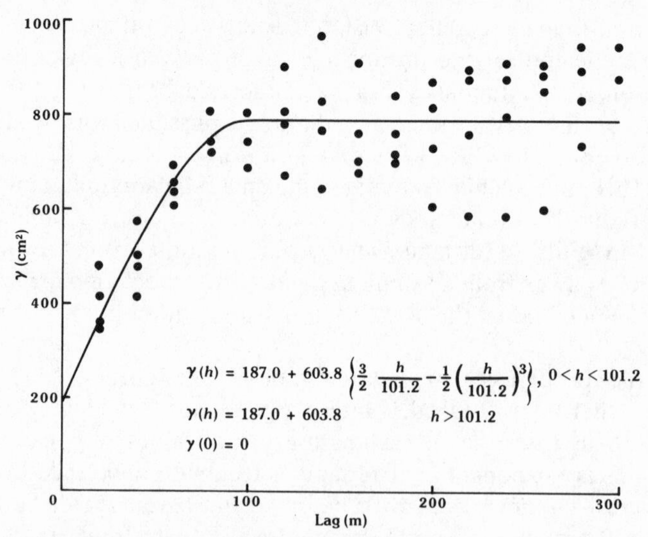

FIG. 14.7 Variogram of the thickness of cover loam at Hole Farm, Norfolk. (From Burgess and Webster 1980)

variances or their square roots at the centres of the grid cells for a range of sampling intervals from equations (14.6) and (14.8). We plot the values against the grid spacing and join them by a line. The optimal spacing corresponds to the specified tolerable error and can be read from the graph.

To illustrate the procedure we use the example from Burgess *et al.* (1981), in which the aim was to map the thickness of cover loam at Hole Farm in Norfolk. Fig. 14.7 shows the variogram obtained by sampling at 20 m intervals, and Fig. 14.8 is the graph derived from it. If we wished to predict the thickness of cover loam at points and were prepared to accept 20 cm as the maximum tolerable error then from this graph we see that we should have to sample at intervals of approximately 35 m. Notice that the difference between the results from the square and triangular grids is not nearly as large as the 12 per cent difference in d_{max} for the two grids. In fact, it is negligible at this spacing. This is partly because of the effect of the nugget variance and partly because a point at the centre of a square grid cell has four neighbouring sampling points whereas one at the centre of a triangle has only three.

The same reasoning and procedure apply to block kriging, though the strategy is not quite so straightforward because of the effect of the block size. For blocks that are very small in relation to the sampling interval the kriging variance will be largest in the centres of the grid cells. As the block is increased in size the estimation variance decreases. Consider, however, a block centred on a grid node. If it is no bigger than the support of the

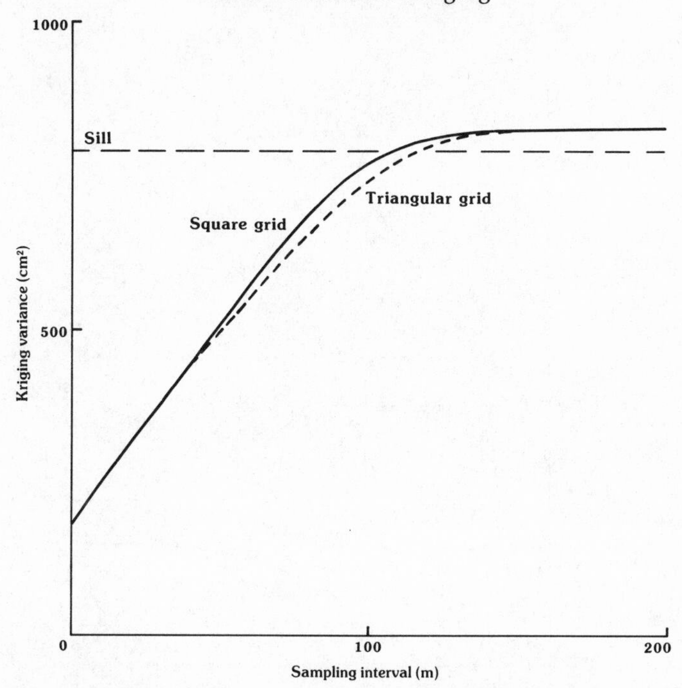

FIG. 14.8 Graph of punctual kriging error against sampling interval to map the thickness of cover loam at Hole Farm. (From Burgess *et al.* 1981)

sample then its estimation variance is zero and we have punctual kriging. As the block increases, its estimation variance initially increases because the dominant effect of the observation at its centre declines. Only when it is big enough for the nearest neighbours to be more influential does the estimation variance start to decline. This difference of configuration has another important effect. As the block increases in size, so the weights of the sampling points nearest its centre decrease while the weight of those further away increase. A block size is eventually reached at which its estimation variance equals that for a block centred in a grid cell. If the block is enlarged further then its estimation variance can be greater than that of a block of the same size centred in a grid cell. Burgess *et al.* (1981) illustrate these effects.

To find the most economical sampling scheme for making an isarithmic map by block kriging graphs must be drawn both for blocks centred on grid grid cells and blocks centred on grid nodes. A horizontal line is drawn through the maximum tolerable error on the ordinate to cut both graphs. Dropping perpendiculars from the points of intersection gives the spacings for the two configurations, and the shorter of the two must be chosen for mapping. Fig. 14.9 shows examples of graphs on which this can be done. In

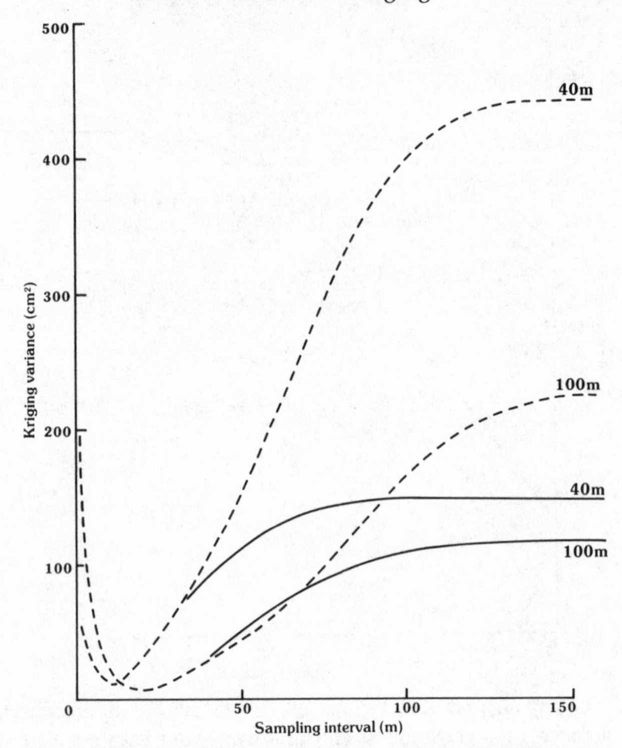

Fig. 14.9 Graphs of kriging error of the thickness of cover loam against sampling interval to map from blocks of 40 m × 40 m and 100 m × 100 m at Hole Farm (— centred on grid nodes, --- centred in grid cells)

it the block-kriging errors are plotted against sampling interval for Hole Farm. For blocks of 40 m × 40 m the configuration giving the largest error is always cell-centred. For 100 m × 100 m blocks, however, blocks centred on the nodes require the smaller spacing if the maximum tolerable error lies between about 5.5 and 8.5 cm.

Fig. 14.8 illustrates two subsidiary features of sampling and estimation. The first is that it is impossible to design a scheme for punctual kriging that will diminish σ^2_{max} to less than the nugget variance. The second feature is that σ^2_{max} increases to a maximum at which it flattens. This maximum is not the sill value of the variogram, as might at first be expected, but a somewhat larger value. It is in fact the sill value plus the Lagrange multiplier, ψ of equation (14.4). Once d_{max} exceeds the range of the variogram—in this case, once the grid interval exceeds $\sqrt{2}$ times the range—all the weights are equal. If there are n observations then the weights are $1/n$. All the semi-variances in the kriging equations equal the sill value, and the additional quantity, ψ, is σ^2/n, representing the additional uncertainty of predicting the value at a place from only local data rather than the whole population.

These essentially simple procedures enable a surveyor to plan sampling for mapping most of a region. Near the margins of the region, however, some modifications may be needed if sampling cannot be extended outside it. In these circumstances the variance tends to increase at the margin (see for example Fig. 14.5), and so sampling would have to be increased near the margin to keep within the tolerance.

Dealing with anisotropy

When the soil varies anisotropically we should feel intuitively that an equilateral grid would not be best. We should want to make some adjustment to the grid spacings whereby we sample more intensively in the direction of maximum change than in other directions. The problem is how to choose these spacings so that we keep within the specified tolerable error for least effort. The optimum solution depends on the precise form of the anisotropy. The one we illustrate is that for strict geometric anisotropy (Burgess *et al*. 1981) and is likely to serve well for surveys of small areas.

Consider the linear variogram

$$\gamma(h, \theta) = \Omega|\mathbf{h}|, \tag{14.11}$$

in which Ω is the sinusoidal function

$$\Omega = \sqrt{\{A^2(\cos^2(\theta - \phi) + B^2 \sin^2(\theta - \phi)\}}. \tag{14.12}$$

In this equation ϕ is the direction of maximum variation, A is the gradient of the variogram in that direction, and B is the gradient in the perpendicular direction, $\phi + \pi/2$. When $\theta = \phi$, (14.11) reduces to

$$\gamma_1(h) = Ah,$$

and when $\theta = \phi + \pi/2$ it becomes

$$\gamma_2(h) = Bh,$$

We can define an anisotropy ratio q that is the ratio of the maximum to minimum gradients:

$$q = A/B = \gamma_1(h)/\gamma_2(h), \tag{14.13}$$

The semi-variance in direction ϕ at any lag h is thus equal to the semi-variance at lag qh in direction $\phi + \pi/2$:

$$\gamma_1(h) = \gamma_2(qh). \tag{14.14}$$

For the common bounded variograms the function Ω replaces the distance parameter. For example, the geometrically anisotropic exponential model is

$$\gamma(h, \theta) = c\{1 - \exp(-h/\Omega)\},$$

and A and B are the distances. The anisotropy ratio is still A/B.

Using (14.4), the most economical sampling scheme is found as follows. The problem is treated as though variation were isotropic with the variogram $\gamma_1(h)$. The sampling interval d is found in exactly the same way as for the square grid. This becomes the sampling interval in direction ϕ. The anisotropy is then taken into account by making the sampling interval in the perpendicular direction, $\phi + \pi/2$, equal to qd.

We illustrate the procedure using the results of the survey of stone content in the soil at Plas Gogerddan (Burgess *et al.* 1981). The variogram is

$$\gamma(h, \theta) = 9.06 + \surd\{0.472^2 \cos^2(\theta - 2.57) + 0.158^2 \sin^2(\theta - 2.57)\}h. \tag{14.15}$$

The variances are in units of (per cent)^2 and the gradients in variance per metre. The direction of maximum variation is 2.57 radians. The graph of maximum kriging error against sample spacing is drawn to find the spacing d in this direction from the variogram

$$\gamma(h) = 9.06 + 0.472h.$$

This is the upper graph in Fig. 14.10. The anisotropy ratio is $0.472/0.158 = 2.99$, so the sample spacing in direction $2.57 + \pi/2$ is $2.99d$. Fig. 14.10 shows the line for this direction too. The ratio of close to 3 in the sampling intervals for any given variance is clear.

This analysis places the familiar logistics of grid surveys on a sound quantitative footing. Observations are made by sampling along parallel transects in the direction of maximum variance. The sampling interval on the transects is d; and the spacing between the transects is qd. Not only is the size of the sample minimized in this way, but so also is the time and cost of travelling. Note also that, if the total sample and therefore sampling intensity is fixed by the money available for this part of a survey, the optimal strategy is to align the sampling grid as above. The sampling interval, t, along transects in the direction of maximum variation is chosen such that the product $t \times qt$ is the reciprocal of the affordable sampling intensity.

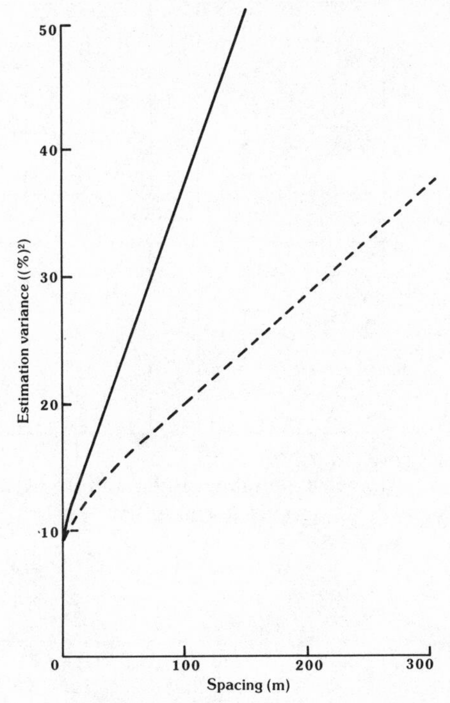

FIG. 14.10 Graphs of the kriging variance of stone content of the soil against sampling interval in the directions of maximum variation (solid line) and minimum variation (dashed line) at Plas Gogerddan, Wales

Designs for isolated regions

Where we wish to estimate the average value in a small isolated region, such as a single field, the situation is somewhat different because we usually do not, or possibly cannot, sample outside the region. We should expect the same general principles to apply, but it is not immediately clear how this constraint on sampling within the region affects the estimation variance.

To illustrate the application of this principle we consider the simple case of isotropic variation in a square field. This is most easily sampled on a square grid. The question is: how should the sampling points be disposed? Fig. 14.11 shows some possible symmetric configurations. In part a nine points are clustered near the centre: in the other parts the points are increasingly dispersed until in Fig. 14.11d eight of the nine points are on the margins of the field.

We can calculate the kriging variances and their square roots for these configurations using (14.4) and (14.5) and plot them against the sample spacing. When we do so we obtain results such as those in Figs. 14.12 and

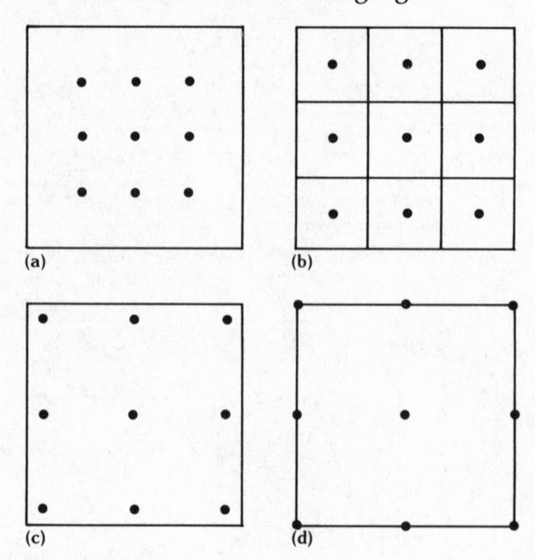

Fig. 14.11 Possible symmetric sampling configurations of nine points on a square grid in a square field

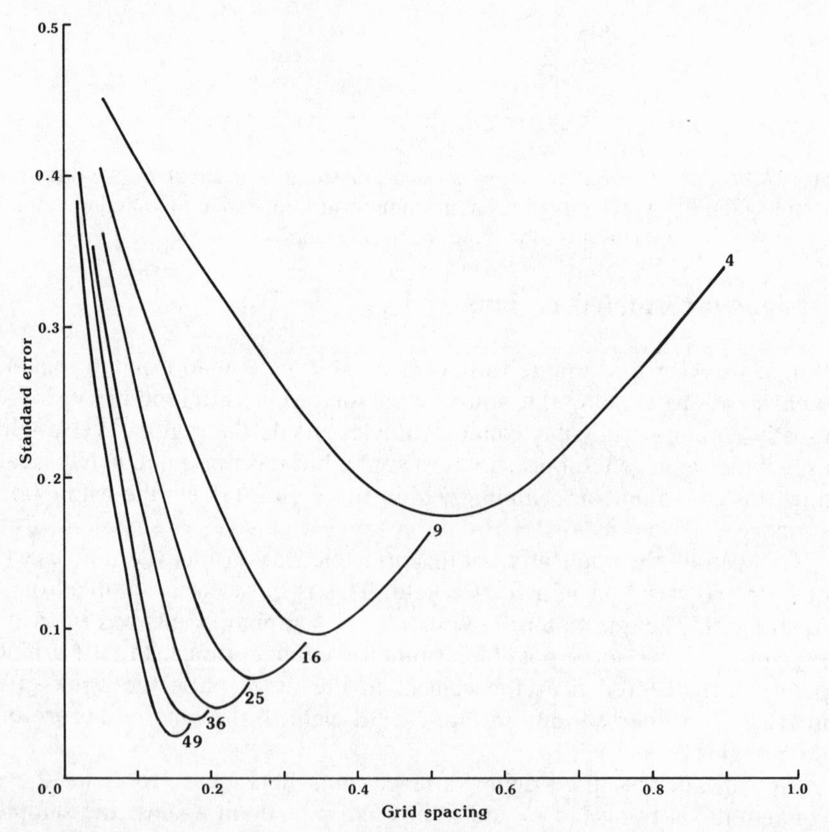

Fig. 14.12 Kriging error for estimating the mean value in a field with samples of different sizes and varying spacing. The variogram is $\gamma(h) = h$ and the field has side 1 unit of distance

14.13. The first of these is for a linear variogram $\gamma(h)=wh$ in a field of side = 1 unit of distance. As the spacing between sampling points is increased from a very small value, the kriging error decreases to a minimum and then increases again. The minimum occurs when the spacing is 1/3. Thus the optimal configuration is the one in which the field is divided into nine smaller squares (tiles) with the sampling points at their centres, Fig. 14.11b. We can do the same for other sizes of sample on a square grid, and we find that the optimum spacing is $1/\sqrt{n}$ where n is the size of the sample. Taking the minimum estimation variances calculated in this way and plotting them against n, a graph, such as the lower one in Fig. 14.13, is obtained. The graph is discontinuous, and the curve connecting the points is drawn to guide the eye only. To determine how large the sample should be to achieve a given precision we can read the sample size for the particular error from the graph.

The upper curve in Fig. 14.13 represents the standard error computed using the classical formula and assuming no spatial dependence. If the sample size were determined from such a graph then the sampling would be very inefficient; it could lead to oversampling by two- or three-fold.

The same procedure can be followed for other forms of variogram. Fig.

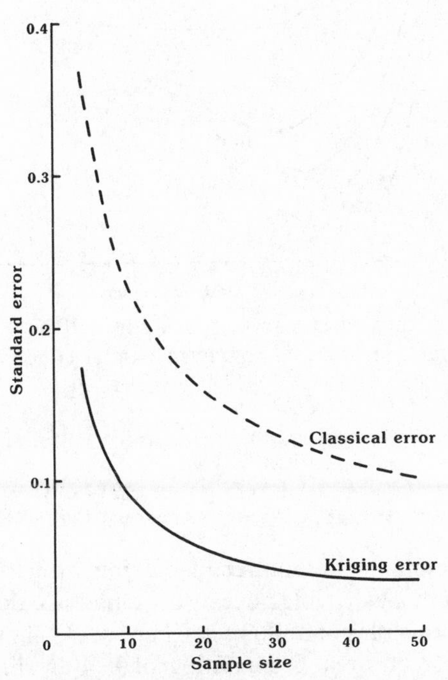

Fig. 14.13 Kriging error against size of sample for optimal configuration and variogram $\gamma(h)=h$, lower graph. The upper curve links the standard errors that would have been calculated assuming no spatial dependence

14.14 shows the kriging error for different sizes of sample and spacings for estimating the average thickness of cover loam in a square 10 ha field. The variogram is spherical and is shown in Fig. 14.7. For four sampling points we find that the optimal configuration is as we stated above. For nine, however, it is not quite so: the optimal spacing is slightly larger than one-third of the side of the field, and for larger samples the optimal spacing is also larger than $\sqrt{(100\,000/n)}$ metres. In general, where the variogram is not linear the optimal spacing will be approximately equal to $1/n$ times the side of the square whose mean is to be estimated, and given that the variogram itself is never known exactly, a square grid will almost certainly be near enough optimal. The lower graph in Fig. 14.15 is the minimum kriging error against sample size in this situation.

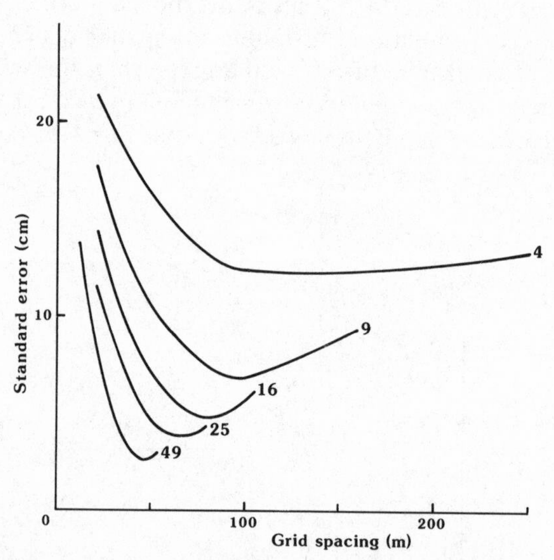

FIG. 14.14 Kriging error against sample spacing in a 10 ha square field for the spherical variogram, for thickness of cover loam at Hole Farm, Norfolk

Bulking

At this point we return to the matter of bulking, which we had to leave unfinished in the previous chapter because we had not dealt with kriging. We described the principles of bulking in Chapter 13. There we considered a bulked sample for an area B consisting of n individual cores and we assumed that the variance within the area was wholly nugget; i.e., we took

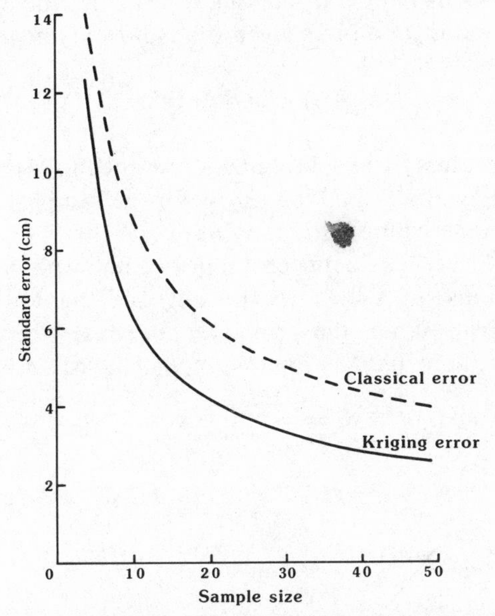

FIG. 14.15 Graph of standard error against sample size for the average thickness of cover loam at Hole Farm

no account of any spatial correlation and took the estimation variance of the mean in B, $z(B)$, as

$$\sigma^2(B) = \sigma^2/n, \tag{14.16}$$

where σ^2 is the variance of the cores. This is the same as if we had measured each core separately and computed a mean from individual measurements. In these circumstances n can be chosen large enough to diminish $\sigma^2(B)$ to the desired threshold. The positions from which the cores are taken are immaterial.

Where there is spatial correlation, however, the true estimation variance of $\hat{z}(B)$ is given by (14.3). This is the same equation as that for kriging except that in kriging the weights, λ_i, found by solving (14.4) are in general different from one another. They are equal only in special circumstances.

With an isotropic linear variogram the kriging weights are equal when a square region is divided into smaller equal squares and the sampling points occupy the centres of the smaller squares, and for $n>4$ these are the only circumstances. Bulking is usually done by mixing cores containing equal amounts of soil. This effectively gives equal weight to all cores regardless of their positions in the field. So, unless the sampling points are arranged on a grid as above, the estimation variance will be larger than it need be.

Fig. 14.16 shows this effect of bulking to estimate the average percentage of clay in the soil in 1 ha plots given the isotropic linear variogram

$$\gamma(h) = 34.76 + 0.839h, \tag{14.17}$$

where h is in metres. The solid lines represent the kriging errors and the dashed lines show the estimation errors for the samples bulked from cores taken on the same configuration. When $n=4$ the lines coincide, whereas for $n>4$ the differences between them become apparent only when the extreme positions are closer to the edge of the square than half the sampling interval. When the grid interval equals the side of the block divided by $\sqrt{n}$, here $100/\sqrt{n}$ metres, $\sigma^2(B)$ is least and equals the kriging

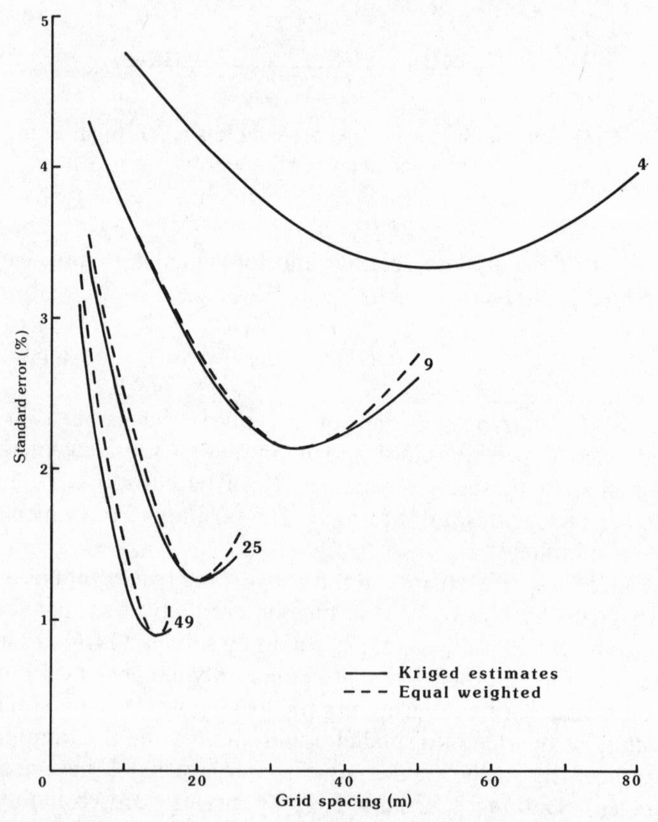

FIG. 14.16 Kriging error against sample spacing for estimating the average clay content of the soil in a 1 ha square given the variogram of equation (14.17). The dashed lines show the bulking errors for the same configurations

variance. This then defines the optimal bulking strategy. The difference between the bulking variance and the kriging variance is more marked with a nonlinear variogram. Fig. 14.17 compares the two for the spherical variogram that describes the variation in clay content over 10 ha. Its coefficients are 30.8(%)^2, 170.8(%)^2, and 274 m for the nugget variance, spatially dependent variance, and range respectively. Except for $n = 4$ the variance of the bulked sample is perceptibly larger than the kriging variance, and for $n = 49$ it is so even with the optimum spacing. We could reduce the bulking variance to the kriging variance by mixing different weights of soil from the different cores. This is hardly practicable and is scarcely worth it. If only a little more precision is desired then a few additional cores will ensure that the bulking variance is no larger than the calculated kriging variance at the best grid spacing.

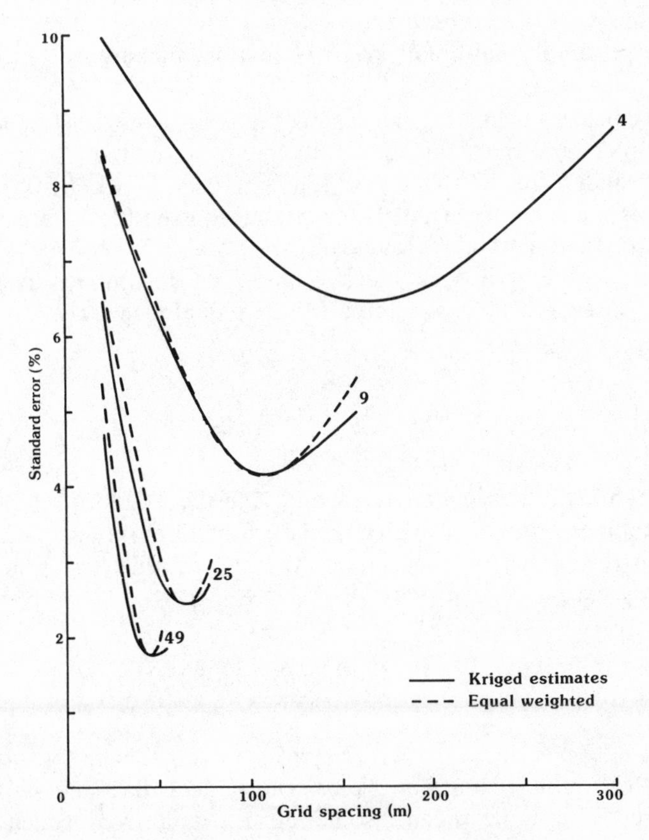

FIG. 14.17 Kriging error against sample spacing for estimating the average clay content in the soil of a 10 ha field with the spherical variogram. The dashed lines show the bulking errors for the same configuration

Global estimation

We have already seen in Chapter 3 that we can estimate the average value of a property in a large region most precisely by sampling on a regular grid. In the example there we obtained the estimation variances by repeated sampling. This is clearly inefficient and impracticable in most routine field surveys. Extending the ideas of the last section to larger regions and using kriging to estimate the error might seem to solve the problem. Unfortunately there are reasons why the approach cannot be pursued.

1. It is unwise to assume that a property that is locally stationary in the mean and semi-variances maintains that stationarity throughout a large region.
2. The variogram is usually known accurately only over short distances, and almost certainly will not be well estimated for lags approaching the distance across a large region.
3. A large sample may create kriging equations that are too large to solve.

Nevertheless, estimating the regional mean is simple. The most convenient procedure is to divide the region into small rectangular blocks (strata) and estimate the mean value in each block by kriging. The estimates are then averaged. If for any reason the blocks are not all the same size then a weighted average can be taken with each weight proportional to the area of its block. Thus, for a region, R, divided into n blocks, B_i, $i=1, 2, \ldots, n$, of area H_i, the global mean, $z(R)$, is estimated by

$$\hat{z}(R) = \sum_{i=1}^{n} H_i \hat{z}(B_i) \Big/ \sum_{i=1}^{n} H_i, \tag{14.18}$$

where $\hat{z}(B_i)$ is the estimated average of Z in the ith block.

A problem arises in calculating the estimation variance. Clearly, the error in the global average equals the sum of the errors in the local estimates; i.e.,

$$\hat{z}(R) - z(R) = \sum_{i=1}^{n} H_i\{\hat{z}(B_i) - z(B_i)\} \Big/ \sum_{i=1}^{n} H_i. \tag{14.19}$$

The estimation variance, $\sigma^2(R) = \mathrm{E}[\{\hat{z}(R) - z(R)\}^2]$, is not estimated without bias by a simple sum, however, because the estimates in neighbouring blocks are not independent; some of the data from which they are computed are common. The solution to the problem is to consider the error that results from using the value at a sampling point to estimate the average value over the portion of the region that is nearer to it than to any other; i.e. for its Thiessen polygon or Dirichlet tile. For a rectangular grid each Thiessen polygon is a rectangle with an observation at its centre, $\mathbf{x}_c$, and sides equal to the sampling intervals along the principal axes of the grid. The variance of estimating its average is

$$\sigma^2(B) = 2\bar{\gamma}(\mathbf{x}_c, B) - \bar{\gamma}(B, B), \tag{14.20}$$

where $\bar{\gamma}(\mathbf{x}_c, B)$ is the average semi-variance between the centre and all other points in the rectangle, and $\bar{\gamma}(B, B)$ is the variance within the polygon. So if the estimated values for these rectangles are $\hat{z}(B_i)$, $i = 1, 2, \ldots, n$, then the average for the region is approximately

$$\hat{z}_B(R) = \frac{1}{n}\sum_{i=1}^{n} \hat{z}(B_i). \tag{14.21}$$

The error of this estimate is approximately $z(R) - \hat{z}_B(R)$, and the corresponding variance of the global mean is

$$\begin{aligned} \mathrm{E}[\{z(R) - \hat{z}_B(R)\}^2] &\simeq \frac{1}{n^2}\mathrm{E}[\{z(B_i) - z(\mathbf{x}_i)\}^2] \\ &= \frac{1}{n^2}\sigma^2(B). \end{aligned} \tag{14.22}$$

The approximation improves with increasing n.

As with stratified random sampling, the error in the global estimate depends on the variance within the rectangles, which are effectively strata. This variance is likely to be smaller, in many instances much smaller, than the variance over the whole region. Its advantage over stratified random sampling is that it takes the maximum advantage either of efficiency or precision from using a regular grid.

To illustrate the advantages of this approach Table 14.1 presents the results for two of the surveys that we have used earlier, and have been described by McBratney and Webster (1983a). At Hole Farm the kriged mean thickness of the cover loam, 66.2 cm, is very close to the simple arithmetic mean of the measurements, 66.3 cm. However, the variance of the kriged mean computed as above at 0.5084 is less than one-third of the estimation variance that would have been computed taking no account of the spatial dependence. Similar comparisons can be made for the percentage of stones in the soil at Plas Gogerddan. The two means are very

TABLE 14.1 Estimation variances of the average thickness of cover loam at Hole Farm and stone content at Plas Gogerddan

	Thickness of cover loam (cm)	Stone content of topsoil (%)
Sample size	452	434
Sampling interval on square grid	20 m	15.2 m
Sample mean	66.3	26.5
Sample variance	786.7	78.28
Classical estimation variance	1.7404	0.1827
Standard error	1.32	0.427
Kriged mean	66.2	26.4
Kriged variance	0.5084	0.02767
Kriged standard error	0.713	0.166

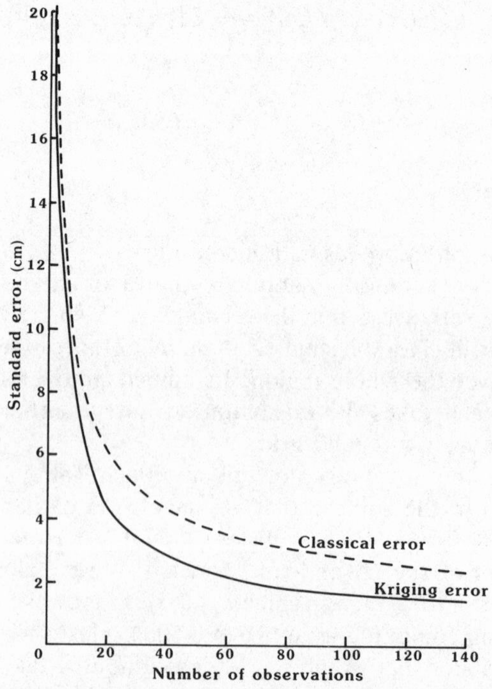

FIG. 14.18 Standard error against sample size for thickness of cover loam at Hole Farm

similar. The ratio of the variances is even wider than at Hole Farm. The kriged variance is less than one-sixth of the classical estimation variance.

Equations (14.20) and (14.22) can be used to determine the precision that would be achieved for any other grid spacing or sampling effort. Alternatively, they can be used to determine the sampling effort needed to estimate the global mean within a particular tolerance. As for small regions, the values of the semi-variance are inserted into the equations which are then solved for a range of grid spacings. The estimation variances, or their square roots the estimation errors, are plotted against the sample size n, and the value of n for a particular error read from the graph. Fig. 14.18 shows such a graph of error against sample size for the cover loam at Hole Farm. The variation was isotropic, and the results derive from estimation on square grids. For stone content at Plas Gogerddan, where variation was anisotropic, the results are for grids elongated 2.99 times in the direction of least variation (Fig. 14.19). In Figs. 14.18 and 14.19 we have drawn on the same axes the standard errors that we should have calculated from classical theory disregarding or not knowing of any spatial dependence. The figures show that, for all but very sparse sampling, kriged estimates are much more precise than those that

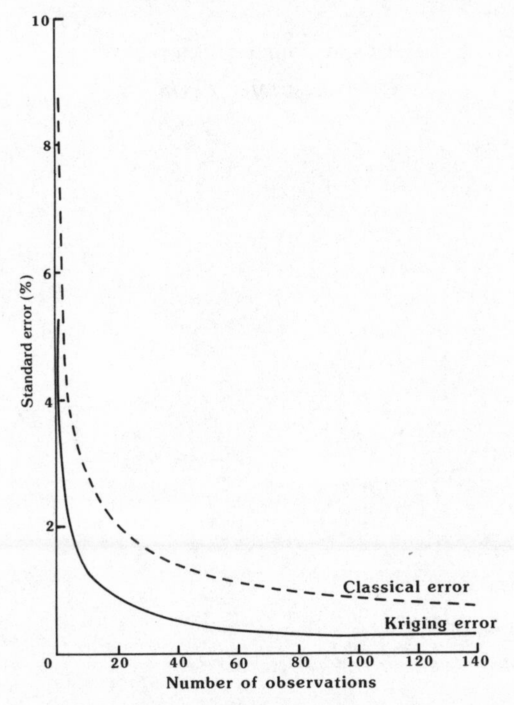

FIG. 14.19 Graph of standard error against sample size for the average stone content at Plas Gogerddan, Wales

classical theory would predict. The difference is largest at Plas Gogerddan, mainly because the nugget variance is very small in relation to the spatially dependent variation. At Hole Farm the nugget variance is about a quarter of the total. Note also that at Hole Farm the variogram is spherical with a range of only 100 m. With small samples the spacing between sampling points can exceed this so that the data are effectively independent. The result is that with small samples the two curves approach one another.

The Future

This book's predecessor, *Quantitative Methods* (Webster 1977), closed by encouraging soil scientists to apply regionalized variable theory to survey. It predicted a rich future, and it has not disappointed. There is undoubtedly much still to discover. The theoreticians continue to develop the subject to handle an ever wider range of situations. Environmental scientists and resource surveyors in all fields will do well to keep abreast of these developments and to apply them where they are advantageous.

And so in our research we shall be limited only by our lack of imagination.

What's to come is still unsure.
W. Shakespeare, *Twelfth Night*

APPENDIX: MATRIX METHODS AND NOTATION

She died because she never knew
These simple little rules and few.
H. Belloc, *The Python*

Matrix algebra provides a convenient and succinct notation for expressing simultaneously large numbers of relationships and the operations that can be performed on them. The subject is reasonably easy, and once the conventions are learned many of the ideas of statistics and specifications for analysis can be presented in matrix form. The purpose of this appendix is to describe just sufficient of matrix algebra to make the material in the main text intelligible.

A matrix is a rectangular array of numbers, or symbols to represent numbers, such as

$$\begin{bmatrix} 2.1 & 5.0 & -6.1 \\ 3.7 & 7.9 & 3.5 \end{bmatrix} \quad \text{and} \quad \begin{bmatrix} 3 & 6 & 8 \\ -9 & 5 & 4 \\ 9 & 6 & 5 \end{bmatrix}, \tag{A1}$$

and in general

$$\begin{bmatrix} a_{11} & a_{12} & a_{13} & \cdots & a_{1n} \\ a_{21} & a_{22} & a_{23} & \cdots & a_{2n} \\ \cdot & \cdot & \cdot & \cdots & \cdot \\ \cdot & \cdot & \cdot & \cdots & \cdot \\ \cdot & \cdot & \cdot & \cdots & \cdot \\ a_{m1} & a_{m2} & a_{m3} & \cdots & a_{mn} \end{bmatrix}. \tag{A2}$$

It is usual to enclose the array in square brackets, though sometimes parentheses, ( ), or double bars, ‖ ‖, are used. A matrix can have any number of rows and columns, and a matrix with m rows and n columns is said to be of order $m \times n$. However, if a matrix has only one row then it is known as a *row* vector; if it has only one column it is known as a *column* vector. If both m and n are 1 then we have a single number, which in the language of matrix algebra is a *scalar*.

The numbers in a matrix are called *elements*. They can be referred to individually by two subscripts giving their row and column, in that order. Thus a_{23} in the matrix above is the element in the second row and third column; a_{ij} is the element in the ith row and jth column.

The whole array of numbers contained in a matrix can be conveniently referred to using a single symbol, for example

$$\mathbf{A}=\begin{bmatrix} 2.1 & 5.0 & -6.1 \\ 3.7 & 7.9 & 3.5 \end{bmatrix}. \tag{A3}$$

It is usual to print symbols for matrices in bold face characters, using capitals for matrices proper and lower case letters for vectors.

Some types of matrix

A matrix that has the same number of rows as columns is called a *square matrix*. If there are m rows and columns then it is of order m. Square matrices are especially important in the analysis of multivariate data. In a square matrix, say $\mathbf{A}$, the elements a_{11}, a_{22}, a_{33}, . . . , a_{mm} are called the *diagonal elements*; their sum is known as the trace of $\mathbf{A}$. A square matrix in which the elements $a_{ij}=a_{ji}$ for all i and j is said to be *symmetric*, as for example

$$\begin{bmatrix} 1 & 2 & 6 \\ 2 & 4 & 3 \\ 6 & 3 & 8 \end{bmatrix}. \tag{A4}$$

It is symmetric about the diagonal.

Two special kinds of symmetric matrix are: a *diagonal matrix*, in which the diagonal elements are non-zero and all off-diagonal are zero, for example

$$\begin{bmatrix} a_{11} & 0 & 0 \\ 0 & a_{22} & 0 \\ 0 & 0 & a_{33} \end{bmatrix}, \tag{A5}$$

and an *identity matrix* or *unit matrix*, which is a diagonal matrix with all diagonal elements equal to 1 and usually denoted by **I**; thus

$$\mathbf{I} = \begin{bmatrix} 1 & 0 & 0 \\ 0 & 1 & 0 \\ 0 & 0 & 1 \end{bmatrix}. \tag{A6}$$

Elementary matrix operations

Transposition

If we have an $m \times n$ matrix and interchange its rows and columns we obtain a new matrix of order $n \times m$ called its *transpose*. For example, if **A** is the matrix of (A3) above its transpose, denoted by $\mathbf{A}^T$, is

$$\mathbf{A}^T = \begin{bmatrix} 2.1 & 3.7 \\ 5.0 & 7.9 \\ -6.1 & 3.5 \end{bmatrix}. \tag{A7}$$

Thus each element a_{ij} occupies the position in the jth row and the ith column in the transpose.

If a matrix **A** is symmetric then $\mathbf{A} = \mathbf{A}^T$.

Addition

Two matrices can be summed as in ordinary algebra provided they are of the same size or order, i.e. if they have the same numbers of rows and columns. The result is a matrix whose elements are the sums of the corresponding elements in the two original matrices. Thus

$$\mathbf{C} = \mathbf{A} + \mathbf{B} \tag{A8}$$

means that $c_{ij} = a_{ij} + b_{ij}$ for all values of i and j. Similarly, the matrix **B** might be subtracted from matrix **A**, in which case $c_{ij} = a_{ij} - b_{ij}$. Matrices of the same order are said to be *conformable* for addition and subtraction.

Multiplication

Matrices can be multiplied under certain conditions. For example, if we have two matrices

$$\mathbf{A}=\begin{bmatrix} a_{11} & a_{12} \\ a_{21} & a_{22} \\ a_{31} & a_{32} \end{bmatrix} \text{ and } \mathbf{B}=\begin{bmatrix} b_{11} & b_{12} \\ b_{21} & b_{22} \end{bmatrix}, \tag{A9}$$

then their product **AB** in that order is the 3×2 matrix, say **C**:

$$\mathbf{C}=\begin{bmatrix} a_{11}b_{11}+a_{12}b_{21} & a_{11}b_{12}+a_{12}b_{22} \\ a_{21}b_{11}+a_{22}b_{21} & a_{21}b_{12}+a_{22}b_{22} \\ a_{31}b_{11}+a_{32}b_{21} & a_{31}b_{12}+a_{32}b_{22} \end{bmatrix}. \tag{A10}$$

Thus each element of **C** is the sum of the products of the elements in a row of **A** and the elements in a column of **B**. The operation is *row by column*. In general, if **A** is of order $m \times n$ we may multiply it by an $n \times \mathrm{p}$ matrix **B** to obtain a $m \times p$ matrix **C**. The elements of **C** are then

$$\begin{aligned} c_{ij} &= a_{i1}b_{1j} + a_{i2}b_{2j} + \ldots + a_{in}b_{nj} \\ &= \sum_{k=1}^{n} a_{ik}b_{kj} \end{aligned} \tag{A11}$$

for $i=1, 2, \ldots, m$ and $j=1, 2, \ldots, p$.

Note that for multiplication to be possible there must be the same number of columns in **A** as there are rows in **B**. **A** is then *conformable* to **B** for multiplication. When **A** is conformable to **B** for multiplication **B** is not necessarily conformable to **A**. Even when the two are conformable the product **AB** is generally not the same as the product **BA**.

A matrix can be multiplied by a scalar, which is equivalent to multiplying every element of the matrix by that quantity.

The calculation of variances and covariances is one of the most frequent operations in statistical analysis and is economically represented in matrix form. Suppose we have a set of observations $x_1, x_2, \ldots, x_n$ of some variable variable and we enter them as deviations from their mean in a column vector **x** of order $n \times 1$. We obtain the sum of squares of deviations from the mean by *pre-multiplying* the vector by its transpose:

$$[x_1, x_2, \ldots, x_n] \begin{bmatrix} x_1 \\ x_2 \\ \cdot \\ \cdot \\ \cdot \\ x_n \end{bmatrix} = \sum_{i=1}^{n} x_i^2. \tag{A12}$$

This can be represented therefore as $\mathbf{x}^T\mathbf{x}$. When there is more than one variate we also require sums of products. By convention we hold the data for n individuals on which we have measured p variables in an $n \times p$ matrix. The data are *centred* by subtracting their column means to give a matrix, say $\mathbf{X}$, of deviations from variate means. We *pre-multiply* $\mathbf{X}$ by its transpose to obtain $\mathbf{X}^T\mathbf{X}$, in which the diagonal elements contain the sums of squares and the off-diagonal elements the sums of products. Multiplying by the scalar $1/(n-1)$ then gives the variances and covariances that we require. Both $\mathbf{X}^T\mathbf{X}$, which is the sums of squares and products (SSP) matrix, and the variance–covariance matrix are symmetric.

Inversion

One matrix cannot be divided by another in the ordinary sense of algebra. However, we can make use of the idea that division by a quantity, say, z, is equivalent to multiplication by its reciprocal, $1/z$, and that $z(1/z)=1$. In matrix algebra when we wish to divide by a matrix, say $\mathbf{A}$, which must be square, we first find the matrix $\mathbf{B}$ such that $\mathbf{AB}=\mathbf{I}$, and incidentally $\mathbf{BA}=\mathbf{I}$. Matrix $\mathbf{B}$ is called the *inverse* of $\mathbf{A}$ and is usually written $\mathbf{A}^{-1}$. We may then proceed to multiply by $\mathbf{A}^{-1}$.

Not all square matrices can be inverted, for if any of the rows (or columns) of a matrix are linearly dependent on others there is no unique inverse. The matrix is then said to be *singular*. Sometimes two or more rows or columns of a matrix are almost linearly related. If so the matrix is very nearly singular and is said to be *ill conditioned*. Its inverse is likely to depend heavily on chance variation in the original data and on rounding errors in computing.

Although the inverse of a matrix can easily be defined, its calculation is very tedious for order more than about 5. Inversion of larger matrices is almost always done by computer nowadays. The usual method of inversion for hand calculating is known as *pivotal condensation*, and is described in standard texts. It is generally not used, at least without modification, in computer programs because rounding errors can become serious.

Determinants

Any square matrix $\mathbf{A}$ has associated with it a scalar quantity known as its determinant, denoted by $|\mathbf{A}|$ or sometimes det $\mathbf{A}$. If the order of the matrix is m its determinant is derived from the $m!$ products that can be formed by choosing one and only one element from each column of the matrix. Consider the columns of **A**; there are $m!$ ways of arranging them, and for each arrangement there is a unique set of diagonal elements which gives one of the products required. Each arrangement of the columns can be obtained from the original matrix by interchanging two columns, followed by another two, and so on until the desired arrangement is achieved. The number of interchanges needed to produce the arrangement determines the sign of the product. If the number is even the product has a positive sign; if the number is odd the sign is negative. The determinant is then the sum of these signed products. Thus, if **A** is of order 3 then it is easily seen that its determinant $|\mathbf{A}|$ is

$$\begin{vmatrix} a_{11} & a_{12} & a_{13} \\ a_{21} & a_{22} & a_{23} \\ a_{31} & a_{32} & a_{33} \end{vmatrix} = \begin{matrix} +a_{11}a_{22}a_{33} - a_{12}a_{21}a_{33} - a_{11}a_{23}a_{32} \\ +a_{12}a_{23}a_{31} + a_{13}a_{21}a_{32} - a_{13}a_{22}a_{31} \end{matrix}. \tag{A13}$$

It is worth noting that if **A** is singular its determinant is zero, but not otherwise.

Quadratic forms

Suppose we have a column vector **x** and a square matrix **A** of the same order. The scalar product

$$\mathbf{x}^T\mathbf{A}\mathbf{x} \tag{A14}$$

is then known as a *quadratic form* since it is a quadratic function of the xs. If, for example, **A** is of order 3 then on expansion

$$\mathbf{x}^T\mathbf{A}\mathbf{x} = a_{11}x_1^2 + a_{22}x_2^2 + a_{33}x_3^2 + (a_{12}+a_{21})x_1x_2 \\ + (a_{13}+a_{31})x_1x_3 + (a_{23}+a_{32})x_2x_3.$$

In general, if **x** and **A** are of order n then

$$\mathbf{x}^T\mathbf{A}\mathbf{x} = \sum_{i=1}^{n} a_{ii}x_i^2 + \sum_{j=2}^{n}\sum_{i=1}^{j-1}(a_{ij}+a_{ji})x_ix_j. \tag{A15}$$

Note that if **A** is symmetric then $a_{ij} = a_{ji}$ for all values of i and j, and expression (A14) becomes

$$\mathbf{x}^T\mathbf{A}\mathbf{x} = \sum_{i=1}^{n} a_{ii}x_i^2 + 2\sum_{j=2}^{n}\sum_{i=1}^{j-1} a_{ij}x_ix_j. \tag{A16}$$

Matrix **A** is now unique; it is the only symmetric matrix for which the quadratic form can be expressed as $\mathbf{x}^T\mathbf{A}\mathbf{x}$.

A quadratic form that is positive for all values of **x** other than $\mathbf{x}=\mathbf{0}$, i.e. when all the elements of **x** are real numbers and not all zero, is known as a *positive definite* quadratic form. The associated matrix is similarly known as a *positive definite matrix*. If $\mathbf{x}^T\mathbf{A}\mathbf{x} \geqslant 0$ the quadratic form and its matrix are called *positive semi-definite*, often abbreviated to p.s.d.

Latent roots and vectors

The derivation of latent roots and vectors involves the following question: given a square matrix **A**, do there exist a vector **c** and a scalar λ that satisfy the equation

$$\mathbf{A}\mathbf{c} = \lambda\mathbf{c}? \tag{A17}$$

If so then (A17) is equivalent to

$$\mathbf{A}\mathbf{c} - \lambda\mathbf{c} = \mathbf{0}, \tag{A18}$$

and alternatively,

$$(\mathbf{A} - \lambda\mathbf{I})\mathbf{c} = \mathbf{0}, \tag{A19}$$

where **0** is a null vector and **I** is the identity matrix of the same order as **A**. It can be shown that these equations have non-zero solutions for **c** and λ only if the determinant of $(\mathbf{A} - \lambda\mathbf{I})$ is zero; thus

$$|\mathbf{A} - \lambda\mathbf{I}| = 0. \tag{A20}$$

Equation (A20) defines the conditions under which (A17) is true, and it is known as the *characteristic equation* of **A**. If **A** is of order m then the determinant expands to a polynomial of degree m in λ. The characteristic equation therefore has m solutions, $\lambda_1, \lambda_2, \ldots, \lambda_m$, which are known as *latent roots, characteristic roots,* or *eigenvalues*. Corresponding to each of the m latent roots λ_i, there is a vector $\mathbf{c}_i$ that satisfies (A17)–(A19). These vectors are known as *latent vectors*, *characteristic vectors*, or *eigenvectors*. Unless two or more roots happen to be equal, each vector is unique apart from a scaling factor. In statistical work it is usual to scale a vector so that the sum of the squares of its elements is 1, and this is achieved by dividing each element of the vector by $\sqrt{(\mathbf{c}_i^T \mathbf{c}_i)}$.

When two or more roots $\lambda_j, \lambda_{j+1}, \ldots$, are equal, there are many possible $\mathbf{c}_j$ that will satisfy (A18), and any one of these can be chosen arbitrarily.

A simple example in which the latent roots and vectors of a symmetric matrix are found is given in Chapter 8.

In practice, most variance–covariance matrices and correlation matrices are positive-definite. As such, all their latent roots are real and positive. Occasionally they are p.s.d., in which case at least one latent root is zero. Many types of similarity or dissimilarity matrix are p.s.d. with one zero root and the remainder positive.

REFERENCES

Say, from whence you owe this strange intelligence?
William Shakespeare, *Macbeth*

Ahrens, L. H. (1965). *Distribution of Elements in our Planet*. McGraw-Hill, New York.

Aitchison, J. and Brown, J. A. C. (1957). *The Lognormal Distribution*. The University Press, Cambridge.

Anderson, A. J. B. (1971). A similarity measure for mixed attribute types. *Nature, London* **232**: 416–17.

Avery, B. W. (1973). Soil classification in the Soil Survey of England and Wales. *Journal of Soil Science* **24**: 234–8.

Banfield, C. F. and Bassill, L. C. (1977). Algorithm AS113: a transfer algorithm for non-hierarchical classification. *Applied Statistics* **26**: 206–10.

Bartlett, M. S. (1937). Some examples of statistical methods in agriculture and applied biology. *Journal of the Royal Statistical Society, Supplement* **4**: 137–83.

—— (1947). The use of transformations. *Biometrics* **3**: 39–52.

Beale, E. M. L. (1969). *Cluster Analysis*. Scientific Control Systems, London.

Beckett, P. H. T. and Webster, R. (1965a). *A Classification System for Terrain*. Report no 872, Military Engineering Experimental Establishment, Christchurch.

—— (1965b). *Field Trials of a Terrain Classification—Organisation and Methods*. Report no 873, Military Engineering Experimental Establishment, Christchurch.

—— (1971). Soil variability: a review. *Soils and Fertilizers* **34**: 1–15.

Beckner, M. (1959). *The Biological Way of Thought*. Columbia University Press, New York.

Benzécri, J. P. (1973). *L'Analyse des données*. Vol. 2, *L'Analyse des correspondances*. Dunod, Paris.

Berry, B. J. L. (1962). *Sampling, Coding and Storing Flood Plain Data*. Agricultural Handbook no 237. US Department of Agriculture, Washington, DC.

Bray, J. R. and Curtis, J. T. (1957). An ordination of the upland forest communities of southern Wisconsin. *Ecological Monographs* **27**: 325–49.

Brink, A. B. A., Partridge, T. C. and Williams, A. A. B. (1982). *Soil Survey for Engineering*. Clarendon Press, Oxford.

Buraymah, I. M. and Webster, R. (1989). Variation in soil properties caused by irrigation and cultivation in the Central Gezira of Sudan. *Soil and Tillage Research* **13**: 57–74.

Burgess, T. M. and Webster, R. (1980). Optimal interpolation and isarithmic mapping of soil properties. I: The semi-variogram and punctual kriging. *Journal of Soil Science* **31**: 315–31.

—— (1984). Optimal sampling strategies for mapping soil types. I: Distribution of boundary spacings. *Journal of Soil Science* **35**: 641–54.

Burgess, T. M., Webster, R. and McBratney, A. B. (1981). Optimal interpolation and isarithmic mapping of soil properties. IV: Sampling strategy. *Journal of Soil Science* **32**: 643–59.

Burrough, P. A. (1986). *Principles of Geographical Information Systems for Land Resources Assessment*. Clarendon Press, Oxford.

—— and Webster, R. (1976). Improving a reconnaissance soil classification by multivariate methods. *Journal of Soil Science* **27**: 554–71.

Butler, B. E. (1980). *Soil Classification for Soil Survey*. Clarendon Press, Oxford.

Cain, A. J. and Harrison, G. A. (1958). An analysis of the taxonomist's judgement of affinity. *Proceedings of the Zoological Society of London* **131**: 85–98.

Campbell, J. B. (1978). Spatial variation in sand content and pH within single contiguous delineations of two soil mapping units. *Soil Science Society of America Journal* **42**: 460–4.

Campbell, N. A., Mulcahy, M. J. and McArthur, W. M. (1970). Numerical classification of soil profiles on the basis of field morphological properties. *Australian Journal of Soil Research* **8**: 43–58.

Clark, I. (1976). Some auxiliary functions for the spherical model of geostatistics. *Computers and Geosciences* **1**: 255–63.

—— (1979). *Practical Geostatistics*. Applied Science Publishers, London.

CIE (1971). *Colorimetry: Official Recommendations of the International Commission on Illumination*. Publication CIE no 15 (E–1.3.1), 1971, Bureau Central de la CIE, Paris.

Cochran, W. G. (1977). *Sampling Techniques*. 3rd edition, John Wiley, New York.

Cooke, G. W. (1975). *Fertilizing for Maximum Yield*. 2nd edition. Crosby Lockwood, London.

Cressie, N. (1985). Fitting variogram models by least squares. *Mathematical Geology* **17**: 563–25.

—— and Hawkins, D. M. (1980). Robust estimation of the variogram. *Mathematical Geology* **12**: 115–25.

Crommelin, R. D. and Gruijter, J. J. de (1973). *Cluster Analysis Applied to Mineralogical Data from the Coversand Formation in the Netherlands*. Soil Survey Paper no 7, Soil Survey Institute, Wageningen.

Cuanalo, H. E. de la and Webster, R. (1970). A comparative study of numerical classification and ordination of soil properties in a locality near Oxford. *Journal of Soil Science* **21**: 340–52.

Dale, M. B., MacNaughton-Smith, P., Williams, W. T. and Lance, G. N. (1970). Numerical classification of sequences. *Australian Computer Journal* **2**: 9–13.

Dalenius, T., Hajek, J. and Zubrzycki, S. (1961). On plane sampling and related geometrical problems. *Proceedings of the 4th Berkeley Symposium on Probability and Statistics* **1**: 164–77.

Das, A. C. (1950). Two-dimensional systematic sampling and the associated stratified and random sampling. *Sankhya* **10**: 95–108.

David, M. (1977) *Geostatistical Ore Reserve Estimation*. Elsevier, Amsterdam.

De La Rosa, D. (1979). Relation of several pedological characteristics to engineering qualities of soil. *Journal of Soil Science* **30**: 793–9.

Demirmen, F. (1969). *Multivariate Procedures and Fortran IV Program for Evaluation and Improvement of Classifications*. Computer Contribution no 31, State Geological Survey, University of Kansas, Lawrence.

Draper, N. R. and Smith, H. (1981). *Applied Regression Analysis*. 2nd edition. John Wiley, New York.

Eades, D. C. (1965). The inappropriateness of the correlation coefficient as a measure of taxonomic resemblance. *Systematic Zoology* **14**: 98–100.

Edwards, A. W. F. and Cavalli-Sforza, L. L. (1965). A method for cluster analysis. *Biometrics* **21**: 362–75.

Fisher, R. A. (1936). The use of multiple measurements in taxonomic problems. *Annals of Eugenics* **7**: 179–88.

—— and Yates, F. (1963). *Statistical Tables for Biological, Agricultural and Medical Research*. 6th edition. Longman (Oliver and Boyd), Edinburgh.

Friedman, H. P. and Rubin, J. (1967). On some invariant criteria for grouping data. *Journal of the American Statistical Association* **62**: 1159–78.

Gandin, L. S. (1965). *Objective Analysis of Meteorological Fields*. Israel Program for Scientific Translations, Jerusalem.

Gates, G. E. and Shiue, C.-J. (1962). The analysis of variance of the S-stage hierarchical classification. *Biometrics* **18**: 529–36.

Genstat 5 Committee (1987). *Genstat 5 Reference Manual*. Clarendon Press, Oxford.

Gilmour, J. S. L. (1937). A taxonomic problem. *Nature, London* **139**: 1040–2.

Goodall, D. W. (1954). Vegetational classification and vegetational continua. *Angewandte Pflanzen Soziologie* **1**: 168–82.

Goulding, K. W. T., McGrath, S. P. and Johnston, A. E. (1989). Predicting lime requirements of soils under permanent grassland and arable crops. *Soil Use and Management* **5**: 54–8.

Gower, J. C. (1962). Variance component estimation for unbalanced hierarchical classification. *Biometrics* **18**: 537–42.

—— (1966). Some distance properties of latent root and vector methods used in multivariate analysis. *Biometrika* **53**: 325–38.

—— (1967). A comparison of some methods of cluster analysis. *Biometrics* **23**: 623–37.

—— (1968). Adding a point to vector diagrams in multivariate analysis. *Biometrika* **55**: 582–5.

—— (1971). A general coefficient of similarity and some of its properties. *Biometrics* **27**: 857–71.

—— (1974). Maximal predictive classification. *Biometrics* **30**: 643–654.

—— (1983). Data analysis: multivariate or univariate and other difficulties. In *Food Research and Data Analysis*, ed. H. Martens and H. Russwurm. Applied Science Publishers, London, pp. 39–67.

—— and Ross, G. J. S. (1969). Minimum spanning trees and single linkage cluster analysis. *Applied Statistics* **18**: 54–64.

Gruijter, J. J. de (1977). *Numerical Classification of Soils and its Application in Survey*. Pudoc, Wageningen.

Guarascio, M. (1976). Improving the uranium deposits estimations (the Novazza case). In *Advanced Geostatistics in the Mining Industry*, ed. M. Guarascio, M. David and C. Huijbregts. Reidel, Dordrecht, pp. 351–67.

Halperin, M. (1970). On inverse estimation in linear regression. *Technometrics* **12**: 727–36.

Hammond, L. C., Pritchett, W. L. and Chew, V. (1958). Soil sampling in relation to soil heterogeneity. *Soil Science Society of America Proceedings* **22**: 548–52.

Harman, H. H. (1976). *Modern Factor Analysis*. 3rd edition. University of Chicago Press.

Harvard Computation Laboratory (1955). *Tables of the Cumulative Binomial Probability Distribution*. Harvard University Press, Cambridge, Mass.

Hill, M. O. (1973). Reciprocal averaging: an eigenvector method of ordination. *Journal of Ecology* **61**: 237–49.

—— (1974). Correspondence analysis: a neglected multivariate method. *Applied Statistics* **23**: 340–54.

Hodgson, J. M., Hollis, J. M., Jones, R. J. A. and Palmer, R. C. (1976). A comparison of field estimates and laboratory analyses of the silt and clay contents of some West Midlands soils. *Journal of Soil Science* **27**: 411–19.

Hole, F. D. and Hironaka, M. (1960). An experiment in ordination of some soil profiles. *Soil Science Society of America Proceedings* **24**: 309–12.

Hotelling, H. (1931). The generalization of Student's ratio. *Annals of Mathematical Statistics* **2**: 369–78.

Jaccard, P. (1908). Nouvelles recherches sur la distribution florale. *Bulletin de la Société vaudoise des Sciences naturelles* **44**: 223–70.

Jeffers, J. N. R. (1959). *Experimental Design and Analysis in Forest Research*. Almqvist and Wiksell, Stockholm.

Jenkins, G. M. and Watts, D. G. (1968). *Spectral Analysis and its Applications*. Holden-Day, San Francisco.

Johnston, A. E., Goulding, K. W. T. and Poulton, P. R. (1986). Soil acidification during more than 100 years under permanent grassland and woodland at Rothamsted. *Soil Use and Management* **2**: 3–10.

Journel, A. G. and Huijbregts, C. J. (1978). *Mining Geostatistics*. Academic Press, London.

Kaiser, H. F. (1958). The varimax criterion for analytic rotation in factor analysis. *Psychometrika* **23**: 187–200.

Kantey, B. A. and Williams, A. A. B. (1962). The use of soil engineering maps for road projects. *Transactions of the South African Institution of Civil Engineers* **4**: 149–59.

Kendall, M. G. and Stuart, A. (1967). *The Advanced Theory of Statistics*, Vol. 2. 3rd edition. Griffin, London.

Kolmogorov, A. N. (1941). Interpolirovanie i ekstrapolirovanie statsionarnykh sluchainykh posledovatel 'nostei (Interpolated and extrapolated stationary random sequences), *Isvestia AN SSSR, seriya matematicheskaya* **5** (1).

Krige, D. G. (1966). Two-dimensional weighted moving average trend surfaces for ore-evaluation. *Journal of the South African Institute of Mining and Metallurgy* **66**: 13–38.

Krumbein, W. C. and Slack, H. A. (1956). Statistical analysis of low-level radioactivity of Pennsylvanian black fissile shale in Illinois. *Bulletin of the Geological Society of America* **67**: 739–62.

Kruskal, J. B. (1964). Nonmetric multidimensional scaling: a numerical method. *Psychometrika* **29**: 115–29.

Krutchkoff, R. G. (1967). Classical and inverse regression methods of calibration. *Technometrics* **9**: 425–39.

Kyuma, K. (1973a). A method of fertility evaluation for paddy soils. II: Second approximation: evaluation of four independent constituents of soil fertility. *Soil Science and Plant Nutrition* **19**: 11–18.

—— (1973b), A method of fertility evaluation for paddy soils. III: Third approximation: synthesis of fertility for soil fertility evaluation. *Soil Science and Plant Nutrition* **19**: 19–27.

—— and Kawaguchi, K. (1973). A method of fertility evaluation for paddy soils. I: First approximation: chemical potentiality grading. *Soil Science and Plant Nutrition* **19**: 1–9.

Lance, G. N. and Williams, W. T. (1966). A generalized sorting strategy for computer classifications. *Nature, London* **212**: 218.

—— (1967a). Mixed-data classificatory programs. I: Agglomerative systems. *Australian Computer Journal* **1**: 15–20.

—— (1967b). A general theory of classificatory sorting strategies. I: Hierarchical systems. *Computer Journal* **9**: 373–80.

—— (1967c). Note on classification of multi-level data. *Computer Journal* **9**: 381–3.

Lawley, D. N. and Maxwell, A. E. (1971). *Factor Analysis as a Statistical Method*. 2nd edition. Butterworths, London.

Lindley, D. V. and Miller, J. C. P. (1953). *Cambridge Elementary Statistical Tables*. The University Press, Cambridge.

MacQueen, J. (1967). Some methods for classification and analysis of multivariate observations. *Proceedings of the Fifth Berkeley Symposium on Mathematical Statistics and Probability* **1**: 281–97.

Mahalanobis, P. C. (1927). Analysis of race mixture in Bengal. *Journal of the Asiatic Society of Bengal* **23**: 301–33.

Marcuse, S. (1949). Optimum allocation and variance components in mixed sampling with application to chemical analysis. *Biometrics* **5**: 189–206.

Mardia, K. V. (1972). *Statistics of Directional Data*. Academic Press, London.

Mark, D. M. and Church, M. (1977). On the misuse of regression in earth science. *Mathematical Geology* **9**: 63–77.

Marriott, F. H. C. (1971). Practical problems in a method of cluster analysis. *Biometrics* **27**: 501–14.

—— (1974). *The Interpretation of Multiple Observations*. Academic Press, London.

Matérn, B. (1960). Spatial variation: stochastic models and their application to some problems in forest surveys and other sampling investigations. *Meddelanden från Statens Skogsforskningsinstitut* **49**: 1–144.

Matheron, G. (1965), *Les Variables régionalisées at leur estimation*. Masson, Paris.

—— (1969). *Le Krigeage universel*. Cahiers du Centre de Morphologie Mathématique de Fontainebleau, no 1.

—— (1971). *The Theory of Regionalized Variables and its Applications*. Cahiers du Centre de Morphologie Mathématique de Fountainebleau no 5.

—— (1976). A simple substitute for conditional expectation: the disjunctive kriging. In *Advanced Geostatistics in the Mining Industry*, ed. M. Guarascio, M. David and C. Huijbregts. Reidel, Dordrecht, pp. 221–36.

McBratney, A. B. and Webster, R. (1981). Spatial dependence and classification of the soil along a transect in north-east Scotland. *Geoderma* **26**: 63–82.

—— (1983a). How many samples are needed to estimate the regional mean of a soil property? *Soil Science* **135**: 177–83.

—— (1983b).Optimal interpolation and isarithmic mapping of soil properties. V: Co-regionalization and multiple sampling strategy. *Journal of Soil Science* **34**: 137–62.

—— (1986). Choosing functions for semi-variograms of soil properties and fitting them to sampling estimates. *Journal of Soil Science* **37**: 617–39.

McBratney, A. B., Webster, R. and Burgess, T. M. (1981). The design of optimal sampling schemes for local estimation and mapping of regionalized variables. I: Theory and method. *Computers and Geosciences* **7**: 331–4.

McBratney, A. B., Webster, R., McLaren, R. G. and Spiers, R. B. (1982). Regional variation of extractable copper and cobalt in the topsoil of south-east Scotland. *Agronomie* **2**: 969–82.

McKeague, J. A., Day, J. H., and Shields, J. A. (1971). Evaluating relationships among properties by computer analysis. *Canadian Journal of Soil Science* **51**: 105–11.

Melville, M. D. and Atkinson, G. (1985). Soil colour: its measurement and its designation in models of uniform colour space. *Journal of Soil Science* **36**: 495–512.

Miesch, A. T. (1975). Variograms and variance components in geochemistry and ore evaluation. *Geological Society of America Memoir* **142**: 333–40.

Moore, A. W. and Russell, J. S. (1967). Comparison of coefficients and grouping procedures in numerical analysis of soil trace element data. *Geoderma* **1**: 139–58.

—— and Ward, W. T. (1972). Numerical analysis of soils: a comparison of three soil profile models with field classification. *Journal of Soil Science* **23**: 193–209.

Moroney, M. J. (1956). *Facts from Figures*. 3rd edition. Penguin Books, Harmondsworth.

Müller, V. and Böttcher, J. (1987). Verteilung von Grenzabständen auf Bodenkarten und Ermittlung von Risikofunctionen. *Catena* **14**: 561–70.

National Bureau of Standards (1950). *Tables of the Binomial Probability Distribution*. US Government Printing Office, Washington, DC.

Nelder, J. A. (1975). *Computers in Biology*. Wykeham Publications, London.

Norris, J. M. and Dale, M. B. (1971). Transition matrix approach to numerical classification of soil profiles. *Soil Science Society of America Proceedings* **35**: 487–91.

Norris, J. M. and Loveday, J. (1971). The application of multivariate analysis to soil studies. II: The allocation of soil profiles to established groups: a comparison of soil survey and computer method. *Journal of Soil Science* **22**: 395–400.

Nortcliff, S. (1978). Soil variability and reconnaissance soil mapping: a statistical study in Norfolk. *Journal of Soil Science* **29**: 403–18.

Northcote, K. H. (1971). *A Factual Key for the Recognition of Australian Soils*. 3rd edition. Rellim Technical Publications, Glenside, South Australia.

Olea, R. A. (1975). *Optimum Mapping Techniques Using Regionalized Variable Theory*. Series on Spatial Analysis, no 2. Kansas Geological Survey, Lawrence.

—— (1977). *Measuring Spatial Dependence with Semi-Variograms*. Series on Spatial Analysis no 3. Kansas Geological Survey, Lawrence.

Oliver, M. A. (1984). *Soil Variation in the Wyre Forest: Its Elucidation and Measurement*. Ph.D. thesis, University of Birmingham.

—— and Webster, R. (1986). Combining nested and linear sampling for determining the scale and form of spatial variation of regionalized variables. *Geographical Analysis* **18**: 225–42.

—— (1987a). The elucidation of soil pattern in the Wyre Forest of the West Midlands, England. I: Multivariate distribution. *Journal of Soil Science* **38**: 279–91.

—— (1987b). The elucidation of soil pattern in the Wyre Forest of the West Midlands, England. II: Spatial distribution. *Journal of Soil Science* **38**: 293–307.

—— (1989). A geostatistical basis for spatial weighting in multivariate classification. *Mathematical Geology* **21**: 15–35.

Olson, J. S. and Potter, P. E. (1954). Variance components of cross-bedding direction in some basal Pennsylvanian sandstones of the Eastern Interior Basin: statistical methods. *Journal of Geology* **62**: 26–49.

Osmond, D. A., Swarbrick, T., Thompson, C. R. and Wallace, T. (1949). *A Survey of the Soils and Fruit in the Vale of Evesham, 1926–1934*. Bulletin no 116, Ministry of Agriculture and Fisheries. H. M. Stationery Office, London.

Pearson, E. S. and Hartley, H. O. (ed.) (1966, 1972). *Biometrika Tables for Statisticians*, Vol. I, 3rd edition; Vol. II. The University Press, Cambridge.

Quenouille, M. H. (1949). Problems in plane sampling. *Annals of Mathematical Statistics* **20**: 355–75.

Rand Corporation (1955). *A Million Random Digits with 100 000 Normal Deviates*. Free Press, Glencoe, Ill.

Rao, C. R. (1948). The utilization of multiple measurements in problems of biological classification. *Journal of the Royal Statistical Society* **B10**: 159–93.

—— (1952). *Advanced Statistical Methods in Biometric Research*. John Wiley, New York.

Rayner, J. H. (1966). Classification of soils by numerical methods. *Journal of Soil Science* **17**: 79–92.

—— (1969). The numerical approach to soil systematics. In *The Soil Ecosystem,* ed. J. G. Sheals. Systematics Association Publication, no 8, London, pp. 31–9.

Reyment, R. A., Blackith, R. E. and Campbell, N. A. (1984). *Multivariate Morphometrics*. 2nd edition. Academic Press, London.

Romig, H. G. (1952). *50–100 Binomial Tables*. John Wiley, New York.

Ross, G. J. S. (1969). Algorithms AS 13–15. *Applied Statistics* **18**: 103–10.

—— (1987). *MLP User Manual*. Numerical Algorithms Group, Oxford.

Rubin, J. (1967). Optimal classification into groups: an approach for solving the taxonomy problem. *Journal of Theoretical Biology* **15**: 103–10.

Russell, J. S. and Moore, A. W. (1968). Comparison of different depth weightings in the numerical analysis of anisotropic soil profile data. *Proceedings of the 9th International Congress of Soil Science* **4**: 205–13.

Sampson, R. J. (1978). *The Surface II Graphics System*. Revision one. Series in Spatial Analysis no 1. Kansas Geological Survey, Lawrence.

Shaw, D. M. (1961). Element distribution laws in geochemistry. *Geochemica e Cosmochimica Acta* **23**: 116–34.

Shukla, G. K. (1972). On the problem of calibration. *Technometrics* **14**: 547–53.

Sichel, H. S. (1966). The estimation of means and associated confidence limits for small samples from lognormal populations. *Journal of the South African Institute of Mining and Metallurgy* **66**: 106–23.

Sneath, P. H. A. (1957). The application of computers to taxonomy. *Journal of General Microbiology* **17**: 201–26.

—— and Sokal, R. R. (1973). *Numerical Taxonomy*. W. H. Freeman, San Francisco.

Snedecor, G. W. and Cochran, W. G. (1980). *Statistical Methods*. 7th edition. Iowa State University Press, Ames.

Soil Survey Staff (1975). *Soil Taxonomy*. US Department of Agriculture, Washington, DC.

Sokal, R. R. and Michener, C. D. (1958). A statistical method for evaluating systematic relationships. *Kansas University Science Bulletin* **38**: 1409–38.

Sokal, R. R. and Sneath, P. H. A. (1963). *Principles of Numerical Taxonomy*. W. H. Freeman, San Francisco.

Sprent, P. (1969). *Models in Regression and Related Topics*. Methuen, London.

Thiessen, A. H. (1911). Precipitation averages for large areas. *Monthly Weather Review* **39**: 1082–4.

Thomasson, A. J. (ed.) (1975). *Soils and Field Drainage*. Soil Survey Technical Monograph no 7, Lawes Agricultural Trust, Rothamsted Experimental Station, Harpenden.

Thorndike, R. L. (1953). Who belongs in families? *Psychometrika* **18**: 267–76.

Voltz, M. (1986). *Variabilité spatiale des propriétés physiques du sol en milieu alluvial*. Thèse; Ecole Nationale Supérieure Agronomique de Montpellier.

Ward, J. H. (1963). Hierarchical grouping to optimize an objective function. *Journal of the American Statistical Association* **58**: 236–44.

Webster, R. (1966). The measurement of soil water tension in the field. *New Phytologist* **65**: 249–58.

—— (1968). Fundamental objections to the 7th Approximation. *Journal of Soil Science* **19**: 354–66.

—— (1977). *Quantitative and Numerical Methods in Soil Classification and Survey*. Clarendon Press, Oxford.

—— and Beckett, P. H. T. (1968). Quality and usefulness of soil maps. *Nature, London* **219**: 680–2.

—— and Burgess, T. M. (1980). Optimal interpolation and isarithmic mapping of soil properties. III: Changing drift and universal kriging. *Journal of Soil Science* **31**: 505–24.

—— (1984a). Une approche probabiliste de la cartographie du sol. *Sciences de la Terre, Série Informatique Géologique* **18**: 175–85.

—— (1984b). Sampling and bulking strategies for estimating soil properties of small regions. *Journal of Soil Science* **35**: 127–40.

—— and Burrough, P. A. (1972a). Computer-based soil mapping of small areas from sample data. I: Multivariate classification and ordination. *Journal of Soil Science* **23**, 210–21.

—— (1972b). Computer-based soil mapping of small areas from sample data. II: Classification smoothing. *Journal of Soil Science* **23**: 222–34.

—— (1974). Multiple discriminant analysis in soil survey. *Journal of Soil Science* **25**: 120–34.

—— and Butler, B. E. (1976). Soil survey and classification studies in Ginninderra. *Australian Journal of Soil Research* **14**: 1–24.

—— and Cuanalo de la C., H. E. (1975). Soil transect correlograms of North Oxfordshire and their interpretation. *Journal of Soil Science* **26**: 176–94.

—— and McBratney, A. B. (1987). Mapping soil fertility at Broom's Barn by simple kriging. *Journal of the Science of Food and Agriculture* **38**: 97–115.

—— (1989). On the Akaike Information Criterion for choosing models for variograms of soil properties. *Journal of Soil Science* **40**: 493–6.

—— and Nortcliff, S. (1984). Improved estimation of micronutrients in hectare plots of the Sonning series. *Journal of Soil Science* **35**: 667–72.

—— and Oliver, M. A. (1989). Disjunctive kriging in agriculture. In *Geostatistics,* ed. M. Armstrong. Kluwer, Dordrecht, pp. 421–32.

Wilks, S. S. (1932). Certain generalizations in the analysis of variance. *Biometrika* **24**: 471–94.

Williams, C. and Rayner, J. H. (1977). Variability in three areas of the Denchworth soil map unit. III: Soil grouping based on chemical composition. *Journal of Soil Science* **28**: 180–95.

Williams, E. J. (1959). *Regression Analysis*. John Wiley, New York.

—— (1969). A note on regression methods in calibration. *Technometrics* **11**: 189–92.

Williams, W. T. and Lambert, J. M. (1959). Multivariate methods in plant ecology. I: Association-analysis in plant communities. *Journal of Ecology* **47**: 83–101.

Wishart, D. (1969a). Mode analysis: a generalization of nearest neighbour which reduces chaining effects. In *Numerical Taxonomy,* ed. A. J. Cole. Academic Press, London, pp. 282–311.

—— (1969b). An algorithm for hierarchical classification. *Biometrics* **25**: 165–170.

—— (1987). *Clustan User Manual*. 4th edition. Computing Laboratory, The University, St. Andrews.

Yates, F. (1935). Some examples of biased sampling. *Annals of Eugenics* **6**: 202–13.

—— (1948). Systematic sampling. *Philosophical Transactions of the Royal Society* **A241**: 345–77.

—— (1981). *Sampling Methods for Censuses and Surveys*. 4th edition. Griffin, London.

Yfantis, E. A., Flatman, G. T. and Behar, J. V. (1987). Efficiency of kriging estimation for square, triangular and hexagonal grids. *Mathematical Geology* **19**: 183–205.

Youden, W. J. and Mehlich, A. (1937). Selection of efficient methods for soil sampling. *Contributions of the Boyce Thompson Institute for Plant Research* **9**: 59–70.

AUTHOR INDEX

SUBJECT INDEX